Polymer Fiber Optics

Materials, Physics, and Applications

OPTICAL SCIENCE AND ENGINEERING

1. Electron and Ion Microscopy and Microanalysis: Principles and Applications, *Lawrence E. Murr*
2. Acousto-Optic Signal Processing: Theory and Implementation, *edited by Norman J. Berg and John N. Lee*
3. Electro-Optic and Acousto-Optic Scanning and Deflection, *Milton Gottlieb, Clive L. M. Ireland, and John Martin Ley*
4. Single-Mode Fiber Optics: Principles and Applications, *Luc B. Jeunhomme*
5. Pulse Code Formats for Fiber Optical Data Communication: Basic Principles and Applications, *David J. Morris*
6. Optical Materials: An Introduction to Selection and Application, *Solomon Musikant*
7. Infrared Methods for Gaseous Measurements: Theory and Practice, *edited by Joda Wormhoudt*
8. Laser Beam Scanning: Opto-Mechanical Devices, Systems, and Data Storage Optics, *edited by Gerald F. Marshall*
9. Opto-Mechanical Systems Design, *Paul R. Yoder, Jr.*
10. Optical Fiber Splices and Connectors: Theory and Methods, *Calvin M. Miller with Stephen C. Mettler and Ian A. White*
11. Laser Spectroscopy and Its Applications, *edited by Leon J. Radziemski, Richard W. Solarz, and Jeffrey A. Paisner*
12. Infrared Optoelectronics: Devices and Applications, *William Nunley and J. Scott Bechtel*
13. Integrated Optical Circuits and Components: Design and Applications, *edited by Lynn D. Hutcheson*
14. Handbook of Molecular Lasers, *edited by Peter K. Cheo*
15. Handbook of Optical Fibers and Cables, *Hiroshi Murata*
16. Acousto-Optics, *Adrian Korpel*
17. Procedures in Applied Optics, *John Strong*
18. Handbook of Solid-State Lasers, *edited by Peter K. Cheo*
19. Optical Computing: Digital and Symbolic, *edited by Raymond Arrathoon*
20. Laser Applications in Physical Chemistry, *edited by D. K. Evans*

21. Laser-Induced Plasmas and Applications, *edited by Leon J. Radziemski and David A. Cremers*
22. Infrared Technology Fundamentals, *Irving J. Spiro and Monroe Schlessinger*
23. Single-Mode Fiber Optics: Principles and Applications, Second Edition, Revised and Expanded, *Luc B. Jeunhomme*
24. Image Analysis Applications, *edited by Rangachar Kasturi and Mohan M. Trivedi*
25. Photoconductivity: Art, Science, and Technology, *N. V. Joshi*
26. Principles of Optical Circuit Engineering, *Mark A. Mentzer*
27. Lens Design, *Milton Laikin*
28. Optical Components, Systems, and Measurement Techniques, *Rajpal S. Sirohi and M. P. Kothiyal*
29. Electron and Ion Microscopy and Microanalysis: Principles and Applications, Second Edition, Revised and Expanded, *Lawrence E. Murr*
30. Handbook of Infrared Optical Materials, *edited by Paul Klocek*
31. Optical Scanning, *edited by Gerald F. Marshall*
32. Polymers for Lightwave and Integrated Optics: Technology and Applications, *edited by Lawrence A. Hornak*
33. Electro-Optical Displays, *edited by Mohammad A. Karim*
34. Mathematical Morphology in Image Processing, *edited by Edward R. Dougherty*
35. Opto-Mechanical Systems Design: Second Edition, Revised and Expanded, *Paul R. Yoder, Jr.*
36. Polarized Light: Fundamentals and Applications, *Edward Collett*
37. Rare Earth Doped Fiber Lasers and Amplifiers, *edited by Michel J. F. Digonnet*
38. Speckle Metrology, *edited by Rajpal S. Sirohi*
39. Organic Photoreceptors for Imaging Systems, *Paul M. Borsenberger and David S. Weiss*
40. Photonic Switching and Interconnects, *edited by Abdellatif Marrakchi*
41. Design and Fabrication of Acousto-Optic Devices, *edited by Akis P. Goutzoulis and Dennis R. Pape*
42. Digital Image Processing Methods, *edited by Edward R. Dougherty*
43. Visual Science and Engineering: Models and Applications, *edited by D. H. Kelly*
44. Handbook of Lens Design, *Daniel Malacara and Zacarias Malacara*
45. Photonic Devices and Systems, *edited by Robert G. Hunsberger*
46. Infrared Technology Fundamentals: Second Edition, Revised and Expanded, *edited by Monroe Schlessinger*
47. Spatial Light Modulator Technology: Materials, Devices, and Applications, *edited by Uzi Efron*
48. Lens Design: Second Edition, Revised and Expanded, *Milton Laikin*

49. Thin Films for Optical Systems, *edited by Francoise R. Flory*
50. Tunable Laser Applications, *edited by F. J. Duarte*
51. Acousto-Optic Signal Processing: Theory and Implementation, Second Edition, *edited by Norman J. Berg and John M. Pellegrino*
52. Handbook of Nonlinear Optics, *Richard L. Sutherland*
53. Handbook of Optical Fibers and Cables: Second Edition, *Hiroshi Murata*
54. Optical Storage and Retrieval: Memory, Neural Networks, and Fractals, *edited by Francis T. S. Yu and Suganda Jutamulia*
55. Devices for Optoelectronics, *Wallace B. Leigh*
56. Practical Design and Production of Optical Thin Films, *Ronald R. Willey*
57. Acousto-Optics: Second Edition, *Adrian Korpel*
58. Diffraction Gratings and Applications, *Erwin G. Loewen and Evgeny Popov*
59. Organic Photoreceptors for Xerography, *Paul M. Borsenberger and David S. Weiss*
60. Characterization Techniques and Tabulations for Organic Nonlinear Optical Materials, *edited by Mark G. Kuzyk and Carl W. Dirk*
61. Interferogram Analysis for Optical Testing, *Daniel Malacara, Manuel Servin, and Zacarias Malacara*
62. Computational Modeling of Vision: The Role of Combination, *William R. Uttal, Ramakrishna Kakarala, Spiram Dayanand, Thomas Shepherd, Jagadeesh Kalki, Charles F. Lunskis, Jr., and Ning Liu*
63. Microoptics Technology: Fabrication and Applications of Lens Arrays and Devices, *Nicholas Borrelli*
64. Visual Information Representation, Communication, and Image Processing, *edited by Chang Wen Chen and Ya-Qin Zhang*
65. Optical Methods of Measurement, *Rajpal S. Sirohi and F. S. Chau*
66. Integrated Optical Circuits and Components: Design and Applications, *edited by Edmond J. Murphy*
67. Adaptive Optics Engineering Handbook, *edited by Robert K. Tyson*
68. Entropy and Information Optics, *Francis T. S. Yu*
69. Computational Methods for Electromagnetic and Optical Systems, *John M. Jarem and Partha P. Banerjee*
70. Laser Beam Shaping, *Fred M. Dickey and Scott C. Holswade*
71. Rare-Earth-Doped Fiber Lasers and Amplifiers: Second Edition, Revised and Expanded, *edited by Michel J. F. Digonnet*
72. Lens Design: Third Edition, Revised and Expanded, *Milton Laikin*
73. Handbook of Optical Engineering, *edited by Daniel Malacara and Brian J. Thompson*

74. Handbook of Imaging Materials: Second Edition, Revised and Expanded, *edited by Arthur S. Diamond and David S. Weiss*
75. Handbook of Image Quality: Characterization and Prediction, *Brian W. Keelan*
76. Fiber Optic Sensors, *edited by Francis T. S. Yu and Shizhuo Yin*
77. Optical Switching/Networking and Computing for Multimedia Systems, *edited by Mohsen Guizani and Abdella Battou*
78. Image Recognition and Classification: Algorithms, Systems, and Applications, *edited by Bahram Javidi*
79. Practical Design and Production of Optical Thin Films: Second Edition, Revised and Expanded, *Ronald R. Willey*
80. Ultrafast Lasers: Technology and Applications, *edited by Martin E. Fermann, Almantas Galvanauskas, and Gregg Sucha*
81. Light Propagation in Periodic Media: Differential Theory and Design, *Michel Nevière and Evgeny Popov*
82. Handbook of Nonlinear Optics, Second Edition, Revised and Expanded, *Richard L. Sutherland*
83. Polarized Light: Second Edition, Revised and Expanded, *Dennis Goldstein*
84. Optical Remote Sensing: Science and Technology, *Walter Egan*
85. Handbook of Optical Design: Second Edition, *Daniel Malacara and Zacarias Malacara*
86. Nonlinear Optics: Theory, Numerical Modeling, and Applications, *Partha P. Banerjee*
87. Semiconductor and Metal Nanocrystals: Synthesis and Electronic and Optical Properties, edited by *Victor I. Klimov*
88. High-Performance Backbone Network Technology, *edited by Naoaki Yamanaka*
89. Semiconductor Laser Fundamentals, *Toshiaki Suhara*
90. Handbook of Optical and Laser Scanning, *edited by Gerald F. Marshall*
91. Organic Light-Emitting Diodes: Principles, Characteristics, and Processes, *Jan Kalinowski*
92. Micro-Optomechatronics, *Hiroshi Hosaka, Yoshitada Katagiri, Terunao Hirota, and Kiyoshi Itao*
93. Microoptics Technology: Second Edition, *Nicholas F. Borrelli*
94. Organic Electroluminescence, *edited by Zakya Kafafi*
95. Engineering Thin Films and Nanostructures with Ion Beams, *Emile Knystautas*
96. Interferogram Analysis for Optical Testing, Second Edition, *Daniel Malacara, Manuel Sercin, and Zacarias Malacara*
97. Laser Remote Sensing, *edited by Takashi Fujii and Tetsuo Fukuchi*
98. Passive Micro-Optical Alignment Methods, *edited by Robert A. Boudreau and Sharon M. Boudreau*

99. Organic Photovoltaics: Mechanism, Materials, and Devices, *edited by Sam-Shajing Sun and Niyazi Serdar Saracftci*
100. Handbook of Optical Interconnects, *edited by Shigeru Kawai*
101. GMPLS Technologies: Broadband Backbone Networks and Systems, *Naoaki Yamanaka, Kohei Shiomoto, and Eiji Oki*
102. Laser Beam Shaping Applications, *edited by Fred M. Dickey, Scott C. Holswade and David L. Shealy*
103. Electromagnetic Theory and Applications for Photonic Crystals, *Kiyotoshi Yasumoto*
104. Physics of Optoelectronics, *Michael A. Parker*
105. Opto-Mechanical Systems Design: Third Edition, *Paul R. Yoder, Jr.*
106. Color Desktop Printer Technology, *edited by Mitchell Rosen and Noboru Ohta*
107. Laser Safety Management, *Ken Barat*
108. Optics in Magnetic Multilayers and Nanostructures, *Štefan Višňovský*
109. Optical Inspection of Microsystems, *edited by Wolfgang Osten*
110. Applied Microphotonics, *edited by Wes R. Jamroz, Roman Kruzelecky, and Emile I. Haddad*
111. Organic Light-Emitting Materials and Devices, *edited by Zhigang Li and Hong Meng*
112. Silicon Nanoelectronics, *edited by Shunri Oda and David Ferry*
113. Image Sensors and Signal Processor for Digital Still Cameras, *Junichi Nakamura*
114. Encyclopedic Handbook of Integrated Circuits, *edited by Kenichi Iga and Yasuo Kokubun*
115. Quantum Communications and Cryptography, *edited by Alexander V. Sergienko*
116. Optical Code Division Multiple Access: Fundamentals and Applications, *edited by Paul R. Prucnal*
117. Polymer Fiber Optics: Materials, Physics, and Applications, *Mark G. Kuzyk*

Polymer Fiber Optics

Materials, Physics, and Applications

Mark G. Kuzyk

CRC is an imprint of the Taylor & Francis Group,
an informa business

CRC Press
Taylor & Francis Group
6000 Broken Sound Parkway NW, Suite 300
Boca Raton, FL 33487-2742

Printed in the United States of America on acid-free paper
10 9 8 7 6 5 4 3 2 1

International Standard Book Number-10: 1-57444-706-8 (Hardcover)
International Standard Book Number-13: 978-1-57444-706-4 (Hardcover)

Library of Congress Cataloging-in-Publication Data

Kuzyk, Mar G., 1958-
Polymer fiber optics : materials, physics, and applications / by Mark G. Kuzyk.
p. cm. -- (Optical science and engineering ; 117)
Includes bibliographical references and index.
ISBN 1-57444-7068-8 (acid-free paper)
1. Polymers--Optical propeerties. 2. Optical materials. 3. Fiber optics--Materials. I. Title. II. Series: Optical science and engineering (Boca Raton, Fla.) ; 117.

QD381.8.K89 2006
621.36'92--dc22 2006044000

Visit the Taylor & Francis Web site at
http://www.taylorandfrancis.com

and the CRC Press Web site at
http://www.crcpress.com

Dedication

I dedicate this book to my wife, who embodies all those things that are noble in humanity. I am truly privileged to live in her company.

Preface

> In no other intellectual pursuit is such a small investment rewarded on such a grand scale. The mind's interpretation of experiments here and now enables us to understand everything from the very smallest thread of space to the vastness of the universe this moment and throughout eternity.

This book has taken forever to materialize. Part of the problem with writing about polymer fibers is that they are becoming a technology and yet the richness of phenomena observed makes polymer fibers a perpetually hot topic of research. The field is progressing so quickly that any manuscript will necessarily be obsolete by the time an author prints out a first draft. As such, the contents of this book have settled down into what I believe are the most fundamental issues.

While many of the topics that are covered in this book are standard fare in the field of polymer fibers, such as Bragg gratings and photonic crystals, I have chosen to re-derive many of the common results. In addition, some of the material in this book is, to the best of my knowledge, new. However, since it is often easier to predict the future than to remember the past, these new topics (such as the large chunk of the speculative material on photomechanical effects and smart materials) may not be as unique as I believe. Even so, I hope that my presentation provides a fresh perspective on topics new and old.

With such a complex piece of work, there are many opportunities for typos and errors to go unnoticed even after several editorial passes — especially in the more mathematical sections. So, I apologize in advance and would be grateful to readers who bring errors to my attention. If any major issues surface, I will publish an errata on the Physics and Astronomy website at Washington State University.

The two introductory chapters cover the history of how polymer fibers came into being and the theory of wave propagation of light in fibers. Beyond these perfunctory chapters, I have chosen topics that are of the most interest to me, which I also believe will provide the reader with the basic physics and materials-science issues needed to understand how to make fibers, how fibers work, and the kinds of things that can be done with fibers.

In developing new polymeric materials for making fibers, there will always be challenging materials-processing issues. Instead of trying to produce a compendium of all the newest materials and techniques, I have chosen to present the rudimentary methods for making fibers. Even the most high-tech methods used today still employ drawing techniques that are governed by the viscoelastic properties of polymers. As such, I take the approach of making the reader appreciate the basics, which can be applied to all the newfangled approaches that may be used in the future.

A large part of this book covers characterization techniques since these will always be an important part of making fibers for either applications or basic research. Several examples of devices, such as electrooptic fibers and fiber lasers, follow. I have also decided to devote a whole chapter to smart materials — an application of polymer optical fibers that I am convinced will someday be one of the most important. In particular, a large part of this chapter discusses photomechanical effects, which make possible science-fiction-like applications.

I thank the reviewers for making comments that greatly improved the quality of this book. In particular, I am indebted to Prof. Un-Chul Paek for suggestions that made many improvements. Both the physical insight and rigor resulting from his input were invaluable.

I also thank Steve Vigil for writing the first draft of Chapter 2 — on which the foundation of its present form was built and Bob Olsen for his critique of chapter 8. I am indebted to my faculty position at Washington State University, which gives me the opportunity to interact with students who reinvigorate my intellect, and for working in a department that rewards the generation of knowledge and the flow of ideas. Much of the science described in this book was enabled by financial support from various agencies, including the National Science Foundation, the Air Force Office of Scientific Research, and the Army Research Office, as well as companies such as Allied Signal, AT&T Bell Laboratories, and Spectralux. Finally, I would like to thank my family for excusing my eccentricities and absences — both mental and physical. Without them, life would have little meaning.

Mark G. Kuzyk

List of Tables

List of Figures

Contents

1 History of Polymer Optical Fibers

1.1 INTRODUCTION

Many technologies are the result of bringing together unrelated ideas and discoveries. The development of optical fiber for telecommunications is no exception. In this chapter, we show the historical developments that have lead to polymer optical fiber and devices that have been made using this technology.

1.2 USING LIGHT FOR TELECOMMUNICATIONS

The idea of using light for telecommunications goes back to the 19th century. In 1881, Alexander Graham Bell reported on an experiment that demonstrated what he called a photophone — a device in which a voice or musical tone is encoded on a beam of light, and transmitted to a detector that converts the light to an acoustical signal.

Figure 1.1 shows a schematic diagram of one of the schemes that was reported by Bell for making a photophone. [1] He used the sun as the source of light and a silvered piece of mica as a reflector. The sun was intensified using a collimator and directed to the mica reflector, which was thin enough to allow it to undergo oscillations in response to sound. The lowest mode of vibration is shown as a dashed line and the flat line represents the static mirror. The oscillating mirror results in time variations in its focal length — leading to intensity variations in the beam that illuminates the detector. A dish is used to collect the light and to focus it on a selenium detector, which converts the light to an electrical signal that drives a speaker. The two active parts of the system shown in Figure 1.1 are essentially an optical microphone at the transmitting side (conversion of sound to light modulation) and an optical speaker (conversion of light to sound).

In the first long-range demonstration of this system, the transmitter and receiver were separated by 213 meters. Bell's assistant, Mr. Tainter, operated the transmission part of the system from the top of Franklin School House in Washington, DC, while Bell operated the receiver at his lab at 1325 L Street. Once the experiment was set up, Mr. Tainter spoke into the "microphone:" "Mr. Bell, if you hear what I say, come to the window and wave your hat." Since this experiment was successful, we assume that he waved his hat even though he did not report this part of the story.

A large part of Bell and Tainter's research efforts concentrated on improving the selenium detector to make it sensitive enough to detect small intensity fluctuations. They also investigated more novel detection schemes. For example, they directly illuminated their ears with intensity-modulated light and found that they could also hear sounds. Since the largest modulation of the light is achieved by

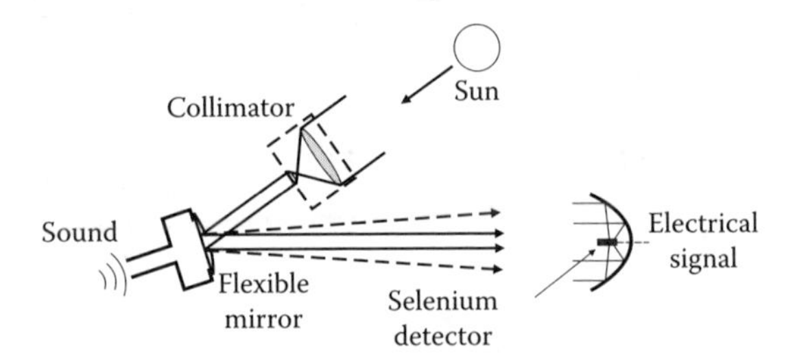

FIGURE 1.1 Alexander Graham Bell's demonstration of the photophone. Adapted from Ref. [1].

periodically blocking and unblocking the beam, they made a rotating perforated disk, as shown in Figure 1.2, to make musical tones at the detector. They used such controlled experiments to study selenium and other materials and found that it was not necessary to convert the light to an electrical signal to get sound. Many materials were found to emit tones. The most efficient light-to-sound generation was found for geometries in which the material was made into a thin membrane. They correctly hypothesized that this geometry lead to the most efficient energy transfer from the material to the surrounding air.

Bell and Tainter also experimented with various light sources and found that a kerosene lamp and even a candle could be used as the light source provided that the distance between the source and the receiver was not too large. They also found that blocking the beam with a hand or other organic materials (such as hard rubber) diminished but did not eliminate the sound generated at the detector end. They postulated that some components of the light energy were able to pass through certain materials.

Many of Bell and Tainter's observations are relevant to modern-day technologies. While the photophone was not extensively used at the time of its first demonstration, the basic principles are used in telecommunications. Bell proposed that light modulation could be used to encode signals, for example, by using Morse code. This foreshadows modern digital signal processing technologies. In addition, the fact that light can be used to excite acoustical waves in a material is of relevance to smart photomechanical materials and devices. These will be

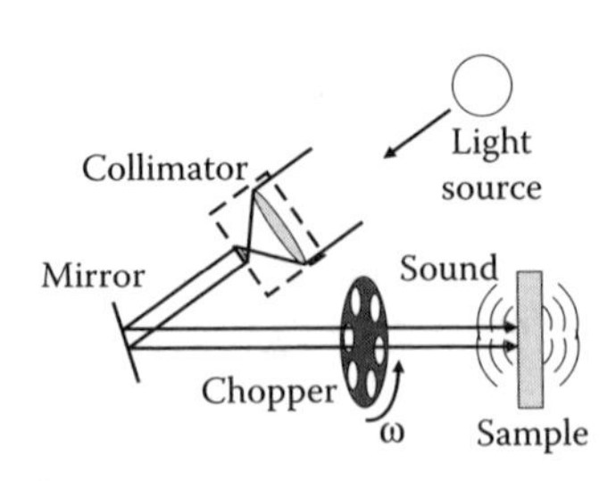

FIGURE 1.2 Bell and Tainter experimented with different light sources and detectors. Based on Ref. [1].

discussed in Chapter 9. It is interesting to read Bell's papers both for their historical significance and for getting an appreciation of the ingenuity of these early pioneers.

1.3 GLASS FIBERS

The disadvantage of the photophone is its requirement of unobstructed line-of-site between transmitter and receiver. A waveguide, however, which confines light inside a flexible structure, allows the light signal to be routed. The road to the development of glass fiber has a long history. In 1897 (shortly after Bell's photophone report), Lord Rayleigh considered waveguides.[2] The theory of waveguides was developed in 1910 by Hondros and Debye[3] and experimentally investigated by Schriever in 1920.[4] By this time, it was understood that long and thin strands of fiber could be used to guide light through total internal reflection.

1.3.1 Optical Imaging

By the 1950s, fibers were being made in glass, quartz, nylon, and polystyrene.[5] At that time, glass was deemed to be the best material due to its tensile strength and transparency. Hopkins and Kapany were the first to demonstrate the use of fiber bundles to carry images.[5] These fiber bundles were made of 25 μm diameter glass cylinders in approximately meter lengths. Figure 1.3 shows a reproduction of the test object whose image was recorded.[5] The idea of using a waveguide to carry an image, however, was patented in Britain in 1927 by Baird.[6]

1.3.2 Multimode Fibers

Optical fibers made before 1960 typically were of large diameter relative to the wavelength of visible light. Such fibers are now called multimode fibers, and ray

FIGURE 1.3 An artistic rendition of the light output from a fiber bundle with a test object "Glas". Adapted from Ref. [5].

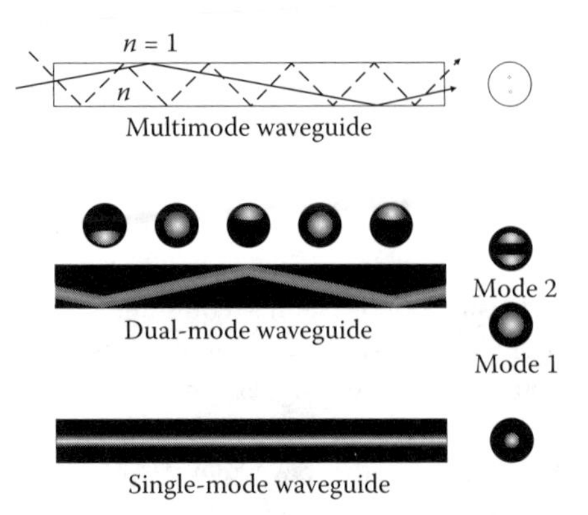

FIGURE 1.4 Schematic diagram of the principle of fiber waveguiding.

optics can be used to describe wave propagation. Figure 1.4 summarizes three different fiber diameters. In a multimode fiber with a large number of modes, any beam that is launched into the guiding region below the critical angle will remain trapped by total internal reflection. Two such rays are shown in the upper figure (the right part of the figures shows the intensity profiles of the rays as they exit the fiber). A multimode fiber is not well suited for high-speed data transmission as can be understood by considering two rays that are launched at different angles. Each of these rays travels a different distance, so the solid ray will arrive at the end of the fiber before the dashed ray (see Figure 1.4). The pulse widths in a data stream sent along an optical fiber and made of many different rays will thus broaden and smear as they propagate. The longer the length, the greater the broadening effect. For a fixed fiber type, its information-carrying capacity, called the bandwidth, depends on the length of the fiber. One may suggest that data sent along a single ray would not suffer the same demise. Unfortunately, some light scatters into different rays upon each reflection, so eventually many rays of nearly equal intensity are generated. For short-range applications, such as connecting computers together in an office, multimode fiber's bandwidth meets today's information transmission needs.

1.3.3 Small Core Fibers

When the fiber diameter is comparable to the wavelength of light, the diameter of the ray is comparable to the size of the waveguide and interference results from the transverse confinement. As such, the light in the waveguide will have an intensity distribution that depends on its shape. Certain intensity distributions, called modes, will propagate down the fiber without change provided that the waveguide's cross-sectional shape is constant along the fiber's length. Each of these modes is a solution of the wave equations and is analogous to the patterns of an oscillating drum head. The number of modes grows with the fiber diameter. If the fiber's cross-section varies (due to a scratch in the surface, for example), one mode can scatter into another mode. Otherwise, modes do not interact. In analogy to the ray picture, each mode propagates at a characteristic speed called the mode velocity.

The middle part of Figure 1.4 shows a fiber with only two modes. In this example, the intensity peaks at the center of the guide for the lowest-order mode. The second mode, however, has a dark region through the center. If light is launched into such a fiber with an intensity profile that matches one of the modes, it will stay in that mode. If light is launched in a way that both modes are excited, each mode will travel independently. (The cross-sectional intensity distribution at different points along the fiber for such a combination is shown above the fiber.) Because each mode travels at a different speed, though, the degree of interference changes at different points along the fiber. As such, the intensity profile looks like a bright spot at each cross-sectional slice. So, it appears to behave like a ray, except that the profile shape oscillates as the light travels in the fiber. In a highly multimode fiber, the number of modes is so large that any allowed ray (i.e., transverse intensity profile) can be constructed from the sum of many modes. In contrast, in the two-mode fiber, only one effective launch angle is supported. Section 2.1.2.1 gives a more rigorous treatment of how the ray picture and mode pictures are reconciled.

A single-mode fiber has such a small diameter that only one mode can be supported. As such, there is no modal dispersion; that is, since there is only one ray (through the middle of the fiber) a signal remains intact. In practice, material dispersion limits the bandwidth of a single mode fiber. Dispersion will be described in more detail in Chapter 3.

The road to single-mode fibers was paved by Snitzer in the early 1960s with a theoretical description of modes in a cylindrical waveguide,[7] with a simultaneously published experimental paper that showed photographs of the modes.[8] Both the theoretical and experimental work focused on cladded fibers (with refractive index 1.52) whose waveguide cores (with refractive index 1.56) were comparable to a wavelength of visible light.

The fibers were made with a draw process, where a larger version of the cylinder (called a preform) is heated to the softening point of the glass, and pulled along its axis to make the fiber. This is similar to the process that is still in use. The cladded fibers so pulled were epoxied into a large glass tube to allow the ends to be polished. A monochromator and carbon arc source were used as a bright tunable light source that was imaged onto the end of the fiber. Individual modes were found to be excited by adjusting the angle of the incident beam relative to the fiber axis. The fact that the angle of incidence determines what mode is excited is intuitively clear when comparing the mode velocity to the axial component of the velocity of a ray.

Figure 1.5 shows a sampling of modes that were measured by Snitzer and Osterberg. In this example, the images correspond to the modes $HE_{1,m+1}$, $EH_{1,m}$, and $HE_{3,m}$ for $m = 1$ and their combinations (Figure 1.5 a, b, and c) and $m = 2$ (Figure 1.5 d, e, and f). In particular, Figure 1.5a is the $HE_{1,2}$ mode, Figure 1.5b is a mixture of the $EH_{1,1}$ and $HE_{3,1}$ modes, and Figure 1.5c is a mixture of the $HE_{1,2}$ with either the $EH_{1,1}$ mode or the $EH_{3,1}$. The modes observed are consistent with the theory.

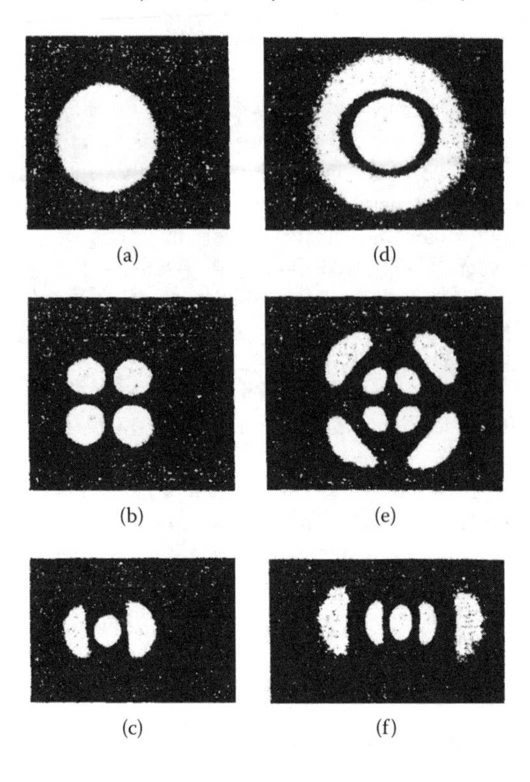

FIGURE 1.5 Examples of modes observed by Snitzer and Osterberg. Adapted from Ref. [8] with permission of the Optical Society of America.

1.4 POLYMER FIBERS

1.4.1 EARLY HISTORY OF POLYMER FIBERS

While glass fibers are used extensively in long-haul telecommunications, they are not without problems. [9] In the field, fiber optic cables need to be flexible so that they can be laid to follow the terrain. Because glass is inherently rigid, a diameter of about $125\mu m$ is needed to make the fiber flexible. It is ironic that glass has a very high tensile strength, yet is very brittle. At these small diameters, glass fibers are difficult to handle and are prone to impact and abrasion damage. In addition, the propensity for their surface to develop Griffith [10] cracks, which lead to fracture, requires that their surface be coated with an elastic polymer. To protect the fragile single mode glass fiber, a multilayer cable is used as shown in Figure 1.6.

Polymers are more flexible and less brittle than glass. As such, it is straightforward to make a large-diameter polymer optical fiber with only minimal protective layers. If a large-diameter core is used, it is also much simpler to couple light into such a fiber. In contrast, connecting single-mode fibers together requires precise positioning and adds to product cost. The difficulty in coupling light into a small fiber core, aside from its size, lies in the angle at which the light needs to be

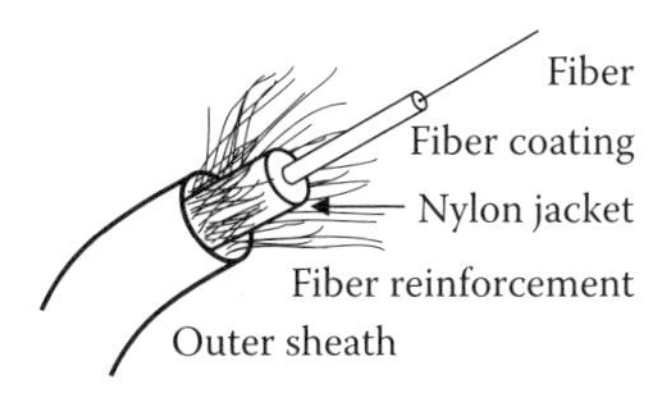

FIGURE 1.6 A single-mode fiber cable.

launched into the fiber. Figure 1.7 shows a ray in a large-core multimode fiber that leads to waveguiding at the critical angle, θ_C. If the incident angle is greater than θ_A, the light will not guide. The numerical aperture is defined by the sine of the largest incident angle that will support total internal reflection at the cladding core interface, or

$$NA = \sin\theta_A. \tag{1.1}$$

Using Snell's Law for the ray entering the core ($\sin\theta_A = n_1 \sin(90 - \theta_C)$), and the condition for the critical angle ($n_1 \sin\theta_C = n_2$), we can solve Equation (1.1) for the refractive index in terms of the refractive indices of the core and cladding,

$$NA = \left(n_1^2 - n_2^2\right)^{1/2}. \tag{1.2}$$

In small-core single-mode fibers, the refractive index difference is small so the numerical aperture is small, requiring very precise launch conditions.

In applications where high bandwidth is not required, or, if fiber is needed to connect devices that are in close proximity, multimode polymer optical fiber is ideal. Polymer fibers are therefore being used for optical interconnection of consumer electronics and are being considered for fiber-to-the-home applications.

By virtue of polymers' low mass density and robustness, polymer fibers have an advantage over glass in applications that require lightweight components or that have space limitations. Indeed, polymers are used as light plates for illuminated displays in commercial aircraft. Polymer fibers are also being considered for placard illumination on planes, and for optical interconnects between electronic components in automobiles.

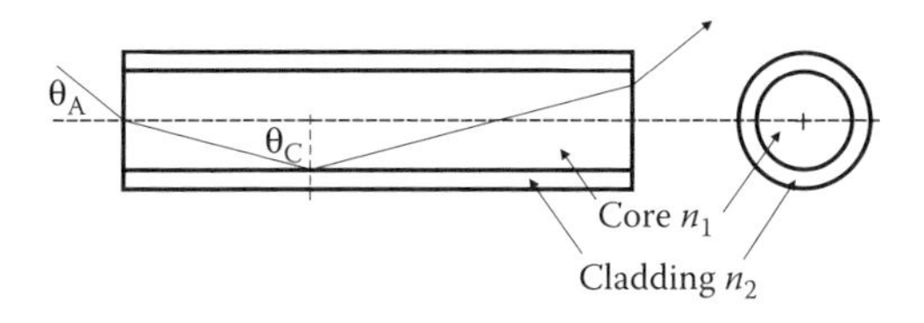

FIGURE 1.7 The numerical aperture of a fiber.

1.4.2 Multimode Fibers

Multimode polymer fibers were first commercially developed in the 1970s by duPont in the United States and Mitsubishi Rayon in Japan. Dupont patented and commercialized a series of fibers under the trade name Crofon^{TM} [11] but has since slowly withdrawn from the market. Mitsubishi, on the other hand, developed a fiber under the trade name Eska,^{TM} which has a poly(methyl methacrylate) (PMMA) core and poly(fluoroalkylmethacrylate) cladding. This lower refractive index cladding is rugged and acts as a barrier to moisture. Eska Extra is currently a product with 125 $dB\ km^{-1}$ loss at 567nm.

In the early 1980s, Kaino [12] and coworkers pioneered studies that determined the loss mechanisms in multimode fibers. This body of knowledge built upon the prior work of Oikawa. [13] Both used an extrusion process to make the core. In Kaino's method, the monomer is polymerized in-situ, extruded, and coated with the cladding in one apparatus. Oikawa, on the other hand, made the polymer separately in an ampoule and extruded it, followed by a coating with a silicon resin. Oikawa studied polystyrene core materials while Kaino concentrated on PMMA.

Figure 1.8 summarizes their results. The series of peaks in the loss spectra originate in harmonic oscillations of hydrogen atoms in the polymer chain. The most common modes are typically associated with carbon. These modes are called C-H stretch modes. While the natural frequencies of such modes are in the infrared part of the spectrum, the overtones continue throughout the visible. The Rayleigh scattering contributions, which are determined separately, are included in the figure. The hydrogen stretch modes become more intense in the infrared and the Rayleigh scattering grows as the inverse of the fourth power of the wavelength. As such, the minimum loss in most polymers is found in the visible.

The optical loss can be decreased by substituting a more massive atom in place of hydrogen, thus shifting the modes further into the infrared. Indeed, deuterated

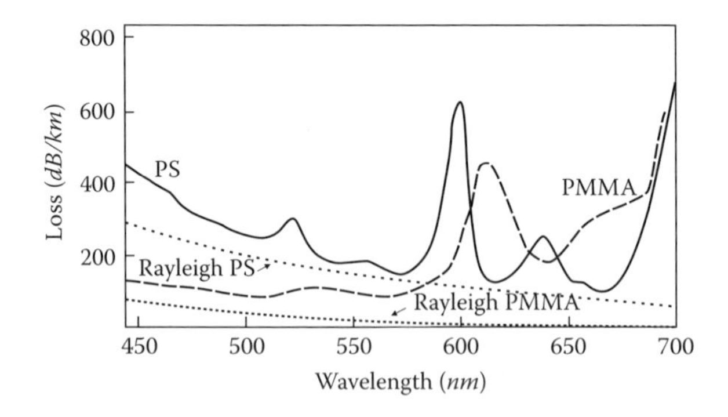

FIGURE 1.8 Measured loss and contribution of Rayleigh scattering loss for PMMA and polystyrene. Adapted from Refs. [12] and [13] with permission of the Optical Society of America.

and fluorinated polymers have a much lower measured loss in the visible. Other loss mechanisms considered by Kaino in this early work included loss due to waveguide imperfections. Typically, the lowest loss is near 650*nm*. This is the "transparency window" that is most often used in polymer fiber devices that require low loss. The issue of loss will be described in greater detail in Sections 4.2 and 4.3.

1.4.3 Gradient-Index Fibers

Gradient-index rods and spheres have been used as lenses for many years. Because any plastic rod can be heated and pulled into a fiber, a gradient-index polymer rod can be turned into a gradient-index fiber. The advantage of a graded index fiber is that it can have both a large numerical aperture (easy to couple light) and a high bandwidth. Section 2.4 shows rigorously how a parabolic refractive index profile leads to the highest bandwidth.

A more intuitive understanding of the bandwidth in a parabolic refractive index profile can be developed by considering the prorogation of rays, as seen in Figure 2.12 in Chapter 2. Because the refractive index is higher on axis, a ray that propagates along this axis will travel more slowly than a ray in the lower refractive index region. Rays that are not launched parallel to the axis will oscillate sinusoidally from the axis as they propagate down the fiber (see Figure 2.12). Even though the sinusoidal rays travel further, excursions far from the axis are in regions of lower refractive index, so the velocity component along the axis is the same as for the ray along the axis.

Bulk gradient-index polymer rods were first made by Ohtsuka and coworkers using a two-step process in which a partially polymerized rod was placed in a different polymerizing monomer. [14,15] Diffusion of the monomer into the partially polymerized rod yielded the gradient refractive index profile. These rods were shown to focus light. [16] Since then, Koike and coworkers have been making gradient-index fibers from rods and have shown that they have a high bandwidth, as expected for a parabolic profile. [17]

1.4.4 Single-Mode Polymer Fibers

Single-mode polymer optical fibers were first demonstrated by Kuzyk and coworkers in the early 1990s. [18] These fibers had dye-doped cores that were responsible for the elevated refractive index as well as potentially having a large intensity-dependent refractive index. Dyes that were used included Disperse Red 1 Azo dye, Squarylium dyes, and Pthalocyanine dyes. The core is typically about 8 μm in diameter with a 125 μm diameter cladding. Figure 1.9 shows a diagram of the fiber with the refractive index profile and single-mode intensity profile labeled below.

Many of the linear properties of these fibers, such as loss and waveguide mode, were extensively characterized. [19] Figure 1.10 shows the intensity profile

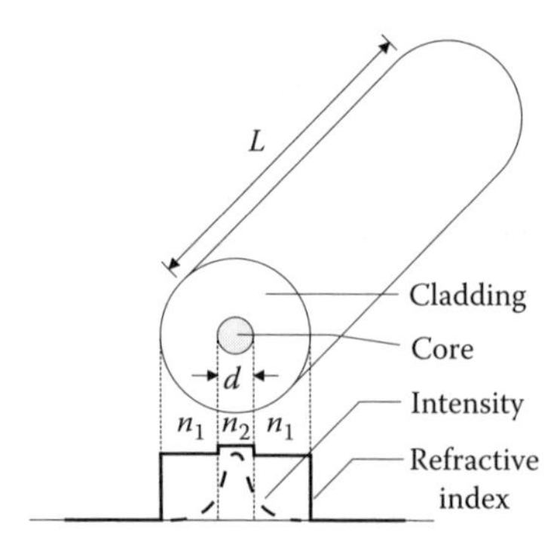

FIGURE 1.9 Cross-sectional view of a single mode. Adapted from Ref. [18] with permission of the American Institute of Physics.

in the core and the theory for a single-mode waveguide. In addition, the intensity-dependent refractive index was measured with a Sagnac interferometric technique and confirmed to be large. [20] Single-mode fibers with copolymers for both core and cladding were later reported by Bosc and Toinen. [21] In this work, the compositions of the copolymer in the core and cladding were adjusted to get a refractive index profile that yielded a single-mode guide. This is in contrast to the work of Kuzyk [18] and Garvey, [19] in which the dye itself is used to control the refractive index of the core. Note that both dye and copolymer composition can be used to control the refractive index and nonlinear-optical properties of single-mode fibers. [22]

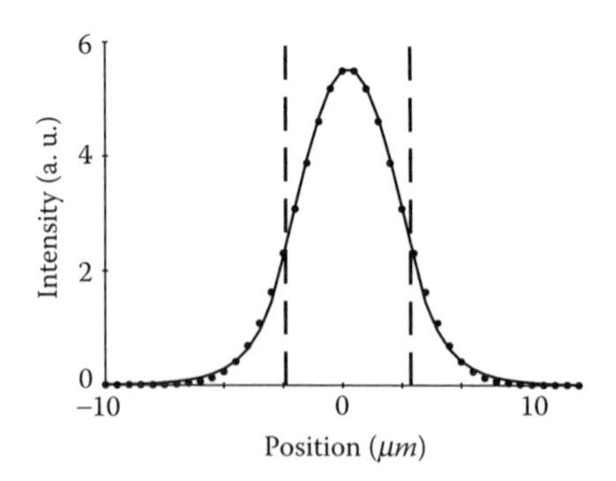

FIGURE 1.10 Light intensity in the cross-section of a fiber and theory for a single mode. Adapted from Ref. [19] with permission of the Optical Society of America.

1.4.5 Electrooptic Fiber Devices

Single-mode polymer fibers show the greatest promise for making in-line devices that are compatible with single-mode silica fiber. For example, at telecommunications wavelengths, the loss in PMMA is greater than 0.1 *dB/cm* — which is much larger than typical losses in silica. The intensity dependent refractive index, however, is much larger. The linear figure of merit, defined as the ratio of the intensity dependent refractive index divided by the optical loss, turns out to be similar for both silica and polymer. The larger nonlinear susceptibility of the polymer, however, makes 10-*cm* long devices possible. In contrast, glass devices require in excess of a kilometer of fiber to make a comparable device. Polymer fibers thus have the advantage of shorter latency times and take up much less space than glass fiber devices.

Dye-doped poled polymers were first demonstrated by Havinga and van Pelt. [23,24] In these studies, the dye chromophores are aligned with an electric field when the polymer is softened at an elevated temperature. When the material is cooled and hardened, the molecules remain aligned. Havinga and van Pelt measured the degree of alignment using linear absorption spectroscopy.

Singer and coworkers first demonstrated that poled polymers have a large second harmonic tensor [25] and electrooptic coefficient. [26] These demonstrations stirred the device community because of the advantages a polymer has over crystalline materials for making an electrooptic device. First, the PMMA polymer used in this work was well known in the electronics industries as a photoresist for making chips. Consequently, the existing know-how to make large area thin films that could be patterned into waveguides made it possible to imagine integrated active optical devices. Furthermore, because such films could be made on silicon wafers, the possibility of making compact devices that included light sources, electronic logic, detectors, and optics on the same chip was an exciting prospect. In addition to fabrication flexibility, the low dielectric constant of most polymers makes it possible to construct traveling wave devices, in which the light and electrical signal can be made to co-propagate in step with each other over longer distances.

Within two years of the demonstration of optically nonlinear poled polymers, Thakara and coworkers made the first electrooptic modulator device. [27] In this device, waveguides were defined by selective poling, where dye alignment leads to a larger refractive index for light that is polarized along the dye molecule's major axis. Today's electrooptic devices, which are made by etching techniques, operate at speeds that exceed 100 *GHz*. [28]

One disadvantage of any device that is based on waveguides with a rectangular cross-section is mode mismatch between the device and the optical fibers to which the device needs to be connected. As such, all-fiber devices are preferable. Electrooptic devices require both a waveguide and electrodes that apply an electric field. The first electrooptic fiber was reported by Welker and coworkers. [29] Figure 1.11 shows a photograph of a preform that is drawn into an electrooptic fiber. The core material is Disperse Red 1 Azo dye and the electrodes are

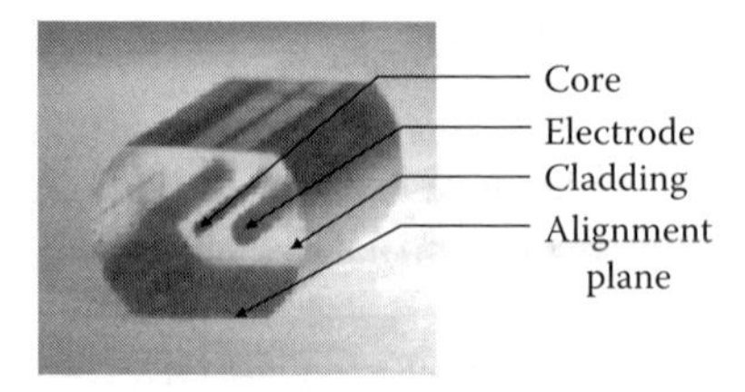

FIGURE 1.11 Electrooptic fiber preform. Adapted from Ref. [29] with permission of the Optical Society of America.

made of an indium alloy that melts near the glass transition temperature of the polymer host. Fiber made from a similar preform was demonstrated to guide and modulate light.

1.4.6 Fiber Amplifiers and Lasers

There are many other applications that could use polymer optical fiber. Tagaya and coworkers showed that fluorescent dyes can be embedded in a gradient-index fiber to make an optical amplifier in the visible. [30] Similarly, such fiber can be used to make a laser. [31] Since graded-index fibers are designed for local area network applications, graded-index fiber amplifiers, and laser sources would be key components in short-haul interconnect and transmission systems.

Telecommunications applications use infrared light sources at 1.3 μm and 1.55 μm. Rare earth atoms, which have emissions at the telecommunications wavelengths, are used as dopants in silica glass for amplifying and lasing applications. In a polymer, the interactions between nearby lanthanide atoms cause aggregation and fluorescence quenching. To get around this problem, the rare earth atoms are surrounded by ligands. Rare earth chelates, as they are called, can then be used as dopants in polymers to make sources and amplifiers. Kobayashi and coworkers have made an amplifier using rare earth chelates as dopants in graded-index polymer fiber. [32] Large-core graded index fiber is not compatible with telecommunications single-mode fiber, but shows the proof of principle. For telecommunications applications, single-mode polymer optical fibers will need to be made.

1.4.7 Dual-Core Couplers and Devices

Parallel waveguides form the basis of splitters, filters, and optical switches. When light is launched into a waveguide with another one nearby, energy is transferred from one guide to the other through the evanescent field. As a result, the intensity oscillates between the two guides over a period called the coupling length. The first demonstration of a dual single-mode-core polymer optical fiber was reported by

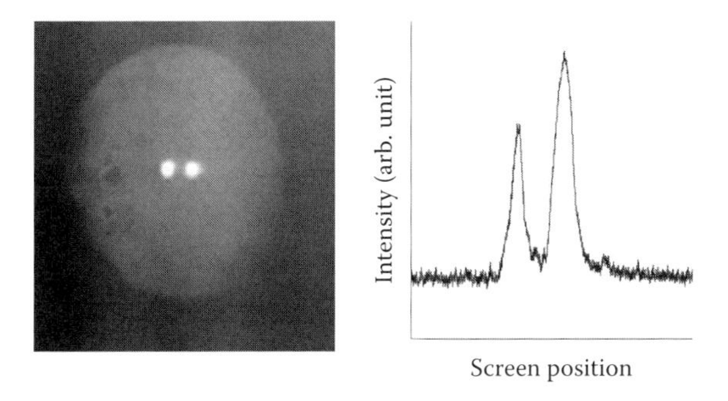

FIGURE 1.12 Dual-core polymer optical fiber. Adapted from Ref. [33] with permission of the Optical Society of America.

Vigil and coworkers. [33] Figure 1.12 shows two single-mode coupled waveguides (left) and the intensity profile (right). The core material is a squaraine dye. The waveguide was cut back, and the mode profile measured as a function of fiber length. This yields a coupling length of about 10 *mm*.

1.5 THE FUTURE

This chapter has highlighted some important developments in polymer optical fibers. One might ask what the future may hold. In a sense, optics — as far as being a technology — is still in its infancy. To get a feeling for its future potential, it is useful to contrast the history of optics with electronics, which has so radically transformed our society. The laser was invented almost a century after the electric generator. The first electrical power generator was the Niagara Falls Power Station that went into service in the summer of 1895. The first consumer product using a laser was the videodisc player released in 1978. That year, the "Digital Audio Disc Convention" was held in Tokyo, Japan, with 35 different manufacturers in attendance. Philips proposed that a worldwide standard be set; and Polygram, a division of Philips, determined that polycarbonate would be the best material for the CD. So, polymer and laser technology were married in the first consumer products using the laser.

In parallel to these developments, the first generation of telephone field trials that used silica optical fibers to transmit light at 850 *nm* using gallium-aluminum-arsenide laser diodes began in 1977. After deregulation of the telecom industry in the early 1980s, fiber optics made it into long distance telecommunication systems. This was exactly a century after electrical wires, carrying telephone conversations, were put into service: In 1877, the first telephone line in Atlanta was a private line connecting the Western and Atlantic Freight Depot with Durand's Restaurant in the Union Passenger Station. At each end of the telephone line was a box phone. At that time, many such private lines began popping up around the United States. By 1880, 30,872 Bell telephone stations were in use in the United States. Conversations were transmitted by overhead lines, and included a 45-mile stretch between Boston, Massachusetts, and Providence, Rhode Island.

Given these century gaps and assuming a linear extrapolation, one would expect the "optical transistor" to be discovered in the middle of the 21st century. Technology, however, is moving so rapidly that the wait will undoubtedly be shorter; and indeed, one might claim that a device that operates as an optical transistor has already been demonstrated. But unfortunately, expensive high power lasers are required to make it work. There is another issue, however, and that is the entrenchment of existing technologies. Even if an optical component's performance far exceeds that of its electronic counterpart, it may be difficult to build such a component into an existing complex system.

Optics will only make inroads where components can be added incrementally. Using telecommunications as an example, a long distance metal wire can be easily replaced with a fiber system that includes the fiber with a source on one end (where an electrical signal is converted to light and launched into the fiber) and a detector on the other end (which converts the light back to the electrical signal). The bandwidth of an optical fiber is so much higher than that of a wire, that it is worth the extra effort and cost to add the light source and detector. Using optics in a complex device such as an optical switching system that handles thousands of independent lines and may include hundreds of thousands of components makes it more difficult to switch components even if there are long-term performance advantages and lower costs.

Keeping these parallels in mind, it is instructive to consider the optical analogs of electrical devices. Table 1.1 shows four device classes that span all electrical devices.

The analogy between wires and optical fibers is clear. Because photons do not interact strongly with each other, information carried on a beam of light, encoded as pulses, remains intact over long distances of propagation. Because electrons interact with each other, the pulses smear over time. Light pulses, however, can also smear out over distances because of material dispersion. Since different

TABLE 1.1
Comparison of electrical and optical devices

Function	Electrical	Optical
Transmission	Metal wires	Optical fibers
	Patterned conductors	Optical waveguides
Sensors	Thermistor	Interferometer
	Stress sensor	Interferometer
Logic	Transistor	Sagnac loop
	Switch	Electrooptic switch
		All-optical switch
Actuator	Piezoelectric	Photomechanical fiber

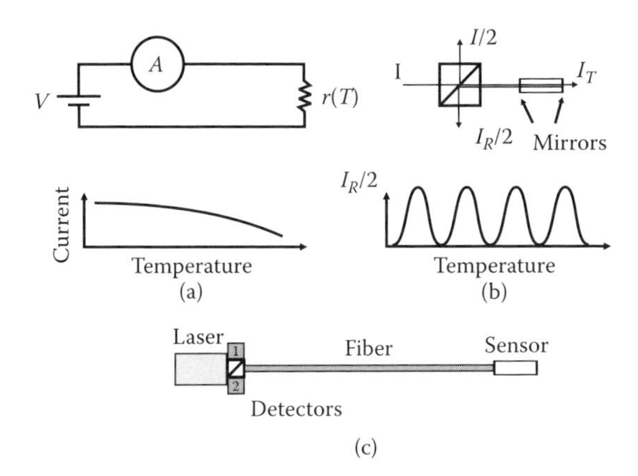

FIGURE 1.13 a) Electrical temperature sensor, b) optical temperature sensor, and c) optical fiber-based sensor.

colors travel at different speeds, and a pulse of light is made of many colors, the degree of spreading depends on the material's refractive index dispersion. Still, the information-carrying capacity of an optical fiber is much higher than of a conducting wire, especially when the fiber material is chosen carefully. Optical fiber is already becoming widespread in long distance telecommunications and local area networks. Furthermore, many consumer electronic products already include fiber connectors, making it possible to interface computers, VCRs, televisions, etc. together. In addition to providing high bandwidth, fewer fiber cables are needed, reducing the dust-laden tangle of wires normally found behind electronic components.

Sensors are devices that convert a physical property into a signal. In electrical devices, the signal is in the form of a voltage or current. For example, a common temperature sensor is made of a material whose resistance depends on temperature. Figure 1.13a shows a temperature sensor based on a thermistor, which has a monotonically decreasing temperature-dependent resistance. A constant voltage is applied and the temperature is determined from the current in the resistor as measured with an ammeter. The temperature-dependence of the resistance must be known *a priori* to calibrate the device.

Figure 1.13b shows the optical version of a temperature sensor, which is made of a transparent material with parallel semitransparent mirrors at each end. The mirrors form a cavity, and the reflected light intensity from the device is a sinusoidal function of the separation between the mirrors (this arises due to interference between the waves that exit the cavity). Changes in temperature cause the material between the mirrors to expand or contract leading to a change in reflected intensity. A beam splitter is used to sample the incident light (I) and the light reflected from the parallel-mirror sensor (I_R). The ratio I_R/I is related to the reflectance, which is a measure of the mirror separation, used to calculate the temperature. Figure 1.13c shows a fiber version of the temperature sensor. A portion of the laser source is sent to detector 1 by way of a beam splitter. The remaining light travels down the fiber, is reflected from the sensor and directed back toward the laser with the fiber. The beam splitter passes this light to detector 2. So, the fiber device works analogously to the bulk device shown in Figure 1.13b.

Such a device can be made very small and the sensor can be placed far away from the light source and detectors. As such, the sensors can be used in harsh electromagnetic environments (with the electrical detection part of the system at a safe distance away) because the light propagating in the fiber is immune to electromagnetic pickup. Another application is in areas where electrical sparks can not be tolerated, such as explosive environments.

The fact that photons don't readily interact with each other may be good for transmitting information; but, just the opposite is needed to make a logic element. In an electronic transistor, a small current applied to one terminal of the device can be used to turn on a much larger current. This same process can be used for amplification, where a small change in the weak current yields a large change in the output current of the device. The transistor is the basis of amplifiers, logic circuits, and electronic switches. A similar device can be made using optical fiber, in which one light beam can be used to control another light beam. Most of these devices are based on nonlinear interferometers, of which the Sagnac interferometer is the most common. These devices will be discussed later in this book.

An actuator is a device that transforms an input into mechanical motion. In a piezoelectric, an applied voltage causes the thickness of a material to change. The optical analog is the photomechanical effect, in which a light beam changes the shape of a material. This allows light to be converted into mechanical energy.

The optical version of each device class has been demonstrated, so in principle, every device made using electronics can be made with an analogous optical system without the need for electronics. Optical fibers and sensors are used widely and can be considered a mature technology. Optical logic and photomechanical effects, on the other hand, have been demonstrated in the lab but have not yet become high-volume commercial products.

This book discusses all the material properties and phenomena in fibers that make these various device classes possible. Chapter 2 covers light prorogation in an optical fiber, including step index, graded index, and dual-core fiber. Chapter 3 shows the details entailed in making polymer optical fibers. Chapter 4 explains how the bulk material properties are related to the microscopic material properties, while Chapter 5 discusses how the properties of the fibers and materials are measured. The remaining chapters discuss applications and devices that take advantage of the unique properties of the various types of polymer optical fibers that can be made.

2 Light Propagation in a Fiber Waveguide

2.1 INTRODUCTION

An optical waveguide is formed from an elevated refractive index that is narrow in at least one dimension and long in the direction of light propagation. The most appropriate description of light confinement in the higher refractive index medium depends on the size of the waveguide compared with the wavelength of the light in the guide. When the smallest dimension of a waveguide is much larger than the wavelength, the ray picture, or geometric optics, is appropriate. When the wavelength is comparable in size to a transverse dimension, the wave nature of the light needs to be taken into account.

Figure 2.1a shows a plane wave. The planes represent surfaces of constant phase, which are separated by a wavelength, λ. The vector that is drawn perpendicular to the phase fronts (planes) is called the wavevector, and when geometrical optics is appropriate, represents the ray. Figure 2.1b shows a ray traveling down a waveguide. In this picture, the rays reflect at the interface between the waveguide and the cladding due to total internal reflection. Figure 2.1c shows a waveguide with a diameter that is comparable to the wavelength of light. Due to the effects of diffraction, a beam cannot be focused to a spot that is smaller than the wavelength. (In contrast, geometrical optics permits light to focus to a point.) When the light couples into the guide, the intensity profile will settle into one mode or a combination of modes. A mode corresponds to a particular intensity profile. If only one mode is excited, the intensity profile will remain unchanged as the light guides down the fiber. If two or more modes are simultaneously excited, each one travels down the waveguide with its own characteristic speed. Thus, at each point along the length of the guide, the phases of each mode will be different — yielding a different pattern of constructive and destructive interference. As such, the intensity profile of light that is made up of multiple modes changes as it propagates.

A mode is stable only under ideal circumstances; i.e., for a perfect waveguide. If there are small perturbations in the diameter or the refractive index of the guiding region, energy can be exchanged between modes. So, if light is launched into one mode, given a long enough propagation length, all modes may end up getting excited. Such mode hopping is common in multimode fibers.

2.1.1 THE RAY PICTURE OF A WAVEGUIDE

Using the ray picture, we can identify several types of waveguiding. Here, we consider only cylindrical waveguides that have a cylindrically symmetric refractive index profile. While this is the most common type of fiber waveguide, this is by

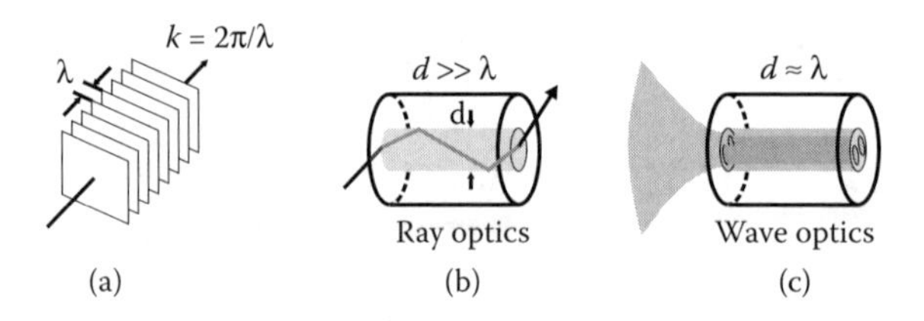

FIGURE 2.1 (a) Plane waves represented by rays; (b) ray optics; and (c) wave optics.

no means an exhaustive review of what is possible. Figure 2.2 shows a multimode, graded index, and single-mode waveguide.

In the multimode fiber, rays can propagate at any angle as long as the ray is within the critical angle. For cores that are large compared to the wavelength, the possible angles appear to form a continuum. If the waveguide size approaches the wavelength, interference effects will only allow certain angles to propagate. (The interference effects are best described in the wave picture, as discussed in Section 2.1.2.) Even in the large core case, the modes are restricted to discrete angles, though they are so close together that this "quantization" is not easily observed. Each distinct ray can be thought of as a mode. Since the modes travel at a different angle relative to the fiber axis, each one has a distinct velocity along the fiber.

Graded index fibers with large cores also support a variety of rays as shown in the top right portion of Figure 2.2. The path of the rays that are launched on axis are sinusoidal with an amplitude that is given by the impact parameter (the distance between the fiber axis and the incident ray). As such, a ray that is launched down the center of the core does not oscillate. While each of these rays are considered a distinct mode, the projection of each mode's velocity vector on the fiber axes are all the same.

Single-mode fibers are best described in the wave picture. The ray explanation, however, can be used to shed some understanding. Because the diameter of single-mode fibers is comparable to the wavelength, the possible angles of propagation

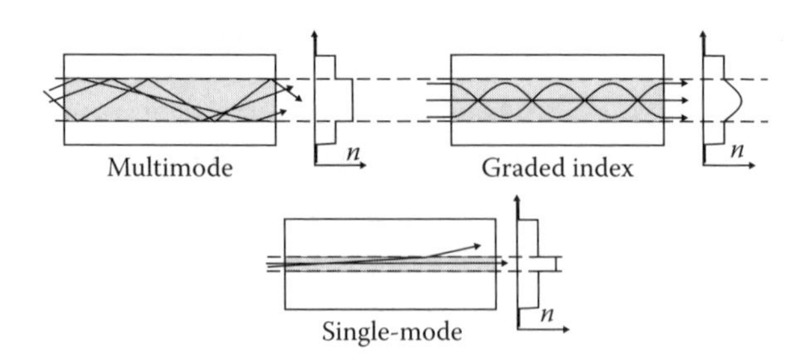

FIGURE 2.2 The rays in a multimode, graded index, and single-mode fiber. The plots to the right of each diagram represent the refractive index profile.

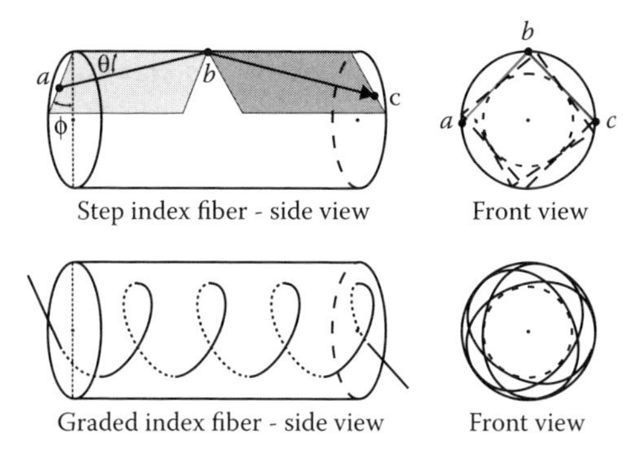

FIGURE 2.3 A ray propagating in the core of a step index (top) and graded index (bottom) fiber; shown from side view and front view. Note that in the graded index fiber, the side view shows a helical mode while the end view shows a precessing helix mode.

are quantized and well separated. The lowest order mode propagates while the second mode is beyond the critical angle, as shown in Figure 2.2. There are many other possible choices for the refractive index profile. Single-mode cores can have a parabolic refractive index profile, two closely spaced parallel cores can be placed in the same cladding, or the refractive index profile can follow an asymmetric profile. There are too many possibilities to discuss all of them in detail.

Clearly, there are many degrees of freedom for how a ray can be launched into a multimode fiber. Consider first the step index case. The top portion of Figure 2.3 shows a ray that is launched into the fiber at an angle ϕ with respect to the vertical; and at an angle θ to the surface of the fiber. In this case, the ray starts in the light gray plane, reflects from the surface of the fiber, and ends up in the dark gray plane. This same pattern repeats for each additional bounce. The angle between two consecutive planes is given by 2ϕ, so the collection of ray segments between reflections forms a helical path. The right part of the figure shows the front view (the gray segments $\overline{ab}$ and $\overline{bc}$ show the first two bounces). The set of rays circumscribes a circle, showing that such rays do not pass through the center cylindrical portion of the fiber core.

There are several special cases as follows:

- **Meridional Ray**: When $\phi = 0$, all rays remain in a plane that contains the axis of the fiber. This is called a meridional ray.
- **Skewed Ray**: When the angle is nonzero, the ray is called a skewed ray.
- **Closed and Open Skewed Rays**: For angles of the form $\phi = \pi/n$, where n is an integer, the path will be closed and the end view of the collection of rays will form a polygon of n sides. If n is a rational non-integer, the path will close after a few finite number of bounces and results in a closed skew ray. If n is irrational, the pattern is open.

The graded index fiber is shown in the bottom part of Figure 2.3. If the light is launched at an angle $\phi = 0$, the ray will be a meridional one as was the case for the step-index fiber. Such a ray will oscillate sinusoidally in a plane, as shown in Figure 2.2. If the ray and axis of the fiber do not fall in the same plane, then a skew ray will result. The propagation of such a ray is similar to the step index fiber, except the trajectories are curved. The front view illustrates how these sets of curves also circumscribe a circle. The skew rays can be open and closed. An example of a closed path is one in which the end view of the ray is a circle. In this case, the trajectory of the ray is a perfect helix. For an open path, the end view would appear as a precessing ellipsoid.

2.1.2 The Wave Picture of a Waveguide

In this section, we start from the ray picture and show how the concept of interference in a scalar wave leads to a mode profile. Then, we apply the Maxwell's wave equation to a transverse electromagnetic field in a waveguide and show the correspondence between the two approaches.

2.1.2.1 Rays and Waves

Before proceeding to the more general calculation, it is instructive to treat a simpler case that illustrates the basic concepts. We will thus consider a thin-film waveguide that confines light in only one transverse direction. Furthermore, we will assume that wave is represented by a scalar and that the two surfaces of the waveguide are perfect reflectors, independent of the angle of incidence. Consider the upper part of Figure 2.4. If the film thickness is d, the wavevector $\vec{k}$ makes an angle θ with respect to the film normal, and a plane wave is represented in complex form as

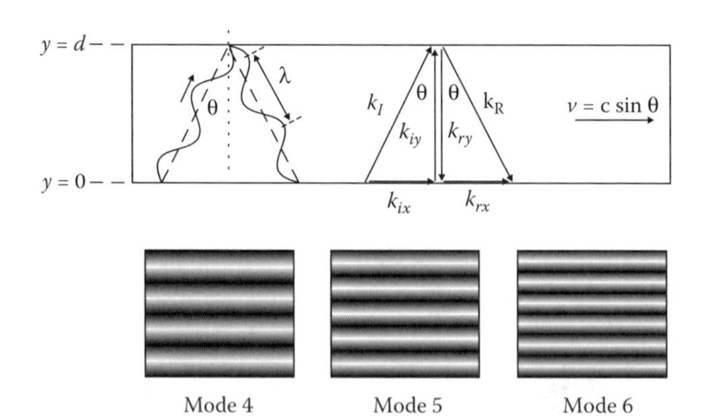

FIGURE 2.4 The upper part of the figure shows the electric field of a plane wave and its reflection at an interface. The bottom figures show the intensity profiles observed for three different modes.

$\exp[i\vec{k} \cdot \vec{r}]$, we can write the total field due to the incident and reflected rays as

$$E = \exp\left[ik\left(y\cos\theta + x\sin\theta\right)\right] - \exp\left[ik\left(-y\cos\theta + x\sin\theta\right)\right], \qquad (2.1)$$

where we have used the fact that $k_x = k\cos\theta$, $k_y = k\sin\theta$, $k = |\vec{k}_I| = |\vec{k}_R|$, $k_{ix} = k_{rx}$, and $k_{ry} = -k_{ix}$.

The intensity is proportional to the absolute magnitude of the field, so using Equation (2.1), we get

$$I \propto |E|^2 = \sin^2[ky\cos\theta]. \qquad (2.2)$$

Since the intensity must vanish at the surfaces of the waveguide, $I(d) = 0$, Equation (2.2) yields

$$\cos\theta_n = \frac{n\lambda}{2d}, \qquad (2.3)$$

where we have used $k = 2\pi/\lambda$, λ is the wavelength, and n is an integer. So, using the ray picture, and associating a plane wave with the incident and reflected waves, we find that only certain angles of propagation are allowed. Equation (2.2) shows that the solutions correspond to light and dark bands along the y direction. (The three images beneath the waveguide in Figure 2.4 show the intensity profile of three different modes.) As the angle of propagation gets larger, n gets smaller, so there are fewer nodes. The lowest-order mode ($n = 1$) has no nodes (dark bands) and mode n has $n - 1$ nodes.

Figure 2.4 shows three modes corresponding to $n = 4$ to $n = 6$. The larger angles correspond to a longer path length so the wave goes through more oscillations; thus, the energy in the mode with the smallest angle travels to the right at the lowest speed (from simple geometry, $v = c\sin\theta$, where c is the velocity of the wave in free space). As such, the lowest-order mode will correspond to the largest velocity.

Figure 2.5a shows the mode profile of the $n = 4$ mode from Figure 2.4 and the associated transverse field. The field vanishes on the boundary when the surfaces are perfectly reflecting. In a real waveguide, the modes are trapped by total internal reflection and the electric field can extend beyond the interface even though energy does not leave the waveguide. (We shall call the region beyond the waveguide

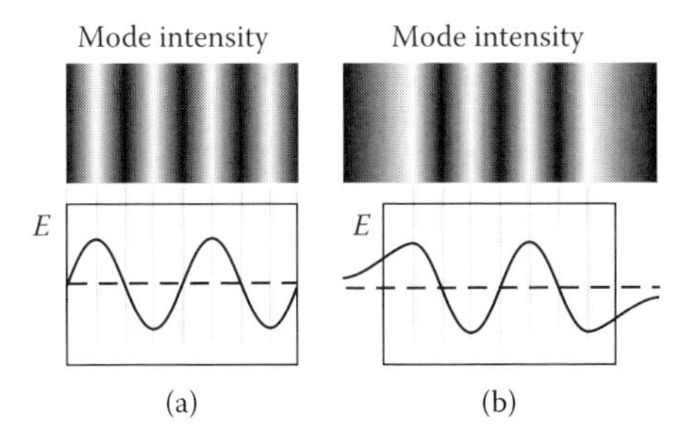

FIGURE 2.5 The lower part of the figure shows the electric field as a function of position of two modes of a planar waveguide. The top figures show the corresponding intensity profiles observed for each mode.

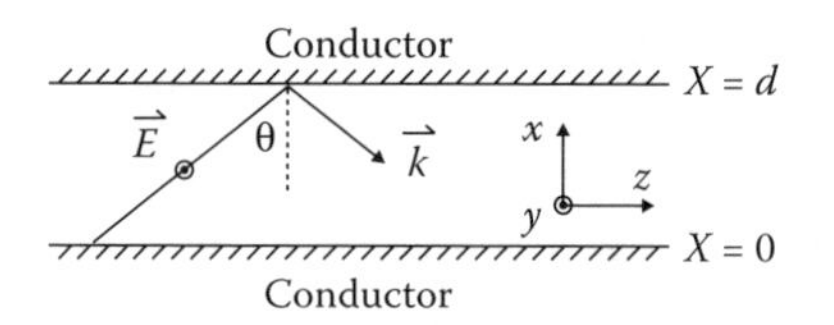

FIGURE 2.6 A waveguide made from two parallel conductors.

a cladding.) This field is called an evanescent electric field, which is shown in Figure 2.5b. As the angle gets smaller, the evanescent field penetrates further into the cladding. As such, if the cladding material has inhomogeneities that interact with the evanescent field, energy can couple out of the waveguide and radiate into the cladding.

Impurities and imperfections in the cladding material are inevitable, so higher-order (low-angle) modes tend to radiate more strongly than the lower-order ones. Once θ falls below the critical angle, most of the light will radiate out of the waveguide, aside from the small fraction of light that is reflected at each bounce due to Fresnel reflections. The Fresnel reflections, however, are so weak (typically a few percent) that most of the light exits the guide after only a few reflections.

2.1.2.2 Solving Maxwell's Wave Equation

Let's consider a waveguide made from two parallel and perfectly reflecting surfaces as shown in Figure 2.6. A beam of light is launched into the waveguide with its wavevector at an angle θ to the conductor's surface. The net energy flow is in the z-direction and the light is polarized in the y-direction. Since the light's polarization is parallel to the conducting surface, the mode is called a transverse electric (TE) mode.

The electric field obeys Maxwell's wave equation:

$$\nabla^2 \vec{E} = \frac{1}{c^2}\frac{\partial^2 \vec{E}}{\partial t^2}. \tag{2.4}$$

Since the electric field propagates along $\hat{z}$, we can guess at the form of the electric field:

$$\vec{E} = A(x)\exp\left[i\left(k_z z - \omega t\right)\right], \tag{2.5}$$

where $A(x)$ is the electric field profile. Recall that since the electric field profile of a mode does not change as the mode propagates, $A(x)$ must be independent of z.

Substituting Equation (2.5) into Equation (2.4), we get an equation for the amplitude A:

$$-k_z^2 A + \frac{\partial^2 A}{\partial x^2} = \frac{\omega^2}{c^2}A, \tag{2.6}$$

which has the solution

$$A(x) = A_0 \sin\left[\left(\sqrt{\frac{\omega^2}{c^2} - k_z^2}\right)x\right]. \tag{2.7}$$

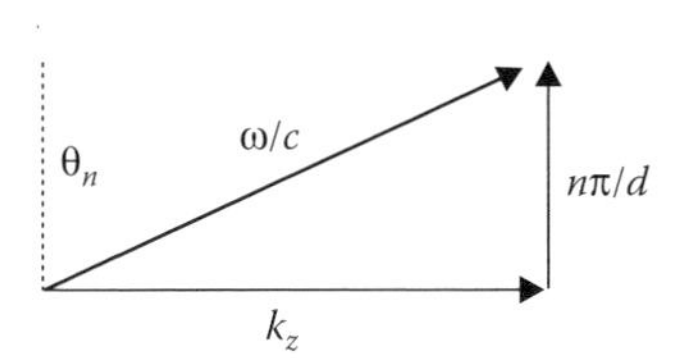

FIGURE 2.7 A graphical relationship between the wavevectors (Equation 2.9).

Note that the general solution is the sum of sines and cosines, but we have used only the sine solution because of the boundary condition that $A(0) = 0$ (the field must vanish at the surface of the conductor). The electric field must also vanish at the surface of the upper conductor, so demanding that $A(d) = 0$ yields

$$\left(\sqrt{\frac{\omega^2}{c^2} - k_z^2}\right) d = n\pi, \tag{2.8}$$

where n is an integer.

We can solve Equation (2.8) for k_z to determine which modes will propagate in the waveguide. We get

$$k_z = \sqrt{\frac{\omega^2}{c^2} - \left(\frac{n\pi}{d}\right)^2}. \tag{2.9}$$

For a propagating wave solution, k_z must be real so the quantity under the square root in Equation (2.9) must be positive. The cutoff frequency, ω_n for the nth mode, is defined by

$$\omega_n = \frac{n\pi c}{d}. \tag{2.10}$$

For the guide to support mode n, the frequency must be at least equal to the cutoff frequency, $\omega > \omega_n$. In a single-mode waveguide (one that supports only one mode), $\omega_2 > \omega > \omega_1$.

The solutions for the mode profile, given by Equation (2.7), along with the boundary conditions given by Equation (2.9), clearly reproduce the intensity profiles shown in Figures 2.4 and 2.5. To understand the correspondence between the ray and wave pictures, consider the modes of a waveguide, which are given by Equation (2.9) and graphically represented in Figure 2.7. From the graph, we get

$$\cos\theta_n = \frac{n\pi/d}{2\pi/\lambda} = \frac{n\lambda}{2d}, \tag{2.11}$$

which is the same as Equation (2.3).

2.2 BOUND MODES OF STEP-INDEX FIBERS

Now that we have shown both the ray and wave pictures of the modes of a slab waveguide, we are ready to derive and discuss the guided modes of both step- and graded-index optical fibers. Because of the cylindrical symmetry of a fiber, the modes will be Bessel functions, in contrast to the sine functions we obtained for the rectangular slab waveguide.

In the following sections, we derive in detail the full vector solution for a homogeneous step-index optical fiber. This can be viewed as either a review of or an introduction to the general solution method for the problem of bound modes of optical fibers. While the full vector solution we will derive is often supplanted by the scalar approximation for weakly guiding fibers, knowledge of it is useful in discussing, deriving, and analyzing the solutions to the scalar approximation problem. We also discuss the problem of energy transfer between two closely spaced waveguides.

The refractive index difference between the core and cladding regions of an optical fiber is what gives rise to total internal reflection, which guides light along the fiber. The confined light must satisfy the boundary conditions occurring within the fiber in order to be a bound (i.e., guided) mode of the fiber. The formal solution method for the bound modes for any type of fiber refractive index profile proceeds as with any problem in classical electromagnetic wave propagation—we solve Maxwell's equations subject to the boundary conditions imposed by the medium in which the electromagnetic field is propagating. We illustrate this by deriving the bound modes for the step-index optical fiber, which is one of the few index profiles having analytic solutions.

2.2.1 Field Relations

We start with Maxwell's equations (in Gaussian units)

$$\nabla \cdot \mathbf{D} = 0 \tag{2.12a}$$

$$\nabla \times \mathbf{E} = -\frac{1}{c}\frac{\partial \mathbf{B}}{\partial t} \tag{2.12b}$$

$$\nabla \cdot \mathbf{B} = 0 \tag{2.12c}$$

$$\nabla \times \mathbf{H} = \frac{1}{c}\frac{\partial \mathbf{D}}{\partial t}, \tag{2.12d}$$

where we have assumed there are no free charges or currents in the dielectric medium, which simplifies Equations (2.12a) and (2.12d). We also recall the constitutive relations between the medium fields (**D**, **H**) and the vacuum fields (**E**, **B**):

$$\mathbf{D} = \epsilon \mathbf{E} = n^2 \mathbf{E} \tag{2.13}$$

$$\mathbf{B} = \mu \mathbf{H}, \tag{2.14}$$

where the electric permittivity ϵ and index of refraction n are related by $n^2 = \epsilon$, and μ is the magnetic permeability of the medium. Unless otherwise noted, we assume the dielectric media are non-magnetic, in which case $\mu = 1$ and $\mathbf{B} = \mathbf{H}$.

We assume the optical fiber has a transverse refractive index profile $n = n(x, y)$ and study the propagation along the fiber's longitudinal axis, which defines the z-direction. Waveguides in which the refractive index profile does not vary along the waveguide axis direction z are *translationally invariant*. The electric and

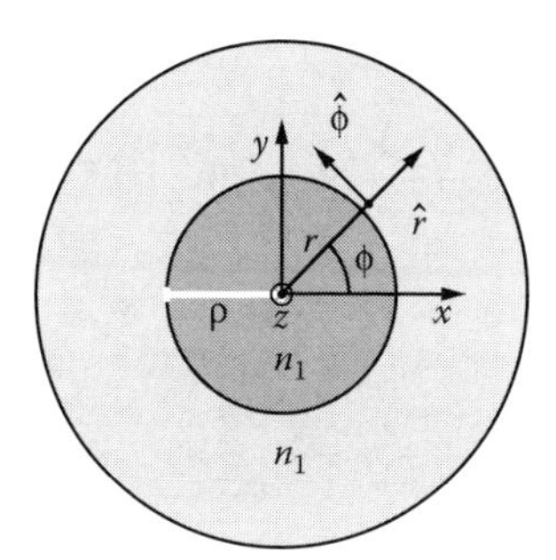

FIGURE 2.8 Transverse cross-section of a step-index optical fiber of radius ρ and the Cartesian and cylindrical-polar coordinate systems used to describe the fiber geometry.

magnetic fields in a translationally invariant waveguide can be expressed in the form [34]

$$\begin{aligned} \mathbf{E}(x, y, z) &= \mathbf{e}(x, y) \exp\left[i(\beta z - \omega t)\right]; \\ \mathbf{H}(x, y, z) &= \mathbf{h}(x, y) \exp\left[i(\beta z - \omega t)\right], \end{aligned} \tag{2.15}$$

in Cartesian coordinates, and

$$\begin{aligned} \mathbf{E}(r, \phi, z) &= \mathbf{e}(r, \phi) \exp\left[i(\beta z - \omega t)\right]; \\ \mathbf{H}(r, \phi, z) &= \mathbf{h}(r, \phi) \exp\left[i(\beta z - \omega t)\right], \end{aligned} \tag{2.16}$$

in cylindrical-polar coordinates. In Equations (2.15) and (2.16) β is the propagation constant of the bound mode, the fields are oscillating at the angular frequency ω, and the coordinate systems are oriented as in Figure 2.8.

As we will show, the transverse fields (i.e., the components of the fields orthogonal to the waveguide axis) can be written in terms of the longitudinal fields (the components of the fields parallel to the waveguide axis). In anticipation of this, we explicitly write the fields in the form

$$\mathbf{E} = (\mathbf{e}_t + e_z\hat{\mathbf{z}}) \exp\left[i(\beta z - \omega t)\right]; \mathbf{H} = (\mathbf{h}_t + h_z\hat{\mathbf{z}}) \exp\left[i(\beta z - \omega t)\right], \tag{2.17}$$

where $\mathbf{e}_t$ is the transverse electric field vector and $\hat{\mathbf{z}}$ is the unit vector parallel to the waveguide axis. We also write the gradient operator in terms of its longitudinal and transverse components:

$$\begin{aligned} \boldsymbol{\nabla} &= \boldsymbol{\nabla}_t + \boldsymbol{\nabla}_z \\ &= \boldsymbol{\nabla}_t + i\beta\hat{\mathbf{z}}, \end{aligned} \tag{2.18}$$

where the second line is a result of the special form of the fields of Equations (2.15) and (2.16).

If we substitute the fields of Equation (2.17) into the source-free Maxwell equations (2.12), we obtain the following relationships between the transverse and longitudinal field components:[1]

$$\mathbf{e}_t = -\frac{1}{k_0 n^2}\hat{\mathbf{z}} \times (\beta \mathbf{h}_t + i\nabla_t h_z) \tag{2.19a}$$

$$\mathbf{h}_t = -\frac{1}{k_0}\hat{\mathbf{z}} \times (\beta \mathbf{e}_t + i\nabla_t e_z) \tag{2.19b}$$

$$e_z = \frac{i}{k_o n^2}\hat{\mathbf{z}} \cdot \nabla_t \times \mathbf{h}_t$$

$$= \frac{i}{\beta}(\mathbf{e}_t \cdot \nabla_t \ln n^2 + \nabla_t \cdot \mathbf{e}_t) \tag{2.19c}$$

$$h_z = -\frac{i}{k_o}\hat{\mathbf{z}} \cdot \nabla_t \times \mathbf{e}_t$$

$$= \frac{i}{\beta}\nabla_t \cdot \mathbf{h}_t. \tag{2.19d}$$

Substituting Equation (2.19b) into Equation (2.19a), some vector algebra allows us to express the transverse electric field $\mathbf{e}_t$ as a function of the longitudinal electric and magnetic fields. This expression for $\mathbf{e}_t$ can be substituted into Equation (2.19b) to obtain the transverse magnetic field as a function of the longitudinal electric and magnetic fields. Applying this prescription we obtain

$$\mathbf{e}_t = \frac{i}{n^2 k_0^2 - \beta^2}(\beta \nabla_t e_z - k_0 \hat{\mathbf{z}} \times \nabla_t h_z), \tag{2.20}$$

and

$$\mathbf{h}_t = \frac{i}{n^2 k_0^2 - \beta^2}(k_0 n^2 \hat{\mathbf{z}} \times \nabla_t e_z + \beta \nabla_t h_z). \tag{2.21}$$

Equations (2.20) and (2.21) show that once the longitudinal fields are known the calculation of the transverse fields is straightforward. We also see from Equations (2.19c), (2.19d), (2.20), and (2.21) that the transverse fields can be either real and the longitudinal fields imaginary, or vice versa. We follow the convention (e.g., [35]) that the transverse fields $(\mathbf{e}_t, \mathbf{h}_t)$ are real and the longitudinal fields (e_z, h_z) are imaginary.

---

[1] The term $\nabla_t \ln n^2$ in Equation (2.19c) is calculated from Equation (2.12a), as follows:

$$\nabla \cdot [n^2(x, y)\mathbf{E}] = 0$$

$$\downarrow$$

$$\mathbf{E} \cdot \nabla n^2 + n^2 \nabla \cdot \mathbf{E} = 0$$

$$\downarrow$$

$$e_z = \frac{i}{\beta}\left(\frac{1}{n^2}\mathbf{e}_t \cdot \nabla_t n^2 + \nabla_t \cdot \mathbf{e}_t\right).$$

Using the chain rule on the first term in the last line above, we can write

$$\frac{1}{n^2}\mathbf{e}_t \cdot \nabla_t n^2 = \mathbf{e}_t \cdot \nabla_t \ln n^2.$$

2.2.2 Longitudinal Fields

We determine the longitudinal electric and magnetic fields by solving Maxwell's equation (2.12). We do this in the usual way by taking the curl of "curl" equations (2.12b) and (2.12d) to derive the vector wave equations for the electric and magnetic fields, respectively. In the case of the electric field,

$$\begin{aligned}\nabla \times \nabla \times \mathbf{E} &= -\frac{1}{c}\nabla \times \frac{\partial \mathbf{B}}{\partial t} \\ &= i k_0 \nabla \times \mathbf{B},\end{aligned} \tag{2.22}$$

where we have used the operator identity

$$\frac{\partial}{\partial t} \equiv -i\omega, \tag{2.23}$$

and the vacuum dispersion relationship

$$k_0 = \frac{\omega}{c}, \tag{2.24}$$

where k_0 is the vacuum propagation constant of the propagating field. Substituting Equation (2.12d) into Equation (2.22) yields

$$\nabla(\nabla \cdot \mathbf{E}) - \nabla^2 \mathbf{E} = n^2 k_0^2 \mathbf{E}, \tag{2.25}$$

where we have used the vector identity $\nabla \times \nabla \times \mathbf{A} = \nabla(\nabla \cdot \mathbf{A}) - \nabla^2 \mathbf{A}$ on the left-hand side of Equation (2.22), and have assumed the refractive index is time-independent. Using Equations (2.12a) and (2.13), we can solve for the first term on the left-hand side of Equation (2.25):

$$\begin{gathered}\nabla \cdot [n^2(x, y)\mathbf{E}] = 0 \\ \downarrow \\ \mathbf{E} \cdot \nabla n^2 + n^2 \nabla \cdot \mathbf{E} = 0 \\ \downarrow \\ \nabla \cdot \mathbf{E} = -\mathbf{E}_t \cdot \nabla_t \ln n^2,\end{gathered} \tag{2.26}$$

where we have used the fact that the refractive index depends only on the transverse coordinate, $n = n(x, y)$, in writing the right-hand side of Equation (2.26). We obtain the wave equation for the electric field by substituting Equation (2.26) into Equation (2.25) which yields

$$\left(\nabla^2 + n^2 k_0^2\right)\mathbf{E} = -(\nabla_t + i\beta\hat{\mathbf{z}})(\mathbf{E}_t \cdot \nabla_t \ln n^2). \tag{2.27}$$

The wave equation for the magnetic field is derived in an analogous fashion, resulting in

$$\left(\nabla^2 + n^2 k_0^2\right)\mathbf{H} = [(\nabla_t + i\beta\hat{\mathbf{z}}) \times \mathbf{H}] \times \nabla_t \ln n^2. \tag{2.28}$$

At this point, we use a property of the vector Laplacian $\boldsymbol{\nabla}^2$ that will allow us to apply it in a simple manner in Equations (2.27) and (2.28). In an arbitrary coordinate system, the vector Laplacian couples the various field components. However, if the field's vector components are referred to fixed Cartesian directions, the components are *not* coupled by the vector Laplacian. In this case, the vector Laplacian can be replaced by the scalar Laplacian ∇^2. (This can be demonstrated in a straightforward manner by expanding the identity $\boldsymbol{\nabla} \times \boldsymbol{\nabla} \times \mathbf{A} = \boldsymbol{\nabla}(\boldsymbol{\nabla} \cdot \mathbf{A}) - \boldsymbol{\nabla}^2\mathbf{A}$ in Cartesian coordinates.) While the field components are referred to fixed Cartesian directions, the spatial variation of the components are expressible in any cylindrical coordinate system.[35] Expressing the field components of Equations (2.27) and (2.28) with respect to Cartesian axes,

$$\mathbf{e}_t = e_x\hat{\mathbf{x}} + e_y\hat{\mathbf{y}} + e_z\hat{\mathbf{z}}, \tag{2.29}$$

$$\mathbf{h}_t = h_x\hat{\mathbf{x}} + h_y\hat{\mathbf{y}} + h_z\hat{\mathbf{z}}; \tag{2.30}$$

the wave equations for the transverse and longitudinal components of the electric and magnetic fields can be written

$$\left[\nabla_t^2 + n^2k_0^2 - \beta^2\right]\mathbf{e}_t = -\boldsymbol{\nabla}_t(\mathbf{e}_t \cdot \boldsymbol{\nabla}_t \ln n^2), \tag{2.31}$$

$$\left[\nabla_t^2 + n^2k_0^2 - \beta^2\right]\mathbf{h}_t = (\boldsymbol{\nabla}_t \times \mathbf{h}_t) \times \boldsymbol{\nabla}_t \ln n^2, \tag{2.32}$$

$$\left[\nabla_t^2 + n^2k_0^2 - \beta^2\right]e_z = -i\beta\mathbf{e}_t \cdot \boldsymbol{\nabla}_t \ln n^2, \tag{2.33}$$

$$\left[\nabla_t^2 + n^2k_0^2 - \beta^2\right]h_z = (\boldsymbol{\nabla}_t h_z - i\beta\mathbf{h}_t) \cdot \boldsymbol{\nabla}_t \ln n^2, \tag{2.34}$$

where the scalar transverse Laplacian ∇_t^2 is expressed in Cartesian coordinates. We can further simplify Equations (2.33) and (2.34) by substituting into them Equations (2.20) and (2.21), respectively, resulting in the coupled wave equations

$$\begin{aligned}\left[\nabla_t^2 + n^2k_0^2 - \beta^2\right]e_z = &\frac{\beta^2}{n^2k_0^2 - \beta^2}\boldsymbol{\nabla}_t e_z \cdot \boldsymbol{\nabla}_t \ln n^2 \\ &- \frac{k_0\beta}{n^2k_0^2 - \beta^2}\hat{\mathbf{z}} \cdot (\boldsymbol{\nabla}_t h_z \times \boldsymbol{\nabla}_t \ln n^2),\end{aligned} \tag{2.35}$$

$$\begin{aligned}\left[\nabla_t^2 + n^2k_0^2 - \beta^2\right]h_z = &\frac{n^2k_0^2}{n^2k_0^2 - \beta^2}\boldsymbol{\nabla}_t h_z \cdot \boldsymbol{\nabla}_t \ln n^2 \\ &+ \frac{k_0n^2\beta}{n^2k_0^2 - \beta^2}\hat{\mathbf{z}} \cdot (\boldsymbol{\nabla}_t e_z \times \boldsymbol{\nabla}_t \ln n^2),\end{aligned} \tag{2.36}$$

for the longitudinal fields e_z and h_z.

2.2.3 Bound Modes

Equations (2.35) and (2.36) represent the complete formal solution for the bound modes (by way of Equations (2.20) and (2.21)) for any translationally invariant waveguide. We now proceed to solve them for a homogeneous step-index optical fiber. The homogeneous step-index fiber is composed of a core region of radius ρ having a uniform refractive index surrounded by a cladding region whose uniform refractive index is lower than that of the core (see Figure 2.8). In an ideal step-index fiber there is a discontinuous step in the refractive index profile at the core-cladding interface. Under these conditions, our approach is to solve the wave equations (2.35) and (2.36) separately in the core and cladding, and then apply the boundary conditions of the fields at the core-cladding interface. Because $\boldsymbol{\nabla}_t \ln n^2 = 0$ within the core and cladding regions (although not at the boundary), this approach results in simplified, uncoupled wave equations for e_z and h_z of the form

$$\left[\nabla_t^2 + n_i k_0^2 - \beta^2\right] f_z = 0, \tag{2.37}$$

where n_i is the index of refraction in region i (core or cladding), and f represents either the electric or magnetic field. Because the fiber is cylindrical, we write f_z as a function of the cylindrical-polar coordinates (r, ϕ) and expand the transverse scalar Laplacian ∇_t^2 in cylindrical coordinates to explicitly express Equation (2.37) in the two regions as

$$\left(\frac{\partial^2}{\partial R^2} + \frac{1}{R}\frac{\partial}{\partial R} + \frac{1}{R^2}\frac{\partial^2}{\partial \phi^2} + U^2\right) f_z(R, \phi) = 0; \quad 0 \le R < 1 \tag{2.38a}$$

$$\left(\frac{\partial^2}{\partial R^2} + \frac{1}{R}\frac{\partial}{\partial R} + \frac{1}{R^2}\frac{\partial^2}{\partial \phi^2} - W^2\right) f_z(R, \phi) = 0; \quad 1 < R < \infty, \tag{2.38b}$$

where the normalized radius $R = r/\rho$, and we have introduced the core and cladding parameters

$$U = \rho\left(n_1^2 k_0^2 - \beta^2\right)^{1/2}, \tag{2.39}$$

$$W = \rho\left(\beta^2 - n_2^2 k_0^2\right)^{1/2}, \tag{2.40}$$

where n_1 is the core refractive index, and n_2 is the cladding refractive index.

Expressing $f_z(R, \phi)$ in the separable form

$$f_z(R, \phi) = \psi(R)\xi(\phi), \tag{2.41}$$

and substituting this into Equation (2.38) results in ordinary differential equations for the radial and azimuthal components ψ and ξ. The angular differential equation in both the core and cladding is simple harmonic, having even and odd solutions

$$\xi_\nu(\phi) = \begin{cases} C_\nu \cos(\nu\phi) \\ D_\nu \sin(\nu\phi) \end{cases}, \quad \nu = 0, 1, 2, \ldots. \tag{2.42}$$

For the radial solution $\psi(R)$, we obtain Bessel's differential equation in the core, having the general solutions

$$\psi_\nu(UR) = A_\nu J_\nu(UR) + B_\nu Y_\nu(UR), \quad \nu = 0, 1, 2 \ldots \tag{2.43}$$

where J_ν is the Bessel function of the first kind of order ν, and Y_ν is the Bessel function of the second kind of order ν. The physical requirement that the field must be finite at $R = 0$ precludes any physically meaningful solution containing Y_ν in the core. In the cladding we obtain Bessel's modified differential equation for $\psi(R)$, having general solutions

$$A_\nu I_\nu(WR) + B_\nu K_\nu(WR), \quad \nu = 0, 1, 2, \ldots, \tag{2.44}$$

where I_ν is the modified Bessel function of the first kind of order ν and K_ν is the modified Bessel function of the second kind of order ν. The physical requirement that the field vanish at $R = \infty$ precludes any physically meaningful solution containing I_ν. We summarize the general solutions for the longitudinal field components in the core and cladding regions:

$$e_z = A_\nu \frac{J_\nu(UR)}{J_\nu(U)} \begin{Bmatrix} \cos(\nu\phi) \\ \sin(\nu\phi) \end{Bmatrix}; \quad h_z = B_\nu \frac{J_\nu(UR)}{J_\nu(U)} \begin{Bmatrix} -\sin(\nu\phi) \\ \cos(\nu\phi) \end{Bmatrix}, \quad 0 \le R < 1, \tag{2.45a}$$

$$e_z = A_\nu \frac{K_\nu(WR)}{K_\nu(W)} \begin{Bmatrix} \cos(\nu\phi) \\ \sin(\nu\phi) \end{Bmatrix}; \quad h_z = B_\nu \frac{K_\nu(WR)}{K_\nu(W)} \begin{Bmatrix} -\sin(\nu\phi) \\ \cos(\nu\phi) \end{Bmatrix}, \quad 1 < R < \infty, \tag{2.45b}$$

where we have written the constants and angular portions in a way that makes the eigenvalue equation calculation somewhat less cumbersome.[35]

2.2.4 HE and EH Modes

We are now in a position to apply the boundary conditions of the fields [36] to obtain the eigenvalue equation that will allow us to determine a bound mode's profile and propagation constant β. For dielectric, source-free media, all components of the relevant fields are continuous across the boundary. Substituting the fields from Equation (2.45) into Equations (2.20) and (2.21) yields the explicit forms of the transverse fields. We describe how to derive the eigenvalue equation from the boundary conditions and leave the algebraic details to the reader.

Initially, all equations contain the unknown constants A_ν and B_ν. The boundary condition equation for one of the transverse field components, for example e_ϕ, can be used to solve for one of these constants in terms of the other. The result can be substituted into any of the three remaining transverse field boundary condition equations, for example the h_ϕ, to obtain the eigenvalue equation

$$\left[\frac{J_\nu'(U)}{UJ_\nu(U)} + \frac{K_\nu'(W)}{WK_\nu(W)}\right]\left[\frac{J_\nu'(U)}{UJ_\nu(U)} + \frac{n_2^2}{n_1^2}\frac{K_\nu'(W)}{WK_\nu(W)}\right] = \left(\frac{\nu\beta}{k_0 n_1}\right)^2 \left(\frac{V}{UW}\right)^4, \tag{2.46}$$

for the bound-mode propagation constant β_ν, where the fiber parameter

$$\begin{aligned} V &= (U^2 + W^2)^{1/2} \\ &= k_0\rho\left(n_1^2 - n_2^2\right)^{1/2}. \end{aligned} \tag{2.47}$$

Equation (2.46) is the eigenvalue equation for the so-called $HE_{\nu m}$ and $EH_{\nu m}$ bound modes (sometimes called hybrid modes).[7] The electric and magnetic fields of these modes contain both transverse and longitudinal components. A mode is designated HE or EH depending on its relative amounts of e_z and h_z.[7] We can see that Equation (2.46) results in *two* eigenvalue equations by letting

$$\xi \equiv \frac{J'_\nu(U)}{UJ_\nu(U)}, \tag{2.48}$$

in Equation (2.46) and solving the resulting quadratic equation to obtain

$$\xi(U) = -\frac{\left(n_1^2+n_2^2\right)}{2n_1^2}\frac{K'_\nu(W)}{WK_\nu(W)} \pm \frac{1}{2}\left\{\left[\frac{K'_\nu(W)}{WK_\nu(W)}\frac{\left(n_2^2-n_1^2\right)}{n_1^2}\right]^2 + 4\left(\frac{\nu\beta}{k_0 n_1}\right)^2\left(\frac{V}{UW}\right)^4\right\}^{1/2}, \tag{2.49}$$

where Equations (2.39) and (2.47) can be used to write (2.49) solely in terms of the indices of refraction, U, and V. The upper root of Equation (2.49) corresponds to the *HE* bound modes and the lower root gives the *EH* bound modes. The subscript indices ν and m indicate the order of the Bessel function and the root number (1,2,3,...) of one of the eigenvalue equations of Equation (2.49), respectively. *The most important mode is the HE_{11} mode, which is the only bound mode having no cutoff and hence is the only mode that propagates in single-mode fibers.* Equation (2.49) is transcendental and must be solved numerically. The *HE* and *EH* curves in Figure 2.9 show the numerical solution of Equation (2.49) and indicate the cutoff values (i.e., the values below which a mode will not propagate as a bound mode) for the first few *HE* and *EH* modes.

It is worthwhile to explicitly determine the relationship between refractive index and core diameter for a single-mode waveguide. Graphically from Figure 2.9

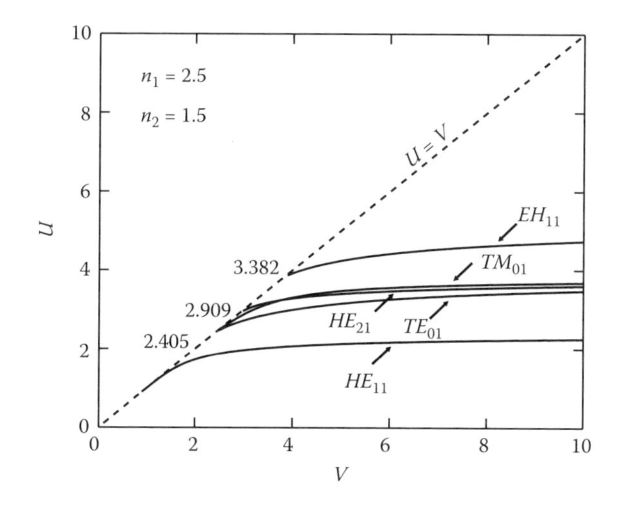

FIGURE 2.9 Numerical solution of the eigenvalue Equations (2.49), (2.57), and (2.58) for the first five bound modes of a step-index optical fiber.

we see that the single-mode condition is $V < 2.405$. In terms of the refractive index of the core, n_1, and the cladding, n_2, this condition is

$$\boxed{\pi \frac{d}{\lambda_0}\sqrt{n_1^2 - n_2^2} < 2.405,} \tag{2.50}$$

where d is the diameter of the core.

2.2.5 TE and TM Modes

Cylindrically symmetric index profiles, i.e., profiles for which $n = n(r)$, permit longitudinal field solutions having no angular dependence. In this case, the transverse Laplacian operator in cylindrical coordinates can be written

$$\nabla_t = \hat{\mathbf{r}}\frac{\partial}{\partial r}, \tag{2.51}$$

and Equations (2.20) and (2.21) yield the simple expressions

$$e_r = i\left(\frac{\beta}{n^2k_0^2 - \beta^2}\right)\frac{\partial e_z}{\partial r}, \tag{2.52}$$

$$e_\phi = -i\left(\frac{k_0}{n^2k_0^2 - \beta^2}\right)\frac{\partial h_z}{\partial r}, \tag{2.53}$$

$$h_r = i\left(\frac{\beta}{n^2k_0^2 - \beta^2}\right)\frac{\partial h_z}{\partial r}, \tag{2.54}$$

$$h_\phi = i\left(\frac{k_0 n^2}{n^2k_0^2 - \beta^2}\right)\frac{\partial e_z}{\partial r} \tag{2.55}$$

for the transverse field components. In the case of a transverse electric (TE) field, $e_z = 0$ and the result of solving Equations (2.38a) for h_z is

$$h_z = \begin{cases} A_0\dfrac{J_0(UR)}{J_0(U)}, & 0 \le R < 1 \\[2ex] A_0\dfrac{K_0(WR)}{K_0(W)}, & 1 < R < \infty. \end{cases} \tag{2.56}$$

Substituting this result into Equations (2.53) or (2.54) yields the eigenvalue equation

$$\frac{J_1(U)}{U J_0(U)} = -\frac{K_1(W)}{W K_0(W)} \tag{2.57}$$

for the TE_{0m} modes. The bound modes resulting from a transverse magnetic (TM) field can be derived in a completely analogous fashion by setting $h_z = 0$, solving Equation (2.38) for e_z, and using Equation (2.52) or (2.55) to derive the eigenvalue equation

$$n_1^2\frac{J_1(U)}{U J_0(U)} = -n_2^2\frac{K_1(W)}{W K_0(W)} \tag{2.58}$$

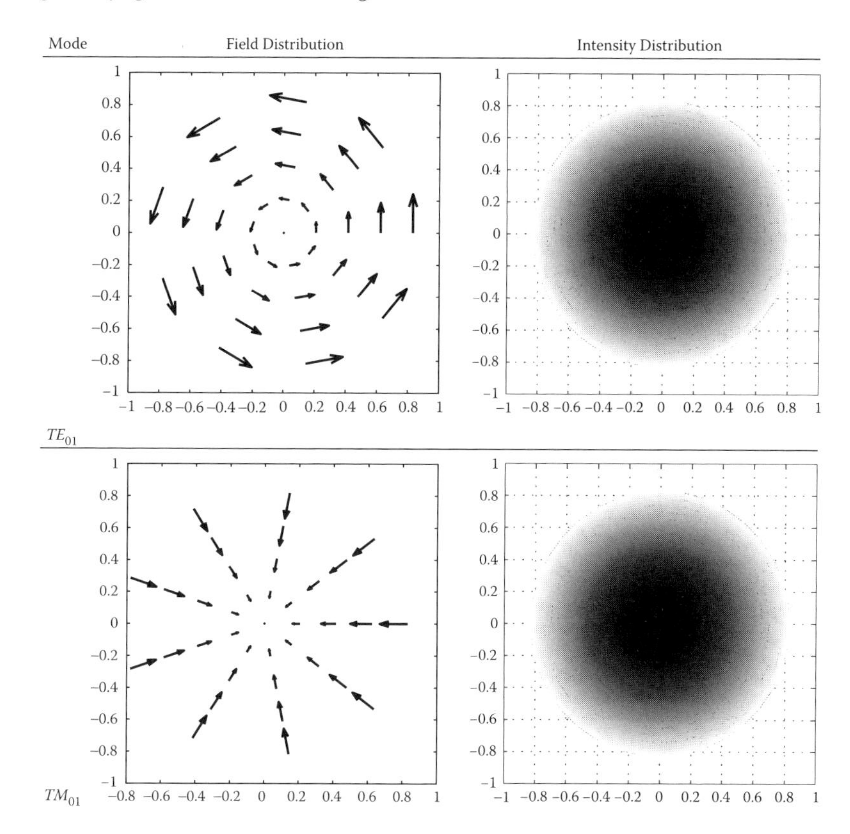

FIGURE 2.10 Transverse electric field and intensity profiles of the modes whose dispersion curves are plotted in Figure 2.9. The vector plots of the electric field are schematic representations only.

for the TM_{0m} modes. We can see that both Equations (2.57) and (2.58) are consistent with Equation (2.46) for $\nu = 0$. Like Equation (2.49), Equations (2.57) and (2.58) must also be solved graphically. Figure 2.9 includes the graphical solutions for the first two TE_{0m} and TM_{0m} modes. Figure 2.10 shows the transverse electric field and intensity profiles for the modes whose dispersion curves are plotted in Figure 2.9. In general, for fibers that are able to support more than the HE_{11} mode the observed intensity profiles can look quite different than the individual profiles of Figure 2.10 as the imperfections in real fibers generally give rise to coupling between all the modes the fiber is capable of supporting, resulting in observed modes that are linear combinations of the fundamental modes.[8]

2.3 MULTIMODE WAVEGUIDES

When the wavelength of light becomes small with respect to the dimensions of the medium in which it propagates, its propagation characteristics can be described in terms of geometrical optics, i.e., in terms of the propagation of rays.[37]

We consider a plane wave propagating in free space incident on an optical fiber. Since we are concerned with the confinement of light within the core, we consider the core to be an aperture and derive the conditions under which bounded light propagation in the core can be described by the ray description provided by geometrical optics in lieu of the modal description obtained by the direct solution of Maxwell's equations. Indeed, when a waveguide can support more than a few modes, it is difficult to observe the individual modes directly, and they become more of abstraction when discussing the measurable properties of the waveguide. If a plane wave having wavelength λ is incident on an aperture, its far-field diffraction pattern is the Fourier transform of the aperture function

$$U(x, y) = \exp(ikz)\exp\left[i\frac{k}{2z}(x^2+y^2)\right] \times \iint U(\xi, \eta)\exp\left[-i\frac{2\pi}{\lambda z}(x\xi + y\eta)\right]\mathrm{d}\xi\,\mathrm{d}\eta, \qquad (2.59)$$

where the wave is propagating in the z-direction, (x, y) are the transverse coordinates in the far field, $U(p_1, p_2)$ is the field amplitude at point (p_1, p_2), k is the propagation constant of the wave in the medium behind the aperture, and ξ and η are the coordinates measured in the aperture plane.[38]

In the case of a circular aperture having radius w, Equation (2.59) yields the diffraction pattern

$$U(r, z) = \frac{\exp(ikz)}{i\lambda z}\exp\left(i\frac{kr^2}{2z}\right)A\left[2\frac{J_1(kwr/z)}{kwr/z}\right], \qquad (2.60)$$

where $r = (x^2 + y^2)$, $A = \pi w^2$, and the aperture is located at $z = 0$. For small core-cladding refractive index differences

$$\frac{r}{z} = \tan\theta_z \approx \sin\theta_z, \qquad (2.61)$$

where θ_z is the angle the ray makes with the z-axis. Thus, rays having $\theta_z < \theta_c$ are bound by total internal reflection at the core-cladding interface. In order for most of the rays to be bound, the divergence θ_z must be much smaller than the critical angle. The number of rays that meet this condition can be described by counting the number of rays that fall within the $1/e^2$ envelope of the intensity diffraction pattern, resulting in

$$J_1(kwr'/z) = \frac{kwr'/z}{2e}. \qquad (2.62)$$

Solving Equation (2.62) graphically yields

$$k\sin\theta_d \approx \frac{2.5}{w}, \qquad (2.63)$$

where $\sin\theta_d = r'/z$ is the divergence angle of a ray. Writing the waveguide parameter V in the form

$$V = kwn\sin\theta_c, \qquad (2.64)$$

where the aperture radius w defines the waveguide radius, and θ_c is the critical angle, we see that by solving Equation (2.63) for w and substituting into Equation (2.64) that

$$V = 2.5\frac{\sin\theta_c}{\sin\theta_d}. \tag{2.65}$$

Thus, for a ray bundle for which $\theta_d \ll \theta_c$ we see that $V \gg 2.5$ is the condition for treating bound light propagation using ray tracing versus a modal description.

2.4 RAY PROPAGATION IN A GRADED-INDEX MEDIUM

In graded-index media, rays follow a trajectory that is described by the eikonal equation[39]

$$\frac{\mathrm{d}}{\mathrm{d}s}\left[n(\mathbf{r})\frac{\mathrm{d}\mathbf{r}}{\mathrm{d}s}\right] = \nabla n(\mathbf{r}), \tag{2.66}$$

where s is the distance measured along the trajectory and $\mathbf{r}$ is the position vector of a point on the trajectory, as shown in Figure 2.11. Although our main focus is on cylindrical optical fibers, our discussion will concern ray propagation in planar graded-index waveguides. This is done for simplicity, and because the result for the transit time of a ray through a waveguide having the optimum profile, i.e., one in which the difference in ray transit times is minimized, is the same for planar and cylindrical waveguides.[35] We assume a semi-infinite planar waveguide, i.e., a waveguide having one transverse dimension much larger than the other, having a translationally invariant refractive index profile $n(x)$ (Figure 2.12). Writing Equation (2.66) in terms of its components

$$\frac{\mathrm{d}}{\mathrm{d}s}\left[n(x)\frac{\mathrm{d}x}{\mathrm{d}s}\right] = \frac{\mathrm{d}n}{\mathrm{d}x} \tag{2.67a}$$

$$\frac{\mathrm{d}}{\mathrm{d}s}\left[n(x)\frac{\mathrm{d}z}{\mathrm{d}s}\right] = 0, \tag{2.67b}$$

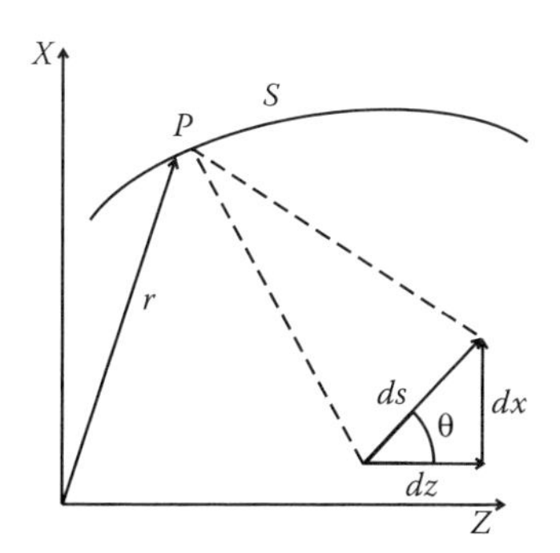

FIGURE 2.11 Ray trajectory and geometry for ray propagation in a graded-index medium.

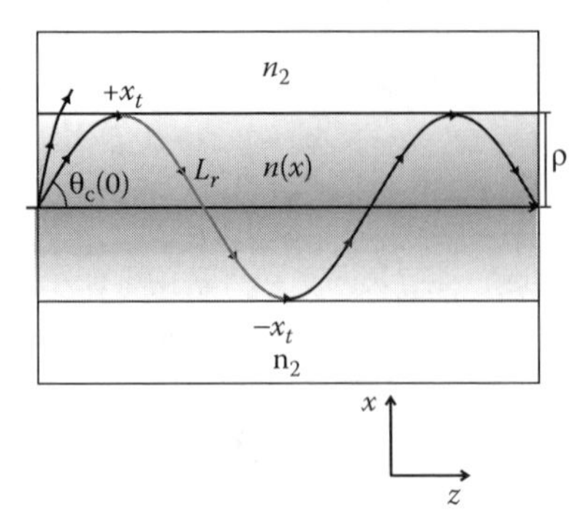

FIGURE 2.12 Ray paths in the waveguiding region of a graded-index planar waveguide having radius ρ for rays launched at greater than the critical angle, the critical angle, and parallel to the fiber axis at $x = 0$. The turning points x_t for rays launched at the critical angle are at $|x_t| = \rho$. The ray path maxima for rays launched at less than the critical angle do not reach the core-cladding interface and thus are bound to the waveguiding region. Rays launched at greater than the critical angle are not totally eliminated from the waveguiding region at the first refracting point, but if the waveguide is many half-periods long their contribution at the output is negligible. L_r denotes a half-period unit.

we see that

$$n(x)\frac{dz}{ds} = \zeta, \tag{2.68}$$

where ζ is a constant.

From Figure 2.11 we have

$$\frac{dz}{ds} = \cos\theta_z(x), \tag{2.69}$$

and combining Equations (2.68) and (2.69), we obtain the *ray invariant*

$$\begin{aligned}\zeta &= n(x)\cos\theta_z(x) \\ &= n(0)\cos\theta_z(0),\end{aligned} \tag{2.70}$$

valid for all values of x.

We note that Snell's Law is a direct consequence of the ray invariant, which must be the same on either side of an interface. Consider an interface between two media of refractive indices n_1 and n_2 and rays in these media that make an angle θ_1 and θ_2 with respect to the interface. From the ray invariant, we get $n_1 \cos\theta_1 = n_2 \cos\theta_2$. However, Snell's law is usually expressed in terms of the angle relative to the surface normal, θ' instead of the angle with respect to the interface, θ. Clearly, $\cos\theta = \sin\theta'$, so we get $n_1 \sin\theta_1' = n_2 \sin\theta_2'$ — the familiar form of Snell's Law.

The condition for bound rays can be derived from Equation (2.70). Figure 2.12 shows that as a ray propagates along a trajectory in the core, it bends according to Equation (2.70). Depending on the launching angle $\theta_z(0)$ and the rate of diminution of the refractive index, there could be a point in the core where $\theta_z(x_t) = 0$. The ray cannot propagate in the x-direction beyond the turning point x_t defined by

$$n(x_t) = n(0)\cos\theta_z(0); \quad 0 \le x_t \le \rho. \tag{2.71}$$

We can also use Equations (2.70) and (2.71) to define the relationship between the ray invariant ζ and the turning point:

$$n(x_t) = \zeta. \tag{2.72}$$

Assuming the index profile is symmetric about the waveguide axis, i.e., $n(-x) = n(x)$, a ray whose initial launching angle is $\theta_z(0)$ will proceed along a sinusoidal-like trajectory, as in Figure 2.12, bounded by the lines $\pm x_t$ if the index profile is such that there is a solution to Equation (2.71). The maximum turning point for a bound ray in a clad profile having radius ρ is at $x_t = \rho$. In this case, Equation (2.71) yields

$$\begin{aligned}\cos\theta_c(0) &= n(\rho)/n(0)\\ &= n_2/n_1,\end{aligned} \tag{2.73}$$

where the second line assumes that $n(0) = n_1$ is the maximum value of $n(x)$. Thus, the launching condition for bound rays is

$$0 \le \theta_z(0) \le \cos^{-1}(n_2/n_1). \tag{2.74}$$

2.4.1 Transit Time

For high-speed optical fiber networks it is of utmost importance to know the transit time of rays launched into the fiber at different angles. The different transit times for rays launched into the fiber at different angles is a function of the refractive index profile as it affects the path the ray takes in the fiber. This difference in transit times is referred to as *ray dispersion* (or intermodal dispersion when discussing propagation in terms of the electromagnetic modes of the fiber). Ray dispersion leads to pulse spreading, which limits the maximum transmission rate or bandwidth of the system.

The basic equation for calculating the transit time t of a ray through a waveguide is

$$\begin{aligned}t &= \frac{1}{c}\int n(x)\mathrm{d}s,\\ &= \frac{1}{c\zeta}\int_0^{z_{\text{fib}}} n^2(x)\mathrm{d}z\end{aligned} \tag{2.75}$$

where z_{fib} is the geometrical length of the waveguide, the velocity of light in the waveguide $v(x) = c/n(x)$—where c is the speed of light in vacuum—and we have used Equation (2.68) to write the bottom line of Equation (2.75).

For arbitrary waveguide lengths, the explicit form of t given by Equation (2.75) can be difficult to calculate. The transit time expression can be simplified for waveguides that are long enough to contain a large number of half-periods (there are three shown for the bound ray in Figure 2.12) as follows: In this case, the basic geometric ray path of length L_r (Figure 2.12) is repeated an integer number N times through the fiber. (Note that this analysis holds for any type of ray, i.e., meridional or skewed, provided that the fiber is cylindrical and the refractive index profile is radial. Under these conditions, the optical path length of a ray between any two successive turning points is the same.) The transit time for a ray

$$t = \frac{NL_r}{v}, \tag{2.76}$$

where $v = c/n$. The number of reflections (i.e., number of turning points) can be written in terms of the total axial distance of travel z as

$$N = n'z, \tag{2.77}$$

where n' is the number of 'reflections', i.e., the number of repetitions of the basic path, per unit length of fiber. We can write n' as

$$n' = \frac{1}{z_a}, \tag{2.78}$$

where z_a is the axial distance between reflections. The transit time can now be rewritten

$$t = \frac{L_{or}}{z_a}\frac{z}{c}, \tag{2.79}$$

where $L_{or} = nL_r$ is the optical path length of the basic ray path.

While the basic expression for the transit time looks very simple, there is work to be done in calculating L_{or} and z_a in Equation (2.79). We thus concentrate on meridional rays. (A similar calculation can also be done for skewed rays, but is more complex.) In order to calculate these quantities, we go back to Equation (2.67a) and write it in terms of x and z using Equation (2.68) to obtain

$$\begin{aligned} \zeta^2 \frac{d^2x}{dz^2} &= n(x)\frac{dn}{dx} \\ &= \frac{1}{2}\frac{dn^2}{dx}. \end{aligned} \tag{2.80}$$

Using the chain rule on the left-hand side of Equation (2.80), we can write

$$\zeta^2 \frac{dx}{dz}\frac{d}{dx}\left(\frac{dx}{dz}\right) = \frac{1}{2}\frac{dn^2}{dx}, \tag{2.81}$$

multiply both sides by dx, and integrate to obtain

$$\frac{dx}{dz} = \frac{1}{\zeta}[n^2(x) - \zeta^2]^{1/2}, \tag{2.82}$$

where the constant of integration on the right-hand side is ζ^2 since $\mathrm{d}x/\mathrm{d}z = 0$ and $n(x) = \zeta$ at x_t. Although we have no explicit need to calculate the transit time, it is simple enough to integrate Equation (2.82) to obtain the trajectory

$$z(x) = \zeta \int_0^x \frac{\mathrm{d}x'}{[n^2(x') - \zeta^2]^{1/2}}. \tag{2.83}$$

We are now in a position to calculate the optical path length of a half-period trajectory

$$\begin{aligned} L_{\mathrm{or}} &= \int n(x)\mathrm{d}s \\ &= \frac{1}{\zeta} \int n^2(x)\mathrm{d}z \\ &= \int_{-x_t}^{x_t} \frac{n^2(x)}{[n^2(x) - \zeta^2]^{1/2}}\mathrm{d}x, \end{aligned} \tag{2.84}$$

where we have used Equations (2.68) and (2.82), respectively, to write the second and third lines in Equation (2.84). We can use Equation (2.82) to solve for the axial distance the ray travels along a half-period trajectory

$$z_a = \zeta \int_{-x_t}^{x_t} \frac{\mathrm{d}x}{[n^2(x) - \zeta^2]^{1/2}}. \tag{2.85}$$

With Equations (2.71), (2.79), (2.84), and (2.85) in hand, we are able to calculate the range of transit times for various index profiles. We calculate analytic solutions for the transit time for the following profiles (see Figure 2.13):

$$\text{Homogeneous step-index:} \quad n_1(x) = n_1; \quad |x| < \rho, \tag{2.86}$$

$$\text{Ideal index profile:} \quad n_1(x) = n_1 \left[1 - 2\Delta \left(\frac{x}{\rho}\right)^q\right]^{1/2}; \quad |x| < \rho \tag{2.87}$$

$$\text{Hyperbolic secant:} \quad n_1(x) = n_1 \mathrm{sech}\left[\sqrt{2\Delta}\left(\frac{x}{\rho}\right)\right]; \quad |x| < \infty, \tag{2.88}$$

where the profile height parameter Δ is defined by[35]

$$\Delta = \frac{1}{2}\left(1 - \frac{n_2^2}{n_1^2}\right) = \frac{\bar{n}}{n_1}\left(\frac{n_1 - n_2}{n_1}\right) \approx \frac{n_1 - n_2}{n_1}, \tag{2.89}$$

where $\bar{n} = (n_1 + n_2)/2$ and the right-hand approximation is valid when the refractive index difference is small. In the process we will show that the maximum ray transit time difference—which in this discussion is defined as the difference in transit time between a ray launched at $\theta_z(0) = \theta_c(0)$ and one launched at $\theta_z(0) = 0$, i.e., the "straight-through" ray—is greatest for the homogeneous step profile, partially eliminated for the clad parabolic profile, and totally eliminated for the hyperbolic secant profile.

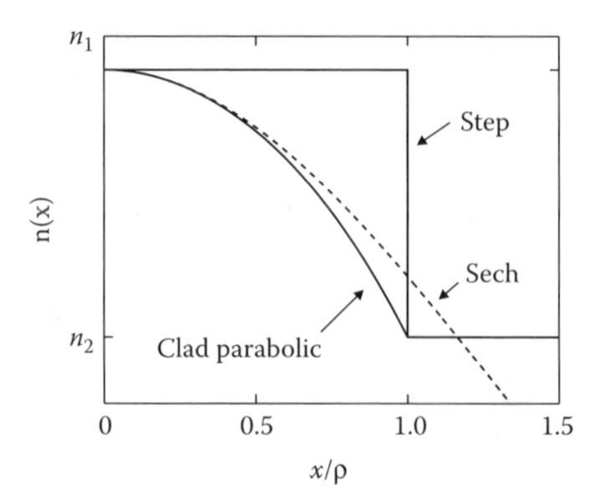

FIGURE 2.13 Step, clad parabolic, and hyperbolic secant refractive index profiles. This plot was computed with a large refractive index difference between the core and cladding ($\Delta \approx 0.2$). For $n_1 \approx n_2$, the clad parabolic and hyperbolic secant profiles are almost indistinguishable within the core region.

2.4.2 Homogeneous Step-Index Profile

2.4.2.1 Transit Time

While this profile is not graded, it is useful to derive its ray propagation characteristics in order to compare them to graded profile waveguides. In this case, the rays travel in straight lines between reflections from the core-cladding interface and the preceding formalism could be replaced by simple geometry. Since the core has no index gradient, all the bound rays have turning points at $|x| = \rho$. The integration of Equations (2.84) and (2.85) is simple, resulting in

$$L_{\text{or}} = \frac{2\rho n_1^2}{[n_1^2 - \zeta^2]^{1/2}}, \tag{2.90}$$

$$z_{\text{a}} = \frac{2\rho\zeta}{[n_1^2 - \zeta^2]^{1/2}}, \tag{2.91}$$

(2.92)

and

$$t = \frac{n_1^2}{\zeta}\frac{z}{c}. \tag{2.93}$$

Equation (2.93) confirms the intuitive notion that rays launched at larger angles (smaller ζ) with respect to the axis take longer to travel through the fiber. For bound rays launched at the maximum angle given by Equation (2.73), $\zeta = n_2$, while straight-through rays have $\zeta = n_1$. Using Equation (2.93), the maximum bound ray transit time difference for a step-index profile

$$\Delta t_{\text{step}} = \frac{n_1}{n_2}(n_1 - n_2)\frac{z}{c}, \tag{2.94}$$

which for small refractive index difference between core and cladding becomes

$$\boxed{\Delta t_{\text{step}} = n_1 \Delta \frac{z}{c}} \tag{2.95}$$

In typical graded-index profiles, where the refractive index decreases from its maximum at the center of the waveguide to the cladding value at the core-cladding interface, the maximum transit time difference will tend to be shorter than that for the homogeneous step-index time difference of Equation (2.94). Figure 2.12 shows that rays launched at an angle with respect to the fiber axis must travel a longer distance to reach the end of the fiber than rays launched parallel to the fiber axis. In the homogeneous step-index fiber the extra time is maximized because all rays travel at the same speed, since the refractive index is the same throughout the waveguiding region. In the graded profile fibers the extra time it takes rays launched at an angle to the fiber axis to travel the extra distance is partially compensated for by the decreasing refractive index away from the fiber axis, thus leading to an increased speed of travel for off-axis rays. This compensation will be shown to be exact for the hyperbolic secant profile in planar waveguides, leading to zero ray dispersion for this profile.

It is instructive to consider a numerical example of transit time difference for a step index fiber. For a fiber of length $z = 1\,km$, core refractive index $n_1 = 1.5$ and $\Delta = 0.01$ (so $n_1 \approx n_2$), we get $\Delta t = 0.05 \mu s$. Thus, the upper limit of the bandwidth of a one kilometer length of typical step-index multimode fiber is about 20 *MHz*.

2.4.2.2 Ray Picture

It is instructive to understand how the ray picture of a step index fiber arrives from Equation (2.83). For a meridional ray launched at an internal angle θ into the core of a step-index fiber of refractive index $n(x) = n_1$, the ray invariant given by Equation (2.70) is $\zeta = n_1 \cos\theta$. With this constant refractive index, the integral in Equation (2.83) simply becomes

$$z(x) = \pm x \tan(\theta). \tag{2.96}$$

Clearly, this corresponds to a ray that reflects from the fiber interface at angle θ. The behavior of the ray at the interface between the core and the cladding is governed by the ray invariant, which as we saw previously is equivalent to Snell's Law.

2.4.3 Ideal Index Profile — The Power Law Profile

2.4.3.1 Transit Time

In this case, the turning points in the core are given by

$$x_t = \pm \rho \left[\frac{n_1^2 - \zeta^2}{2 n_1^2 \Delta} \right]^{1/q}. \tag{2.97}$$

Substituting the turning points into Equation (2.85) and making the substitution $\xi = (x/x_t)^q$ leads to an integral having the form of a beta function, which leads to a result for z_a in terms of a gamma function—[35]

$$z_a = \frac{\zeta}{q}\left[\frac{2\pi}{\Delta}\right]^{1/2}\frac{x_t}{n_1}\left[\frac{\rho}{x_t}\right]^{q/2}\frac{\Gamma(1/q)}{\Gamma(1/q+1/2)}. \tag{2.98}$$

A similar substitution in the optical ray path integral of Equation (2.84) yields

$$L_{or} = \frac{z_a n_1}{q+2}\left[\frac{n_1}{\zeta}q + \frac{2\zeta}{n_1}\right], \tag{2.99}$$

from which we derive the transit time

$$t = \frac{z}{c}\frac{n_1}{q+2}\left[\frac{n_1}{\zeta}q + \frac{2\zeta}{n_1}\right]. \tag{2.100}$$

We can use Equation (2.100) to determine the exponent q that minimizes the transit time difference for rays launched at $x = 0$. The transit time is fastest for the ray launched with $\theta_z(0) = 0$ and slowest for the ray launched at $\theta_z(0) = \theta_c(0)$. The ray invariants ζ for these two rays are $\zeta = n_1$ and $\zeta = n_2$, respectively. Substituting these ζ values into Equation (2.100) yields

$$t = \begin{cases} \dfrac{zn_1}{c}; & \theta_z(0) = 0, \\ \dfrac{zn_1}{c(q+2)}\left[\dfrac{n_1}{n_2}q + \dfrac{2n_2}{n_1}\right]; & \theta_z(0) = \theta_c(0), \end{cases} \tag{2.101}$$

where the first result in Equation (2.101) is reassuring, since it is the transit time for a straight-through ray traveling along the axis. From Equation (2.101) the maximum transit time difference

$$\begin{aligned} \Delta t_{\text{parabolic}} &= t_{\max} - t_{\min} \\ &= \frac{(n_1 - n_2)}{n_2}\left[\frac{qn_1 - 2n_2}{(q+2)}\right]\frac{z}{c}. \end{aligned} \tag{2.102}$$

The denominator is positive-definite, but the numerator yields physically meaningful solutions of $\Delta t \geq 0$ only if

$$q \geq 2\frac{n_2}{n_1}. \tag{2.103}$$

Thus, for the very common case where $n_1 \approx n_2$, the transit time difference is minimized for $q \approx 2$, the clad parabolic profile.

We can use Equations (2.94) and (2.102) to calculate the factor by which the maximum transit times for the homogeneous step and clad parabolic profiles differ:

$$\Delta t_{\text{step}}/\Delta t_{\text{parabolic}} = \frac{2}{1 - \frac{n_2}{n_1}}, \tag{2.104}$$

where we have set $q = 2$ in Equation (2.102). As $n_1 \rightarrow n_2$, while the transit time difference decreases for both profiles, comparatively speaking, the clad parabolic profile exhibits much less ray dispersion. For example, for a maximum refractive index difference between the core and cladding of 1%, which is actually a fairly large value, $\Delta t_{\text{step}}/\Delta t_{\text{parabolic}} = 200$. This example indicates why graded-index waveguides are more desirable than homogeneous step-index waveguides in many applications.

2.4.3.2 Ray Picture

It is instructive to understand how the ray picture of a parabolic-index fiber arrives from Equation (2.83). For a meridional ray launched at an internal angle θ into the center of a parabolic-index fiber of refractive index

$$n(x) = n_0 \left[1 - \left(\frac{x}{a}\right)^2\right]^{1/2}, \tag{2.105}$$

the ray invariant given by Equation (2.70) is $\zeta = n_0 \cos\theta$. With this refractive index function, the integral in Equation (2.83), when solved for $x(z)$ yields

$$x(z) = a \sin\theta \cdot \sin\left(\frac{z}{a\cos\theta}\right). \tag{2.106}$$

This corresponds to a ray that meanders sinusoidally from the fiber axis as it travels down the fiber. A ray that is launched along the fiber axis ($\theta = 0$) does not oscillate, as expected. As the angle of incidence is increased, the amplitude of the ray's oscillation increases as $a \sin\theta$ and the period decreases as $z_0 = 2\pi a \cos\theta$.

2.4.4 Hyperbolic Secant Profile

Because of its infinite extent, all rays launched into a waveguide having a hyperbolic secant index profile are bound rays. The turning points for the hyperbolic secant profile are

$$x_t = \pm\rho(2\Delta)^{1/2}\cosh^{-1}\left(\frac{n_1}{\zeta}\right), \tag{2.107}$$

where $x_t \rightarrow \pm\infty$ as $\zeta \rightarrow 0$. Substituting Equation (2.88) into Equation (2.85) results in the integral

$$z_a = \zeta \int_{-x_t}^{x_t} \frac{\cosh(ax)}{\left(n_1^2 - \zeta^2\cosh^2(ax)\right)^{1/2}}, \tag{2.108}$$

where $a = \sqrt{2\Delta}/\rho$. The substitution

$$\text{sinh}ax = \left(\frac{n_1^2}{\zeta^2} - 1\right)^{1/2} \sin y \tag{2.109}$$

results in the reduction of Equation (2.108) to a simple trigonometric form. After performing a fair amount of algebra to evaluate the result at $\pm x_t$ we obtain the surprisingly simple form for the half-period path length

$$z_a = \frac{\pi \rho}{\sqrt{2\Delta}}, \tag{2.110}$$

which is independent of ζ, indicating it is independent of launching angle. Using the substitution of Equation (2.109) in Equation (2.84) yields

$$L_{or} = n_1 z_a \tag{2.111}$$

for the ray's optical path length. From Equations (2.110), (2.111), and (2.79) we calculate the ray transit time

$$t = \frac{n_1}{c} z, \tag{2.112}$$

which is independent of launching angle.

In the preceding sections, we have derived the ray transit times for a few index profiles in planar waveguides. In fibers, the transit time calculation is complicated by the necessary addition of a second dimension in the transverse direction, which can be avoided in the planar waveguides by making the waveguide semi-infinite in one of the transverse dimensions. Be that as it may, *the calculation of the transit time in fibers yields results identical to the planar waveguide results for the homogeneous step-index and ideal index profile (power law).*[35] The results for the hyperbolic secant profile are identical for meridional rays, i.e., rays that pass through the fiber axis along their path, but there is no analytic solution for skew rays, i.e., rays that do not pass through the fiber axis as they traverse their path. Thus, the bound rays in *fibers* having hyperbolic secant index profiles do *not* all have the same transit time.

2.5 DIRECTIONAL COUPLERS

Optical fiber directional couplers are among the most basic of all-optical devices. Figure 2.14 shows the simplest kind of optical fiber directional coupler—an optical fiber containing two identical waveguiding cores placed in close proximity to each other. Light launched into one core can couple into the other core resulting in a transverse transfer of light energy in addition to the usual longitudinal energy transfer. The efficiency of coupling is determined by how well the cores are matched. The switching properties of directional couplers are fixed once the refractive index profile and fiber geometry are fixed. It is easy to imagine an electrooptic directional coupler in which the refractive index of one (or both) of the cores is changed by application of an external electric field (i.e., not an optical field) in order to change the coupling characteristics of the coupler. These kinds of switches are not all-optical since there must be some electronic component providing the index-changing external field.

We can make the directional coupler into a true all-optical switch by replacing the electronic index-changing fields with an optical field whose intensity is high enough to change the refractive index (due to the intensity-dependent refractive

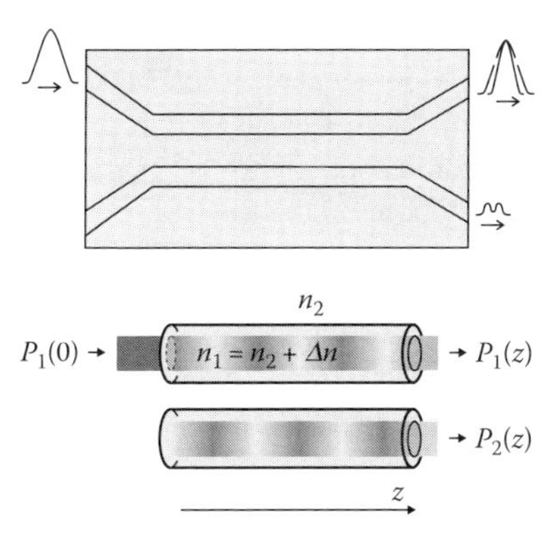

FIGURE 2.14 Schematic of a directional coupler (top) and its principle of operation (bottom). Typically, light is launched into one core and energy is exchanged between the cores. The refractive index profile and geometry of the coupler determine how much light exits from each channel of the coupler. In a nonlinear directional coupler operating via the optical Kerr effect, the light itself changes the switching characteristics of the coupler. The switching characteristics of high-intensity light differ from those of low-intensity light.

index). Using the optical Kerr effect we can switch light with light, using, for example, a strong control beam to switch a weak signal beam, or an intense beam that is able to switch itself. We begin by summarizing the theory of linear directional couplers, i.e., couplers in which the refractive index profile is fixed.

The electromagnetic modes of a coupler system characterized by a refractive index profile ϵ_c are different from those of the individual waveguides, characterized by the refractive index profile ϵ. We can derive the coupler modes from the more familiar isolated fiber modes by expanding the complicated coupler fields in terms of the complete set of fields of isolated fibers. If the modal fields of the isolated fibers are given by

$$\mathbf{E}_m(x, y, z) = [\mathbf{e}_{tm}(x, y) + e_{zm}(x, y)]e^{i\beta_m z}, \tag{2.113a}$$

$$\mathbf{H}_m(x, y, z) = [\mathbf{h}_{tm}(x, y) + h_{zm}(x, y)]e^{i\beta_m z}, \tag{2.113b}$$

where m refers to the mth guided mode, we can write the coupler fields as

$$\mathbf{E}_{ct}^* = \sum_m a_m(z)\mathbf{e}_{tm}^*(x, y), \tag{2.114a}$$

$$\mathbf{H}_{ct}^* = \sum_m a_m(z)\mathbf{h}_{tm}^*(x, y), \tag{2.114b}$$

$$\mathbf{E}_{cz}^* = \frac{\epsilon_c}{\epsilon}\sum_m a_m(z)\mathbf{e}_{zm}^*(x, y), \tag{2.114c}$$

$$\mathbf{H}_{cz}^* = \sum_m a_m(z)\mathbf{h}_{zm}^*(x, y), \tag{2.114d}$$

where Equations (2.114c) and (2.114d) are derived by using Equations (2.114a) and (2.114b) in determining the z-component of the Maxwell curl equations (2.12d) and (2.12b), respectively, for the ϵ_c system.[40] We see from Equations (2.114) that knowledge of the expansion coefficients $a_m(z)$ will result in the knowledge of the energy distribution at any point in the fiber.

The expansion coefficients $a_m(z)$ can be calculated using the conjugated form of the reciprocity theorem for Maxwell's equations.[35] The reciprocity theorem is an integral relationship between the field solutions ($\mathbf{E}$, $\mathbf{H}$) and ($\mathbf{E}'$, $\mathbf{H}'$) of Maxwell's equations for two systems having refractive index profiles characterized by ϵ and ϵ', respectively. If we define a vector function

$$\mathbf{F} = \mathbf{E}\times\mathbf{H}'^{*} + \mathbf{E}'^{*}\times\mathbf{H}, \tag{2.115}$$

then, in dielectric media,

$$\nabla\cdot\mathbf{F} = -ik[(n'^{*})^2 - n^2]\mathbf{E}\cdot\mathbf{E}'^{*}, \tag{2.116}$$

where the fields in both media have a vacuum propagation constant k. The two-dimensional form of the divergence theorem for $\mathbf{F}$ is

$$\int_A \nabla\cdot\mathbf{F}\,\mathrm{d}A = \frac{\partial}{\partial z}\int_A \mathbf{F}\cdot\hat{\mathbf{z}}\,\mathrm{d}A + \oint_l \mathbf{F}\cdot\hat{n}\,\mathrm{d}l, \tag{2.117}$$

where A is the cross-sectional area of the medium, l is the the boundary of A, $\hat{\mathbf{z}}$ is orthogonal to A and parallel to the direction of propagation, and $\hat{n}$ is the unit vector in the plane of A and which is the outward orthogonal to l. For optical waveguides we can take A to be the infinite cross-section A_∞ containing the optical fiber cross-section. In this case the line integral in Equation (2.117) is along a circle with radius $r \to \infty$ and if either or both of the fields comprising F is a guided mode of the isolated system then $\mathbf{F} \to 0$ as $r \to \infty$. Thus for bound modes the line integral makes no contribution to Equation (2.117) resulting in the conjugated form of the reciprocity theorem

$$\int_{A_\infty} \nabla\cdot\mathbf{F}\mathrm{d}A = \frac{\partial}{\partial z}\int_{A_\infty} \mathbf{F}\cdot\hat{\mathbf{z}}\mathrm{d}A. \tag{2.118}$$

Substituting the fields of Equations (2.113) and (2.114) into Equation (2.117) and using the bound mode orthogonality condition

$$\int_{A_\infty} \mathbf{e}_p\times\mathbf{h}_q^{*}\cdot\hat{\mathbf{z}} = \pm\delta_{pq}, \tag{2.119}$$

where the negative sign is for negative q, yields the coupled differential equations for the expansion coefficients

$$\frac{\mathrm{d}a_p}{\mathrm{d}z} + i\beta_p a_p = i\sum_q a_q C_{pq}, \tag{2.120}$$

where

$$C_{pq} = \gamma_p \frac{k}{2} \int_{A_\infty} \left[(n'^*)^2 - n^2) \right] \left(\mathbf{e}_{pt} \cdot \mathbf{e}_{qt}^* + \frac{(n'^*)^2}{n^2} \mathbf{e}_{pz} \cdot \mathbf{e}_{qz}^* \right) \mathrm{d}A, \tag{2.121}$$

and $\gamma_p = 1$ for $p > 0$ and -1 for $p < 0$.

The set of coupled equations embodied in Equation (2.120) quickly becomes analytically intractable if the system supports more than a few modes. For a system supporting only two modes Equation (2.120) reduces to two coupled equations that yield the following solutions for the power in each mode as a function of position

$$\left| a_p(z) \right|^2 = 1 - \left| a_q(z) \right|^2, \tag{2.122a}$$

$$\left| a_q(z) \right|^2 = F_{pq} \sin^2 \beta_b z, \tag{2.122b}$$

where

$$F_{pq} = \left[1 + \left(\frac{\beta_p - \beta_q}{2 \left| C_{pq} \right|} \right)^2 \right]^{-1} \tag{2.123}$$

is the maximum power transfer ratio between the two modes and

$$\beta_b = \frac{\left| C_{pq} \right|}{\sqrt{F_{pq}}}. \tag{2.124}$$

It should be noted that Equations (2.122a and b) are solutions of a system whose initial conditions are $a_p(0) = 1, a_q(0) = 0$. The coupling length l_c is defined as the distance over which the power in a mode goes from its maximum to its minimum, i.e.,

$$l_c = \frac{\pi}{2\beta_b}. \tag{2.125}$$

Thus far mode coupling has been discussed without reference to the details of the refractive index profile. Indeed the previous discussion is perhaps more easily understood in the context of the refractive index profile variations of an imperfect optical waveguide, i.e., a single waveguide whose modes are approximately the modes of an ideal waveguide. In this case, the mode coupling occurs between different modes within the same region of space. Some observations of the nature of this coupling will allow a direct extension to a description of coupling in an ideal linear coupler, i.e., a coupler composed of parallel ideal waveguides. It is evident from Equation (2.123) that the power exchanged between two modes p and q is appreciable only when $\left| \beta_p - \beta_q \right| \ll 2 \left| C_{pq} \right|$. If the perturbation is small in the sense that the modal fields are only slightly different from the modal fields of an ideal waveguide, then it must be that in Equation (2.120) $C_{pq} \ll \beta_p$ for all p and q, i.e., the mode amplitudes don't change very much as they propagate within the waveguide. This condition is contrary to the previously stated condition for strong coupling. Thus, strong coupling can only occur for modes in which $\beta_p \approx \beta_q$. It is reasonable to assume that for a coupler composed of closely spaced, identical waveguides the most efficient coupling occurs between identical modes propagating in the same direction. Ignoring coupling between identical modes traveling in opposite directions and modes for which there exists an "accidental" degeneracy $\beta_p = \beta_q$ the coupled equations describing mode coupling between spatially separated waveguiding regions is described by the coupled waveguide equations

$$\frac{da_p^{(j)}}{dz} + i\beta_p^{(j)} a_p^{(j)} = -i \sum_{s \neq j} a_p^{(s)} C_{pp}^{(j)(s)}, \tag{2.126}$$

where

$$C_{pp}^{(j)(l)} = \frac{k}{2} \int_{A^{(l)}} \left[n^{(l)2} - n^2 \right] \mathbf{e}_p^{(j)} \cdot \mathbf{e}_p^{(l)*} dA \tag{2.127}$$

describes the coupling between the pth mode in the jth waveguide and the pth mode in the lth waveguide.

2.5.1 Nonlinear Directional Couplers

The polarization response of many optically nonlinear media can be described by expanding the polarization as power series in the incident electric field

$$P_i = \chi_{ij}^{(1)} E_j + \chi_{(ijk)}^{2} E_j E_k + \chi_{ijkl}^{(3)} E_j E_k E_l + \cdots, \tag{2.128}$$

where $\chi^{(n)}$ is the nth-order electric susceptibility; $i, j, k = x, y, z$; the Einstein summation convention is assumed for repeated indices, and we are ignoring any static polarization (i.e., $\chi^{(0)} = 0$). A consideration of the inversion symmetry of the polarization shows that all even-order susceptibility terms vanish in centrosymmetric materials (which includes many optical fiber constructions). By assuming an intensity-dependent refractive index

$$n = n_0 + n_2 I, \tag{2.129}$$

where n_0 is the linear refractive index, and considering the polarization response of a medium to third order in the field, it is straightforward to show[41] that

$$n_2 = \frac{12\pi^2}{n_0 c^2} \chi^{(3)}, \tag{2.130}$$

where c is the speed of light. It is important to note that although a switch whose operation is due to the intensity-dependent refractive index is all-optical, the switching speed is determined by the response time of the mechanism responsible for the nonlinearity. This response can range from a thermal response in the millisecond range ($\equiv 10^{-3} s$) to electronic responses in the femtosecond range ($\equiv 10^{-15}\ s$). In order to deploy nonlinear optical devices as switches in telecommunications networks, it is necessary to use materials with "fast" nonlinearities.

As was previously mentioned, the directional coupler can become an all-optical switch if its switching characteristics can be influenced by the light propagating in it. In this case, we have a nonlinear directional coupler (NLDC), a device whose operation is determined by the coherent interaction between the closely spaced optically nonlinear waveguides. The optical nonlinearity that leads to the strongly nonlinear coupling characteristics of the device is due to the intensity-dependent refractive index[42] of the materials used in the device. Of course, the refractive index expressed in Equation (2.129) can also depend on higher-order terms, but the intensities required to excite them to contribute to NLDC effects are currently far beyond the damage thresholds of most optical fibers. The coupling characteristics of the simplest kind of NLDC, one composed of two closely spaced, optically nonlinear, identical waveguides, were first described in the early 1980s by Jensen, who derived an analytic solution for an NLDC composed of two single-mode waveguides.[43,44]

If we assume the modes of the NLDC waveguides are approximately the modes of the linear directional coupler, and substitute the intensity-dependent refractive index of Equation (2.129) into the analysis of directional couplers, we obtain an intensity-dependent term in Equation (2.126). For an NLDC composed of two identical single-mode waveguides, this results in the set of two coupled nonlinear differential equations

$$-i\frac{\mathrm{d}a_1}{\mathrm{d}z} = -(\beta - R_1|a_1|^2)a_1 - C_{12}a_2, \tag{2.131a}$$

$$-i\frac{\mathrm{d}a_2}{\mathrm{d}z} = -(\beta - R_2|a_2|^2)a_2 - C_{21}a_1, \tag{2.131b}$$

where

$$R_n = \frac{n_0 n_2 \omega}{\pi P_0}\int_{A_\infty} |E|^4 \mathrm{d}A, \tag{2.132}$$

and $R_1 = R_2 = R$ for identical, single-mode waveguides. Figure 2.15 shows the intensity-dependent transmission characteristics of an ideal one beat-length

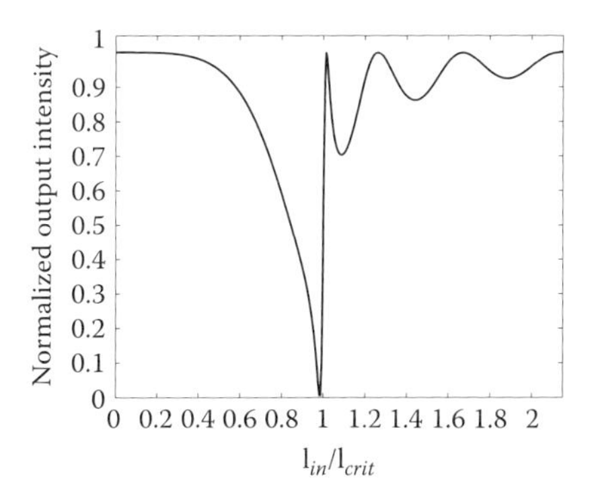

FIGURE 2.15 Intensity-dependent transmission from a one beat length long NLDC. At low intensities, the light exits from the channel into which it was launched. At the critical intensity, the coupling is such that the light exits from the channel into which no light was initially launched. For the ideal coupler, the beat length $L_b = 1/C_{11}$,[45] and the critical power $P_c = 4\pi/RL_b$.[46]

long NLDC in which all the light is initially launched into one of the cores (e.g., $a_1(0) = 1, a_2(0) = 0$).

2.6 CONCLUSION

Light propagation in a waveguide can be described in terms of fundamental modes that each travel at distinct velocities. The shape of a mode depends on the cross-sectional refractive index profile. In a step-index fiber, these modes are Bessel functions. When several modes are simultaneously excited, each mode travels independently of the others. However, because of velocity mismatch, the intensity profile can oscillate along the fiber. The calculations presented here are for ideal fibers made of homogeneous materials. In real applications, imperfections can cause modes to couple. However, over short lengths, the theory presented here is a good approximation of what is observed.

2.7 ACKNOWLEDGMENTS

I thank Steve Vigil for writing most of this chapter.

3 Fabricating Fibers

This chapter investigates two broad methods for making polymer optical fibers: the extrusion process and the drawing process. By far, the largest volume of fiber made to date is multimode step-index fiber, which is most easily mass-produced by an extrusion process. Because the core in such fiber is large, the process is forgiving to variations in the fabrication parameters.

Single-mode fiber and gradient-index fiber, though, are more difficult to make and require stricter fabrication controls. The literature is full of papers on making multimode step index fiber but gradient-index and single-mode fibers with active core materials have only appeared in the literature in the 1990s. While this is not yet a mature technology, the potential for high value-added products makes these high-bandwidth fibers attractive for future development. Potential devices include optical switching, fiber lasers, and optical amplifiers.

The first section of this chapter discusses the process for making fibers using the extrusion method, which is best suited for making multimode fibers. Following is a section that describes how fiber drawing methods can be used to make single-mode, multimode, and gradient-index fibers. In each case, it is possible to incorporate active dyes in the core and cladding of the fiber. This chapter ends with a discussion of how polymer properties, such as birefringence and rheology, are affected by the drawing process.

3.1 MAKING POLYMER FIBERS BY EXTRUSION

Perhaps the most economical method for making polymer optical fiber that is also amenable to high-volume processing is the extrusion process. Figure 3.1 shows a schematic diagram of two versions of such a process. The left part of Figure 3.1 shows a highly schematic diagram of an extrusion process in which the material is not fully polymerized, while the right figure shows a process in which the material is fully polymerized in the reaction chamber. We discuss each below.

3.1.1 Continuous Extrusion

The left part of Figure 3.1 shows a continuous extrusion process. The starting materials are fed into a reaction chamber where the monomer is polymerized. The starting materials can include the chain transfer agent, initiator, monomer, and any other substances required to make the polymer. While not shown in the figure, the monomer and initiator reservoirs can be directly fed from distillators, where the materials are purified. As such, the materials can be continuously fed into one end of the apparatus, yielding fiber at the other end.

The material is partially polymerized into a thick fluid, which is typically 80% polymer and 20% monomer. The temperature of the reactor chamber is

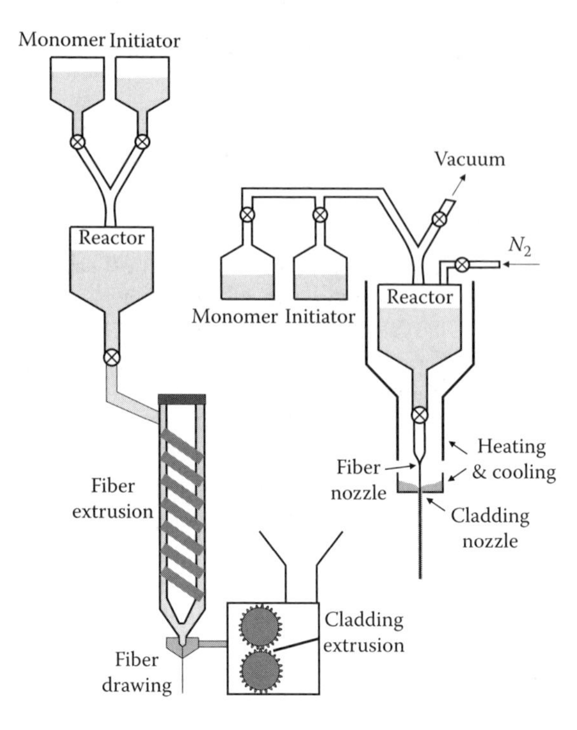

FIGURE 3.1 Extrusion apparatus. Adapted from Ref. [47] with permission of the American Institute of Physics.

typically 180°C. This concentrated solution is directed to the screw extruder, which pushes the material through a small nozzle that creates the fiber. Remaining monomer is evaporated in the core extruder and recovered for further use. Subsequently, the core of the fiber runs into a second chamber and then through a nozzle where the cladding is coated onto the fiber — also using an extrusion process. This leads to the formation of a continuous fiber. Effectively, then, material is fed into one end and the fiber comes out the other end. As such, this is often called a continuous extrusion POF fabrication process.

3.1.2 Batch Extrusion

Because the continuous process is open at the material input end, it may introduce impurities that result in higher optical loss. Closed system designs allow for the starting materials to be cleaned and remain clean during the polymerization process. The right-hand side in Figure 3.1 shows such a design.

The process starts by removing the inhibitor from the monomer by rinsing it with alkali solution. Residual alkali is rinsed away with distilled water and the monomer is dried by adding Na_2SO_4 and CaH_2. The monomer is then distilled under reduced pressure and the middle fraction is collected, which is placed in the monomer vessel. For methyl methacrylate or polystyrene monomer, the initiator azo-*tert*-butane and chain transfer agent *n*-butyl mercaptan are added to the initiator chamber. Both vessels are sealed and evacuated.

The reactor is cooled and the initiator and chain transfer agent are distilled into the reactor and the monomer is added under vacuum and low temperature to eliminate dissolved air. Subsequently, the reaction vessel is heated to 135°C for 12 hours to polymerize the monomer. The temperature is then raised slowly to 180°C and kept at that temperature for another 12 hours to fully polymerize the material.

Once the polymerization process is complete, the reactor vessel is further heated until the polymer becomes fluid. Dry nitrogen gas is used to pressurize the reactor, which pushes the material through a nozzle to form the fiber. The fiber is pulled through another nozzle where it gets coated with a cladding polymer.

In this process, the amount of fiber that can be drawn is limited to the size of the reactor vessel. The materials processing time per batch is in excess of 18 hours, so at most one batch can be made per day.

3.2 MAKING POLYMER FIBER BY DRAWING A PREFORM

This section describes how a preform rod is made and how such a rod can be pulled into a fiber. The preform drawing process is important because it can be used to make single-mode step-index fiber, multimode fiber, and gradient-index fiber. Much of this section focuses on making single-mode fibers with dye-doped cores because this poses the most technically challenging processing and fabrication problems. As such, the process for making a multimode step-index fiber from a preform is not described here because it is similar to the single-mode process but more tolerant to processing variations.

3.2.1 Making Doped Polymers

The most important part of making a single-mode polymer optical fiber is to start with a high optical quality polymer with low levels of impurities. Core materials are typically dye-doped polymers or side chain polymers. As such, the procedure used in making the material must meet stringent requirements for cleanliness throughout the process. The fiber drawing process requires the materials to be free from any remaining volatiles used in the materials synthesis process, impurities, or aggregates of the chromophores used as dopants.

The fiber drawing process relies on the smooth flow of softened polymer. Since thermoplastics are polymers that flow at an elevated temperature, they are well suited as fiber materials. Having a thermoplastic in hand, however, does not ensure that a fiber will draw. An important property of the polymer is its molecular weight, which is a measure of the average chain length. If the molecular weight is too low, the polymer turns into a liquid during drawing that, due to surface tension, breaks apart into drops. If the molecular weight is too high, the chains are too long to permit them to slip and flow. The result is that the polymer remains rubbery even at temperatures that are high enough to cause the material to decompose. The ideal molecular weight allows the material to flow without dripping. The balance is quite delicate and requires the molecular weight to be carefully tuned to the fiber drawing process. In making the polymer, a chain transfer agent is used to control the molecular weight.

The following sections describe the materials fabrication and processing procedure for making a preform, which is drawn into a fiber. In single-mode fiber fabrication, drawing seems to give the best results while long lengths of multimode fibers can be reliably pulled from an extrusion process.

3.2.1.1 Dissolving Polymers

A polymer can be doped with a dye by dissolving the polymer and dye together in a solvent followed by removal of the solvent with evaporation. Indeed, this is a typical process for making thin films using spin coating techniques. While this method results in problems when attempting to make a high-quality material for a fiber preform, we discuss the process briefly and describe the trade-offs.

The polymer and dye, which together are referred to as solids, are dissolved in a solvent. The ratio of polymer to dye chromophores is chosen in proportions that are required in the final solid film. In many common chromophores, such as the azo dyes, the dopant concentration used for nonlinear-optical characterization experiments is about 10 % by weight. At this concentration, the chromophores are evenly dispersed in the polymer matrix. For chromophores such as Disperse Red 1 Azo dye, above 10 weight percent, the dye chromophores begin to aggregate.[48] The degree of aggregation depends on processing conditions and material properties such as concentration, temperature, and rate of evaporation.

To cast a high-quality film, the solution must wet the substrate thoroughly. When spinning the film about its surface normal, a large amount of the material is thrown off, leaving behind a thin film. For spin speeds of about 1,000 to 2,000 revolutions per minute (RPM), a solution with 15% solids by weight yields a film thickness between 1.0 and 3.0μm. Because these films are so thin, the solvents can be baked away in about 20 minutes at a temperature just below the boiling point of the solvent. In fact, just after the 20-second spinning cycle, much of the solvent is gone, leaving a tacky film. For this reason, it is important to make doped polymers in a clean environment to prevent dust from sticking to the surface.

Making a good-quality film often requires a combination of solvents to be used. For example, a mixture of 33% by weight γ-buterolactone and 67% by weight propylene glycol methyl ether acetate are low and high volatility solvents that allow for the tacky film to form quickly, but leaving enough solvent behind so that the film is well planarized. In general, the solution from which thin films are cast needs a viscosity similar to honey. Fifteen percent solids and 85% solvents yield such a solution.

The advantage of using this process for making thin films is the high concentration of dye chromophores that can be dissolved in the polymer in solid solution. The speed with which the solvent evaporates is partly responsible. When making preforms for fiber drawing, however, all residual solvents (and volatiles in general) need to be removed. During the polymer fiber drawing process, the material reaches 250°C. Any volatiles that remain after the baking cycle result in catastrophic boiling. To prevent the preform from deforming during the baking process, baking temperatures just below the glass transition of the polymer are used. Furthermore, the baking temperature can't exceed the boiling point of the solvent because trapped vapor causes bubbling. As such, baking cycles should not exceed 95°C for PMMA polymerization. Solvents that are trapped in the bulk of the polymer will not diffuse enough to leave the material even under impractically long baking times. When the fiber is drawn at 250°C from such a preform, where the polymer is soft, the high vapor pressure of the trapped solvents leads to foaming.

Any process used for making preforms that results in solvent residues will lead to foaming or trapped bubbles. While a large-core multimode fiber with an occasional bubble will still guide light, a bubble in the core of a single-mode fiber will scatter most of the light from the guiding region.

3.2.1.2 Polymerization with Dopants

The best method for making high-quality preforms is to start with a liquid monomer, dissolve the dopant in the monomer, and polymerize the material. Polymerization requires an initiator (which starts the polymerization process), a chain transfer agent (which limits the chain lengths of the polymer formed), and a plasticizer (which controls the rheological properties of the polymer). There are several issues that must be kept in mind when making a preform. First, the process should avoid any steps that might contaminate the monomer with volatiles that get trapped in the bulk of the polymer. Second, the monomer must be pure and kept clean. Finally, the effects of the dopant on the polymerization process as well as the effect of the polymerization process on the stability of the dopant must be understood and controlled. In this section, we describe in detail the process that is used for making the materials that are used to fabricate a high-quality preform that can be pulled into a single-mode polymer optical fiber.

The first step in making the doped polymer is the purification of the monomer. The more pure the polymer, the better the final fiber. The method generally used is similar to the process first described by Kaino.[12]

The purity of the methyl(methacrylate) (MMA) monomer as purchased from any supplier always contains some impurities and residues from the manufacturing process. Toluene is the most common and can account for 5–10% by volume. Such high levels of volatiles would lead to foaming during the draw process. To fabricate a high optical quality polymer, it is necessary to distill the monomer before polymerization. Figure 3.2 shows the apparatus. The monomer is stirred with 1% by weight of CaH_2 in the boiling flask. The calcium hydride removes water that is absorbed from the air by the monomer and removes the polymerization inhibitor that is shipped with the monomer to prevent the dangerous possibility of runaway polymerization. Water, on the other hand, must be removed to prevent optical attenuation in the material in the near IR.[50] This is particularly important in telecom applications, which use $850nm$, $1.3\mu m$, and $1.5mu$m laser sources. The joint above the Claisen flask prevents the undistilled monomer from boiling up into

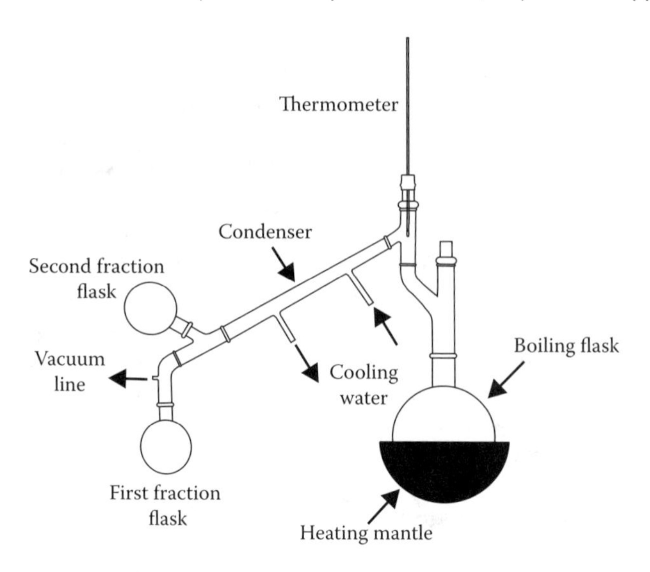

FIGURE 3.2 Monomer distillation apparatus. Adapted from Ref. [49].

the condenser. The bulb of a thermometer is placed at the mouth of the condenser to monitor the temperature of the vapor. The condensed vapor that contains the lower boiling point impurities is then collected by the first fraction flask at the bottom of the condenser. The angled joint between the condenser and the first fraction flask allows another flask to be swung down to collect the second fraction, which contains the purified monomer. (All glassware joints that come in contact with the condensed monomer vapor are sealed using Teflon sleeves, preventing contamination — which leads to increased vacuum leakage. All other joints are sealed using vacuum grease.)

The process for distillation is as follows: The distillation apparatus is evacuated with a mechanical roughing pump with a cold trap to prevent the monomer from clogging the pump. The monomer takes approximately 30 minutes to degas under vacuum. The most common gas products are absorbed air leaving the monomer and hydrogen, which is produced in the reaction between the water and the calcium hydride. After degassing is complete, the boiling flask is heated with a heating mantle. The heating rate depends on the pressure of vacuum. The boiling point scales with the pressure so less heat is required for high vacuums. Typically the voltage to the heating mantle is about 20 *volts* AC and the monomer starts to boil about 10 minutes after the heating process begins. At this point in the process, condensates from the vapor can be observed as the vapor rises up through the Claisen joint and to the mouth of the condenser. As soon as vapor is observed, the temperature as read by the thermometer rises abruptly to the boiling temperature of the volatile impurities, which typically boil at about 31°C under roughing pump vacuum. The temperature remains stable as long as the majority of the impurities are boiling. To get a clean separation of fractions the heat level should be adjusted to the point where there is no more than one drop per second falling into the collection flask. The temperature read by the thermometer rises again after approximately 5–10% of the total amount of monomer has condensed in the first fraction flask. At this point, pure monomer can be collected once the temperature stabilizes at its boiling point — at about 34°C. The second fraction flask is swung down so that the pure monomer condenses and collects there. The second fraction flask is wrapped in aluminum foil to prevent ultraviolet-induced polymerization. The distillation process is continued until only about 10% of the original amount of monomer is left in the boiling flask, which contains the higher boiling point impurities. The heating is then discontinued, the apparatus cooled, and the purified monomer removed and used as soon as possible. Note that to make ultra-pure monomer, this process can be repeated several times inside a sealed system.

In the next step of the process, dye molecules are dissolved in the monomer. The solubility depends on the temperature of the monomer, so higher concentrations can be reached by heating the material. Care must be taken to ensure that the mixture's temperature remains well below the polymerization temperature while the dyes are being mixed with the monomer. For methyl methacrylate, monomer and dyes can be mixed together at temperatures up to 60°C provided that the volume of monomer is less than 15*cc*. **(The reader should be warned that it is dangerous to heat large quantities of monomer because polymerization is a highly exothermic reaction.)** Table 3.1 shows several dyes and their maximum concentrations as determined from solubility studies.[49] Note that the refractive index depends on concentration (see Section 4.1.1).

The procedure for making dye-doped polymers for high-quality preforms was first reported by Garvey.[19] In that process, the dopant dye is added to 15*ml* of purified monomer and mixed in a capped container in an ultrasonic bath for 30 minutes to make a saturated solution. Figure 3.3 shows the structures of dopant dyes used by Garvey to make the core material,[49] with the exception of DR1 dye, which is described later. The dye solubility depends on the monomer, dye structure, and temperature. The table shows the maximum solubilities for several dyes at

TABLE 3.1
Solubilities of dyes in MMA and the resulting concentrations in PMMA[49]

	Maximum Solubility in MMA at 25°C		Maximum Solubility in MMA at 60°C	
Dye	**weight %**	N (cm^{-3})	**weight %**	N (cm^{-3})
ISQ	0.26	5.1×10^{18}	0.52	9.9×10^{18}
BSQ	0.81	11.0×10^{18}	1.9	26.1×10^{18}
TSQ	0.08	1.1×10^{18}	0.32	4.2×10^{18}
PSQ	0.12	1.4×10^{18}	0.42	5.0×10^{18}
HSQ	0.15	1.6×10^{18}	0.52	5.7×10^{18}
DR1	1.17	26.4×10^{18}	2.91	66.0×10^{18}

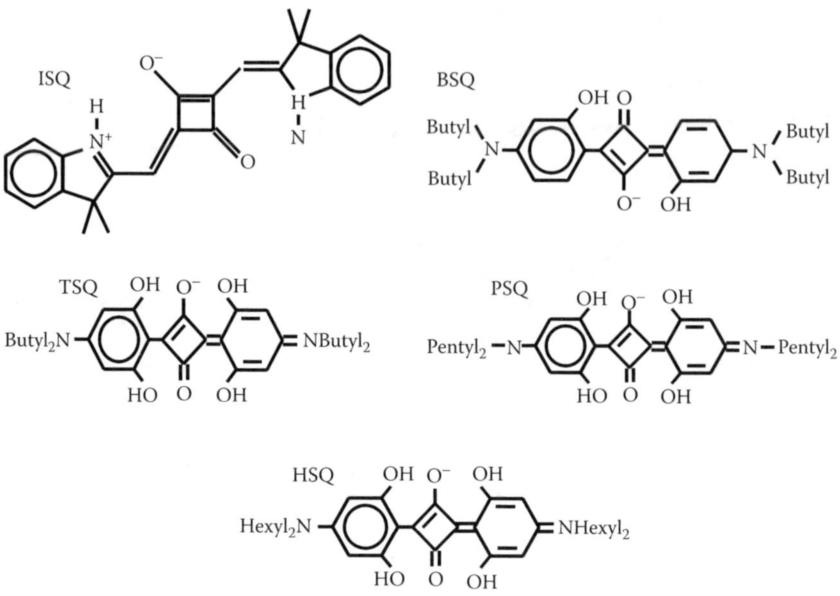

FIGURE 3.3 Structures of dyes used as dopants in fabricating core material.

room temperature and at 60°C. Maximum solubility is found by incrementally adding dye to the solution until the metallic glint of the precipitants can be seen to persist in the bottom of the container. It is found experimentally that for squaraine dyes a number density of approximately $5 \times 10^{18} cm^{-3}$ is needed to produce the required refractive index difference between the core and cladding of $\Delta n = 0.002$. This produces single-mode fiber at $\lambda = 1064nm$ with a $9\mu m$ core diameter — the standard single-mode core size used in glass fiber. The table shows that the solution must be heated to 60°C for several of the squaraine chromophores during the mixing process in order to achieve the required concentration.

After mixing, the solution is passed through a 0.2 micron syringe filter into a 125x16*mm* test tube — thus removing any undissolved dye that might otherwise form scattering centers in the core of the fiber. Subsequently, initiator and chain transfer agent is mixed into the solution, and it is heated to induce polymerization.

It is important to match the initiator, the agent that causes the chemical reaction that forms the polymer, with the dopant dyes. Certain initiators cause the dyes to decompose while others cause bubbling and foaming. The initiator that appears to be optimum for many chromophores that are polymerized in PMMA polymer is tert-butyl peroxide. Typically, $40\mu l$ of this initiator is added to $15ml$ of monomer by using a micropipette. The bond between the two oxygens in the peroxide is very weak and breaks apart from thermal activation with a half-life that decreases with temperature. The two radicals of the peroxide are highly chemically active. Each one bonds to one side of a monomer molecule and opens a chemically active site on the other side of the monomer molecule. This site then bonds to a second monomer molecule to begin chain formation and is followed by sequential chain growth until the process is terminated by encountering a molecule of the chain transfer agent. The average chain length of the polymer is therefore determined by the chain transfer agent concentration.

For the polymer to have the correct mechanical properties for fiber drawing, it is important to control the average chain length during the polymerization process. If the chain length is too short, the fiber is brittle and cracks upon handling. If the chain length is too long, the polymer does not flow even at temperatures well beyond the decomposition temperature of the dyes. The chain transfer agent that is found compatible with many chromophores is butanethiol. Some $50\mu l$ of the chain transfer agent is added to $15ml$ of monomer to make a polymer with an average molecular weight of about 70,000. This is the ideal molecular weight for uniform flow in the drawing process.

3.2.2 Making a Core Preform

A preform is the object from which a fiber is drawn. Garvey and coworkers introduced a two-step process in which a thick fiber (called a core fiber) is made by pulling a solid cylinder of dye doped polymer (called a core preform). This core fiber is then placed into a cladding to make the preform from which the fiber is made.

After adding the chain transfer agent and initiator to the monomer/dye solution, the test tube is capped and placed into a 95°C oven to initiate polymerization. (Note that at room temperature, no polymerization is observed even over a period of weeks.) After 30 minutes, the test tubes are uncapped for 5 minutes to allow degassing of any volatiles that may remain in the hot solution. The test tubes are then recapped and allowed to polymerize for 24–36 hours. Subsequently, the test tubes are removed from the oven and immediately placed into a freezer to cause the polymer to shrink and to separate from the glass. The test tubes are then wrapped in a cloth and shattered with a hammer; and the dye-doped polymer cylinders are removed. The core preforms are then placed back into the 95°C oven for five days to remove any remaining volatiles such as the chain transfer agent.

After this baking process, the core preforms may contain bubbles with a broad distribution of sizes. The bubbles can be microscopic and so are not always visually observable; but nevertheless, they interfere with the fiber drawing process by causing catastrophic foaming when heated in the draw oven. The size and number of bubbles increases with the concentration of dye in the solution and is strongly dependent on the type of initiator used. The bubbles can contain gases that are a by-product of a reaction between the dye and initiator.

To minimize the effects of the gas pockets, the core preforms are placed into a mechanical squeezer after they are degassed. Figure 3.4 shows a photograph of the squeezer used by Garvey and coworkers.[49] The preform is heated to 120°C and compressed with the squeezer assembly to accelerate diffusion that removes the trapped gas and collapses the remaining voids. Typically three preforms are processed in one squeezer. The squeezers are made of two halves, each with half

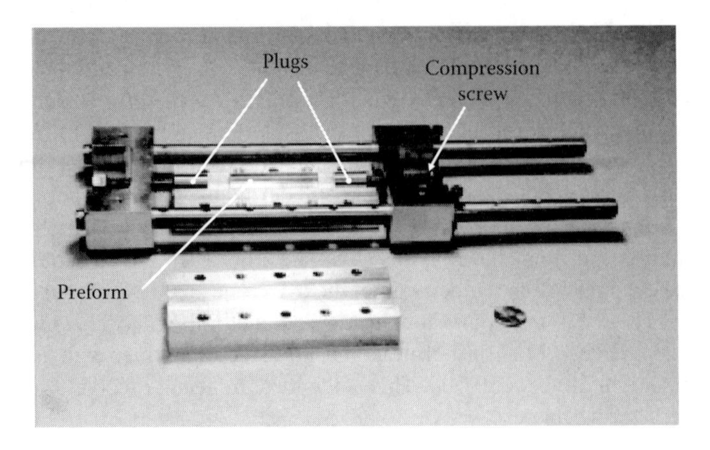

FIGURE 3.4 Squeezer used to compress preform. Adapted from Ref. [19] with permission of the Optical Society of America.

cylinders that fit the preform. By design, these halves make a tight fit around the preform. Subsequently, close-fitting aluminum plugs are pushed into either side of the channel by means of screws through both ends of a rail. The squeezer is kept at 120°C for approximately ten days and the screws are tightened daily until no additional compression is possible by hand-tightening with a wrench.

3.2.3 Making Core Fibers by Drawing

Each of the core preforms is used to make a core fiber, which will be used to make the core of the preform from which the final single mode fiber is drawn. The core fiber is also made by drawing approximately 15 meters of $790\mu m$ diameter fiber that will be used in assembling core fiber preforms as discussed in Section 3.2.4.

Figure 3.5 shows a typical drawing tower,[49] which has an oven that heats the polymer, a stepper motor that feeds the preform into the oven, and a take-up spool. The ratio between the feed speed and spooling speed determines the diameter of the fiber. The oven is made of glass so that the preform can be viewed during drawing.

A glass tube that is wound with nichrome heating wire acts as the oven, which radiatively heats the core preform. A second concentric glass cylinder provides insulation in the air between the cylinders. In some oven designs, a stream of air is passed between the two cylinders to stabilize the temperature from drafts. The inset in Figure 3.5 shows a perspective view of the oven. The optimal heating conditions are determined by the density of the heating coils and need to be empirically determined by trial and error. To control the temperature profile, the ideal pitch of the coil is not uniform along the length of the oven.

The preform forms a neck-down region that starts at the hottest point in the preform. Post-drawing heating below the neck-down region helps reduce stress, making the final fiber more homogeneous and mechanically stable [21,51] (see Section 3.4). Current is supplied to the heating coil with a Variac and the core preform is held by a collet mounted to the feed translation stage. Another collet

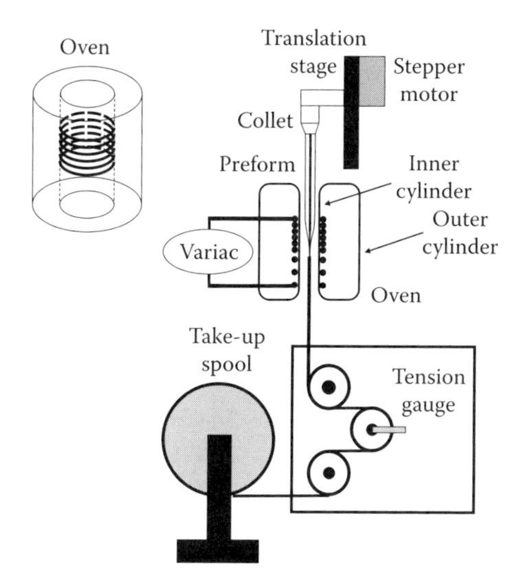

FIGURE 3.5 Fiber drawing tower in cross-section. Inset shows perspective view of the oven. Note that distances are not to scale. Adapted from Ref. [19] with permission of the Optical Society of America.

acts as a weight on the bottom of the preform. When the polymer softens, the weight falls slowly and a neck-down region is formed at the hot spot in the preform.

The fiber that forms below the neck-down region is pulled around a three-pulley stress gauge system. The middle pulley's axis is mounted to a force gauge, which measures twice the tension in the fiber when the fiber entering and leaving the middle pulley are parallel to each other. If the fiber segments around the middle pulley are not parallel, as shown in Figure 3.6, the force measured by the gauge, F, is related to the tension, T, using Newton's Second law:

$$T = \frac{F}{\cos\theta_1 + \cos\theta_2}, \tag{3.1}$$

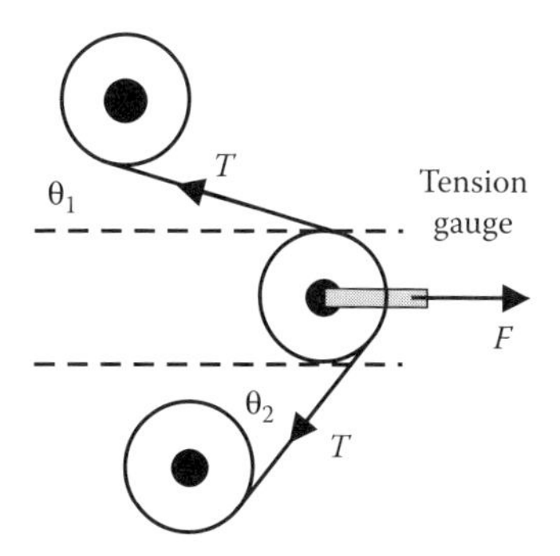

FIGURE 3.6 The tension gauge.

where θ_1 and θ_2 are the angles between the two fiber segments on the middle spool and the horizontal. Note that there are many possible tension gauge designs where the fiber segments are not horizontal.

After the fiber exits the tension gauge pulley system, the fiber is attached to the take-up spool, at which point the feed and take-up motors are started. (Note that an additional motor slowly moves the spool back and forth along the spool axis to allow the fiber to be evenly wrapped on the spool.) The diameter of the fiber is controlled by varying the ratio of the feed and take up speeds. The diameter is calculated from mass conservation,

$$D_f = D_0 \sqrt{\frac{V_{feed}}{V_{takeup}}}, \tag{3.2}$$

where D_f is the diameter of the fiber, D_0 the diameter of the preform, V_{feed} the feed velocity of the preform, and V_{takeup} is the velocity of the fiber, which is determined from the speed of a point on the circumference of the take-up spool.

Figure 3.7 shows the diameters of several core fibers of DR1 dye in PMMA polymer as a function of drawing length from the leading fiber end that leaves the oven.[49] The diameter is measured with a micrometer after the draw. Each core fiber is pulled at the same take-up speed, but the feed rates are different. The symbols represent the measurements and the series of solid lines are the diameters predicted by Equation (3.2) for the different pulling conditions. The data shows transients from equilibrium during early times of the draw process. Furthermore, the thinner the core fiber, the longer the time to equilibrium. In an industrial process where many kilometers of fiber are made, only a small part of the core fiber will be wasted. Note that the core preform is usually made of an insulating material while the metal collets conduct heat. As such, it is expected that both ends of a spool of fiber will suffer from nonequilibrium diameter fluctuations.

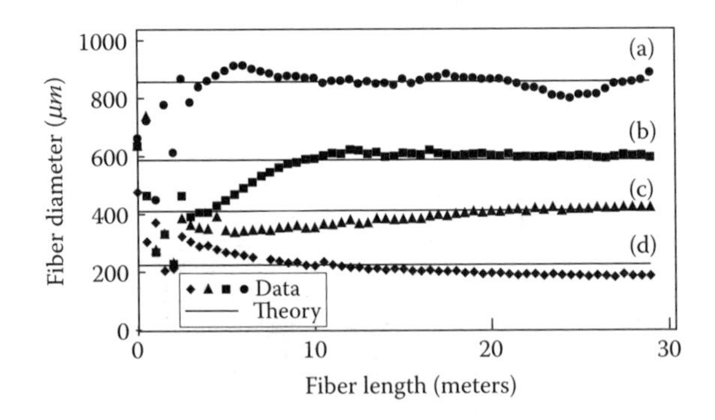

FIGURE 3.7 Core fiber diameter as a function of its length for various drawing parameters. The preform feed speed in these studies are a)160μm/s, b)75μm/s, c)37μm/s, and d)11μm/s. The take-up speed is fixed at 3.33cm/s. The symbols represent data, and the lines represent the diameters predicted by Equation (3.2). Adapted from Ref. [19] with permission of the Optical Society of America.

There are several external influences that can affect the core fiber's optical and mechanical properties. The core fiber, with or without dye chromophores, is sensitive to moisture — which can affect the optical properties of the material. Oxygen, on the other hand, can react with the chromophores and lead to decomposition. UV light can photochemically destroy the chromophore and cause the polymer to become brittle and yellow. As such, before the core fiber is used, it should be stored in an evacuated bell jar covered with a dark cloth to prevent material degradation. Since the core fiber's surface will eventually become the waveguide's interface, the surface must be kept clean from particulate matter and organic grease (fingerprints). It is therefore preferable that the core fiber be prepared in a clean room and should not otherwise be handled on sections that are intended to be used in making a preform.

While the above description makes the fiber appear overly fragile, the final single-mode fiber is quite robust. The typical dimensions of the core are about $9\mu m$ while the cladding is at least $125\mu m$. The cladding thus protects the fiber core. In practical systems, the fiber is jacketed and placed in a cable, which further protects it from the elements. One advantage of polymer fiber, however, is that it is more durable than glass and will not easily crack. As such, it can be used without a jacket in demanding applications such as optical interconnects or optical backplanes.

A good diagnostic of the fiber-making process is the drawing tension of the drawing process. The tension is monitored with a pulley system, as shown in Figures 3.5 and 3.6, where the middle wheel axle is mounted on the sensing mechanism of an electronic scale. It is found that high drawing tension (above 10,000 dynes in a $200\mu m$ diameter fiber) produces fiber with a high degree of chain alignment.[51] When the core fiber is subsequently heated, the alignment of the chains relaxes into a more randomized configuration, which leads to contraction in length and an increase of the core fiber diameter. (The volume of the core fiber is approximately unchanged.) When such a stressed core fiber is placed inside a cladding made of two half-cylinders and heated to produce a preform (the process of making the preform is discussed later in detail), the core shrinks and breaks into several longitudinal segments. The core thus needs to be annealed in an oven before assembling it into a preform. This annealing process must be done in a clean environment. To do so, short pieces of core fiber are removed from the drawing spool. The stress imparted during the drawing process leads to coiling in the fiber, so the degree of stress in the fiber can be estimated from the degree of coiling. Such fiber is placed between two lint-free cloths in a 95°C oven for three days to relieve the stresses. Immediately after the core fiber is removed from the oven, it is used to fabricate the preform.

In the drawing process, it is possible to adjust the temperature and draw rates to control the stress that is imparted to the fiber.[51] Under optimum conditions, the core fiber still needs to be annealed to remove the small residual stresses.

We discuss PMMA polymer draw conditions as an illustration. Other polymers may require different parameters. A drawing tower needs to be empirically adjusted for each new material to make good fiber. The most important condition to determine is the relationship between the tension of the fiber during the drawing process and the temperature of the fiber in the oven. In the work of Garvey and coworkers,[19] their system was calibrated in terms of the Variac voltage rather than the temperature since the temperature of the preform is difficult to measure directly. They found that a Variac voltage of 57V across the nichrome heating element resulted in a fiber draw tension of about 5,000 dynes. This condition produces fiber that will not undergo significant relaxation shrinkage. Preforms made of PMMA polymer with a molecular weight above 70,000 were found to draw at even higher tension. As the molecular weight of the preform is increased, the tension of the draw must be lowered to about 5,000 dynes by increasing the heater voltage. This works only when the molecular weight is not too large. At some point, the required temperature to lower the tension exceeds the decomposition temperature of the polymer.

Similarly, a low molecular weight PMMA polymer (short chain lengths) tends to pull at lower tension. Good fiber cannot be drawn at a tension below 3,000 dynes, where the fiber flows under its own weight. Under these conditions, the surface tension is comparable to the drawing tension, which leads to instabilities in the fiber diameter. This problem can be avoided by decreasing the heating voltage until the tension rises above 5,000 dynes. Low molecular weight polymers, however, are brittle and the glass transition range becomes so narrow that good drawing conditions cannot be found. Molecular weights for PMMA near 70,000 are thus ideal.

Note that the drawing process can change drastically over the temperature changes associated with 1 or 2 volt changes in the oven coil. In fact, the ideal voltage range for drawing a wide range of molecular weights of polymer is reported by Garvey and coworkers to be about 5 volts. Since applications demand highly uniform fibers, an active control system can be used that monitors the tension and fiber diameter, and adjusts the temperature and the draw speeds. A complication in the design of an active feedback control system is the large lag between a temperature change and the corresponding change in the tension. Since this delay can be several minutes, the control system can become chaotic. As such, not only must the sensitivity of the drawing parameters on the tension be determined, the effects of the lag time must be carefully taken into account.

It is difficult to determine the temperature of a preform because the process that gives the best fiber is based on radiative heating. In order to determine the temperature distribution inside a preform, Garvey and coworkers use a thermistor that is embedded in the center of a core preform.[19] The core is drawn into $790\mu m$ fiber at a heating voltage of $57V$, a feed speed of $160\mu m/s$, and a take-up velocity of $4.0 cm/s$. The temperature of the thermistor is recorded as the core is drawn into fiber until the thermistor flows into the heating zone and causes the fiber to break. The measured temperatures are shown as open circles in Figure 3.8. It is found that the temperature of the polymer at the neck-down region is 240°C.

The air temperature of the oven is measured under the same conditions, except the thermistor — which is mounted on the end of a thin glass rod — is lowered into the oven. The air temperature is shown as the solid circles in the graph. The hottest point of the air in the oven without the preform is found to be at the top (as one

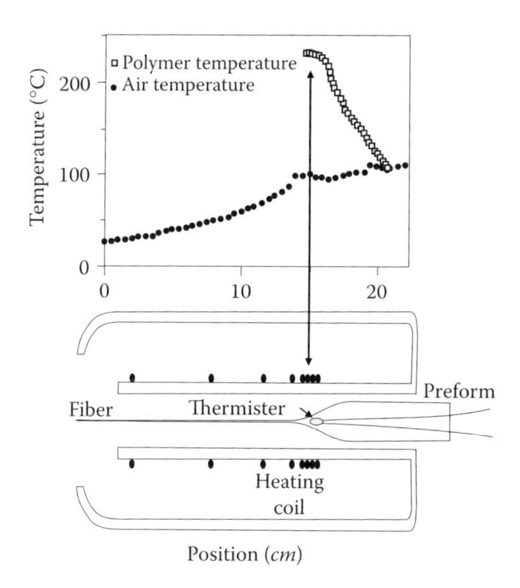

FIGURE 3.8 Temperature profile of the preform and the air in the oven. Adapted from Ref. [19] with permission of the Optical Society of America.

would expect due to convection). The temperature measured at the drawing point in air (the point around which the heating coils are at maximum density) is found to be significantly lower than those measured inside the polymer at the same point. This indicates that the heating of the preform is primarily due to absorption of infrared by the polymer. Indeed, the optical absorption spectrum of polymers is dominated by C-H stretch modes throughout the IR (see Section 5.2.4); so, absorption of the IR by these stretch modes is responsible. As such, drawing conditions for making deuterated or fluorinated polymer will be different (see Section 5.1.2.).

3.2.4 Making Core/Cladding Preforms

There are several methods for making a preform with a core. They include the "wet process" of *in-situ polymerization* for making graded-index fibers, and the "dry processes" in which the core and cladding are made separately and subsequently assembled.

3.2.4.1 Dry Processing

3.2.4.1.1 Squeezing at Elevated Temperature

One method for making a preform with a core is to assemble it from two half-rounds that make the cladding. Two half-inch diameter cylindrical thermoplastic rods of 70,000 molecular weight and a minimum length of 4″ are milled to make

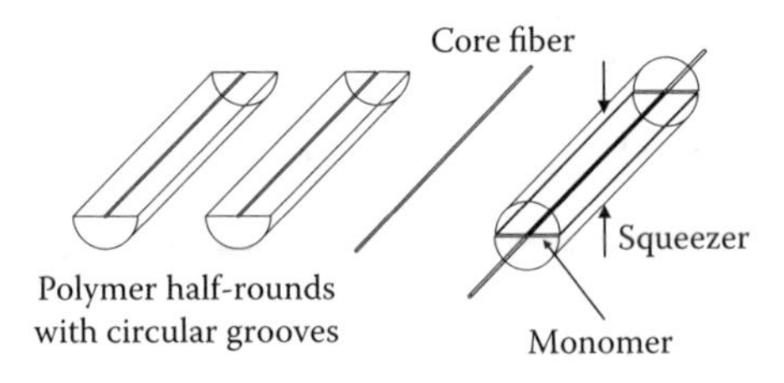

FIGURE 3.9 Two half-rounds are squeezed around the core fiber to make a preform. Adapted from Ref. [19] with permission of the Optical Society of America.

half-rounds. The milling process assures that the two flat halves can be placed together without an appreciable gap. A semicircular groove is machined along the center axis of each with a 1/32″ ball end mill. It is important to clean the milled surfaces with detergent and bake the half-rounds for an hour to remove water from the surface. If the half-round is baked too long, it can warp. The fiber preform is then assembled by placing a length of the core material into one of the grooves and then covering it with a second half-round. The assembly process is shown in Figure 3.9. When the two halves are held together by hand, the core fiber should be snug and difficult to pull out. The grooves, however, must be large enough to allow the two milled planes to be in intimate contact. Monomer can be placed between the two halves to fill any remaining gaps, but often leads to a poor core-cladding interface after the processing described below.

The assembly is placed into one of the squeezers (see Figure 3.4), clamped, and placed into an oven at a temperature near the polymer's glass transition for about 2 days. (For PMMA or polystyrene, 95°C works best.) At this point, the cladding material is observed to flow into the gaps between the squeezer halves. The preform is then left in the oven for an additional 5 days without further tightening. Excessive tightening results in the core meandering away from the preform axis.

Note that it is essential that the half-rounds and core material be kept clean at all times before assembly, so a class 100 clean area is advisable. The half-rounds should be machined using only water for coolant in order to avoid contamination. Even with these precautions a fairly common problem encountered is the formation of bubbles at the core-cladding interface while the preform is drawn. There is a correlation between bubbling during drawing and humidity on the day of preform assembly.

Figure 3.10 shows two half-rounds made of PMMA, the remanent of a core preform after it is drawn with part of the core fiber attached, a core fiber in a half-round, a single- and dual-core preform, and a preform from the end of a draw of single-mode polymer optical fiber with PMMA cladding and a dye-doped core. Also shown is a cross-sectional view of the dual-core fiber.

Figure 3.11 is an example of a preform that was prepared according to the prescribed process but still foamed during drawing. The core material is DR1 dye at about 1 weight percent doping. Extreme bubbling is found near the neck-down region where the temperature is the highest. Most likely, some volatiles remained

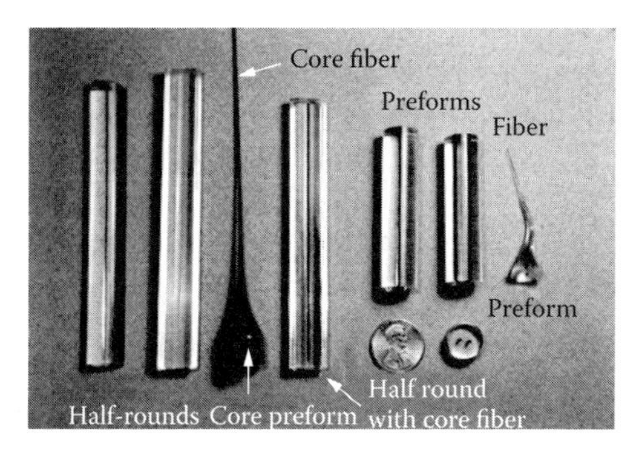

FIGURE 3.10 Various steps in the procedure used in making single-mode fiber. Adapted from Ref. [19].

after the cleaning process, the preform draw temperature was too high, or moisture diffused into the preform. Even when great care is taken to make clean preforms, a large fraction of them may bubble. These effects can have a seasonal effect in climates where humidity swings are large. A highly controlled environment is therefore important in any manufacturing process.

3.2.4.1.2 Stacking and Drilling

Sometimes called holey fibers, photonic crystal fibers use arrays of holes to define the waveguide and confine the light. In essence, these periodic structures result in destructive interference in the cladding region, causing the light to propagate in the core region. The periodicity of the holes are comparable to the wavelength of light. The underlying physics of the light's behavior is analogous to crystals and defects, where the electron wavelength is comparable to the atomic spacing of the crystal. Indeed, many of the properties and phenomena seen in atomic crystals are duplicated in photonic crystals.

There are several common ways of making holey fibers, which rely on making a periodic or quasi-periodic array of holes in a preform. Clearly, the simplest method

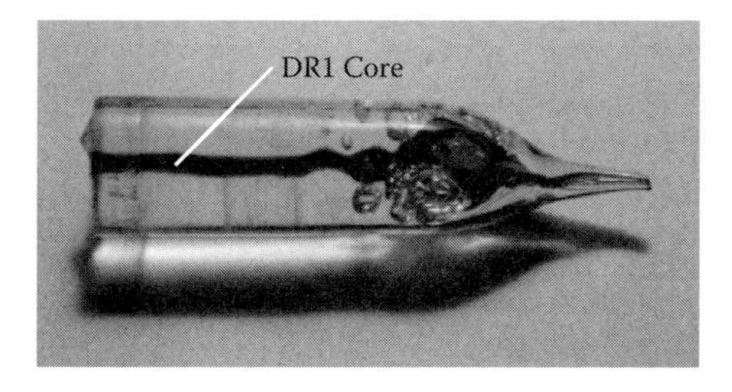

FIGURE 3.11 A preform with even a small amount of volatiles will bubble while drawing. Pictured is a preform that was carefully prepared without solvents but still bubbled. This photograph was supplied by Dennis Garvey.

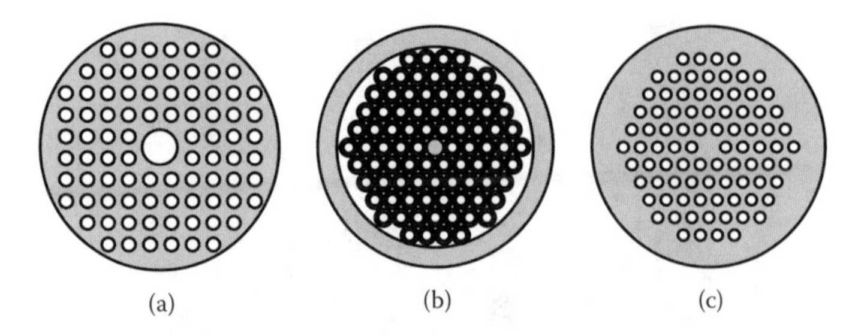

FIGURE 3.12 (a) A preform made by drilling holes; (b) a preform made with a bundle of tubes; and (c) a fiber drawn from the bundled preform.

is to drill holes directly in a solid preform. The ease with which holes can be drilled in a polymer is a great advantage over silica-glass-based fibers, where drilling deep holes is not possible. With this method, preforms with holes of arbitrary sizes and spacing can be mass-produced using a computerized drill press.

Another method to make a preform with periodic holes is to use a bundle of tubes. In the drawing process, the tubes merge, leaving the holes intact. This is the most common method for making glass holey fibers and has also been used to make polymer holey fibers. Figure 3.12 reviews these two methods. When making bundles, only limited types of arrays can be made, such as the hexagonal pattern shown in Figure 3.12b, which is drawn into a fiber with the cross-section shown in Figure 3.12c. Any large mismatch between the individual tubes in a bundle can lead to an additional hole. Furthermore, the bundles are usually placed inside a cladding, which can lead to misalignment of the tubes. In contrast, an arbitrary pattern can be drilled, as shown in Figure 3.12a. In this example, a larger hole is drilled in the middle so that most of the guided mode travels in air. This allows intensities that normally are above the damage threshold of the material to be guided. While it is possible to make tubes of varying sizes to extend the practicality of using bundles of tubes, there are many types of structures that can only be made with drilling.

3.2.4.2 Wet Processing

There are several methods for making a preform that include liquid monomers. We define a wet process as one in which a liquid monomer is in contact with a polymer. Some key methods are described below.

3.2.4.2.1 Liquid Monomer Core

Kuzyk[18] and coworkers first used this process to make a dye-doped single-mode fiber. In this work, the central axis of a polymer cylinder is drilled to make a tube, which is filled with liquid monomer and dye, and polymerized to make the core. Even though the drilling process makes a rough surface, if the core monomer dissolves some of the surrounding polymer on the inner walls of the tube, machining scratches are dissolved and a diffuse boundary is made during core polymerization.

Such a diffuse boundary makes for a better waveguide and decreases scattering losses.

If the cladding is made of the same polymer as the core, as is usually the case for dye-doped polymer fibers, the process of polymerizing the core in a polymer tube is optimum. Even if the core monomer does not dissolve the cladding, one might speculate that the machine marks might lead to excessive scattering losses in the final fiber. This, as it turns out, is not necessarily the case. The reason can be understood by considering the geometrical changes of the preform during the drawing process. If the core of the preform is, for example $1mm$ ($10^{-3}m$) in diameter, and is pulled into a single-mode fiber with a $10\mu m$ ($10^{-5}m$) core, the cross-sectional area decreases by a factor of 10^4. By volume conservation, the length of the final fiber is 10^4 longer than the preform. If the original machining marks are $0.1mm$ in diameter, they will be stretched into $1m$ long features. Furthermore, because of surface tension in the draw process, the fiber will have a natural tendency to smooth out imperfections. Both these effects result in a smoother surface.

Dirk and coworkers[52] and Peng, Chu and coworkers[53] independently developed a method for making preforms using a wet process that makes tubes with smooth walls. Figure 3.13 shows a schematic of the process. A taut Teflon wire is centered in a glass tube and the monomer is polymerized around the wire. Because the polymer does not stick to the Teflon, it is relatively simple to remove the wire. We note that in the technique developed by Dirk, two "plugs" are placed inside a specialized custom test tube in a spring-loaded assembly. In Peng and Chu's apparatus, the Teflon wire comes through the bottom assembly so that a polymer sealing layer needs to be used to prevent leaking. Both techniques, however, yield similar results.

The polymer tube is then removed from the glass tube, the bottom plugged, and the monomer (with initiator, chain transfer agent, etc.) is poured into the hole. The core is then polymerized by heating. The core must have a higher refractive index than the cladding so the core must be made of another polymer, a copolymer, or dye-doped polymer.

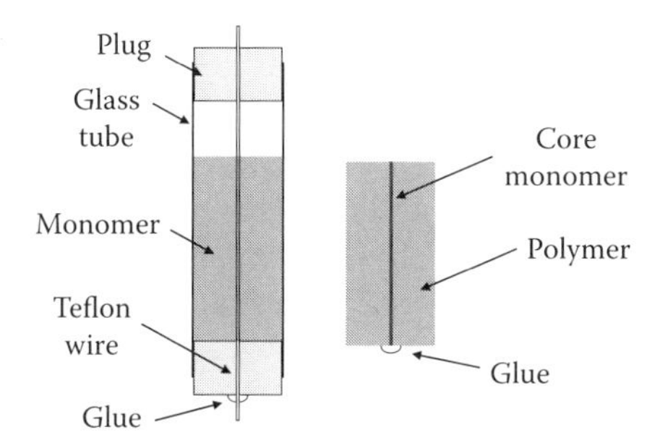

FIGURE 3.13 A tube is made by polymerizing monomer around a Teflon wire. The core is made by polymerizing a liquid monomer in the polymer tube.

The advantage of the Teflon wire method is that it is possible to make small-diameter and long holes along the axis of the polymer cylinder. Indeed, Peng and Chu report that his group has made such preforms in excess of 300*mm* lengths. In contrast, drilling is limited to 25*mm* (drilling depth is smaller for smaller-diameter holes) while the half-round method can be used to make arbitrarily long preforms.

There are trade-offs in using a wet process. The fact that the core diffuses somewhat into the cladding yields a gradient index near the surface. If the core is polymerized slowly, it can diffuse into the polymer cladding, and some of the cladding can dissolve in the monomer core — leading to a smooth, slowly varying gradient index. If the polymerization process is quick, there is little diffusion and the refractive index profile is rectangular. In both cases, this can lead to a better waveguide because surface imperfections are "smoothed" out. If the processing conditions are carefully controlled, the index gradient can be adjusted to meet application needs. The major disadvantage is that core polymerization leads to more bubbling than the dry process. Even though reasonably good cores can be made in the wet process, single-mode waveguides can most likely not be made reliably enough for a manufacturing process. This is in contrast to the dry process, which is a highly reproducible manufacturing process that has led to the production of high-quality single-mode fibers. Indeed, several companies sell single-mode polymer fibers, for example, Paradigm Optics (www.paradigmoptics.com). Therefore, it is more common for the wet process to be applied to making multimode fibers, which have larger cores and larger refractive index differences between core and cladding than do single-mode fibers.

A more serious problem with the wet process is the reactivity at the surface between a monomer and polymer. For example, when liquid styrene monomer is brought into contact with PMMA polymer and polymerized, a cloudy polystyrene polymer is formed. This is independent of the initiators used in the polymerization process. Figure 3.14a shows styrene polymerized in a PMMA rod. The observed scattering is so severe that the polystyrene appears cloudy white. Clearly, there are so many scattering centers that the light is multiply scattered.

Figure 3.14b shows a rod made by injecting a partially polymerized styrene mixture into a PMMA rod. When it is fully polymerized, it becomes cloudy but is still transparent. The cloudiness has a violet tint that is indicative of Rayleigh scattering. Furthermore, when the liquid is viscous, it is difficult to inject it into the rod without creating bubbles.

The quality of the polystyrene improves when the degree of polymerization is increased before placing it in contact with PMMA. Figure 3.14c shows a bead of highly viscous pre-polymerized polystyrene that is polymerized to completion on a PMMA substrate. The quality of the polystyrene bead appears good but a slight bluish tint is still noticeable. Scattering centers are therefore clearly still present. In all these cases, the scattering is observed through 2*mm* of material. Thus, in the limit of dry processing, the optical quality of the material is best.

When making a preform, the core is often so small that such cloudiness may not be easy to observe. After pulling a fiber, the core is even smaller, so scattering will not be visually observable. Since most fiber devices require the light to

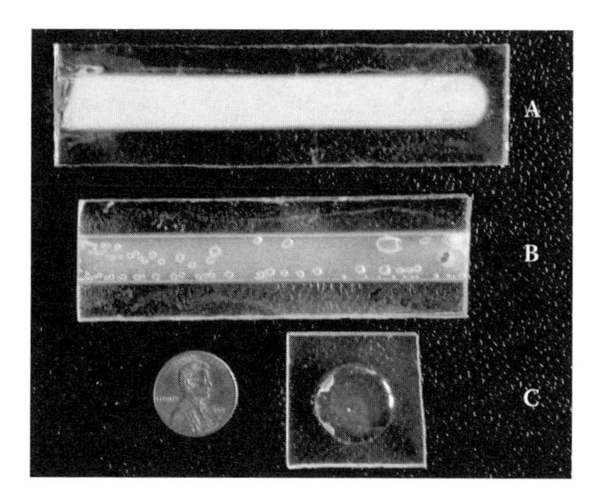

FIGURE 3.14 a) A preform made by polymerizing styrene inside a PMMA rod (left); b) partially polymerizing the polystyrene before injecting it into the rod and then fully polymerizing it; and c) a drop of highly pre-polymerized syrup of polystyrene polymerized to completion on a PMMA substrate. These photographs were supplied by Brent Howell.

propagate from centimeters to tens of meters, the scattering process will lead to an overwhelming amount of loss. Furthermore, since the guiding condition of a single-mode fiber is sensitive to the k-vector of the light, even small amounts of scattering can result in the light leaving the core and entering the cladding. In many materials, the wet process may not be suitable for making good fibers if low loss or single-mode guiding is required by the application.

Dirk and coworkers developed a hybrid method to make a preform in which the cladding is made by polymerizing around a Teflon wire, but the core is made with a dry process. The Teflon wire is removed as in the wet process, but a core fiber is placed in the hole. Even with a highly uniform core fiber, it is difficult to pull it through the hole unless there is sufficient clearance. If such a preform is pulled into a fiber, the uneven gap between the core and the cladding can result in voids. The resulting fiber core makes a poor waveguide. However, experimental tricks can be used to improve the process, such as applying a vacuum at elevated temperatures to cause the cladding to uniformly collapse around the core.

To force intimate contact between the core and the cladding, Dirk uses a series of squeezers that press radially inward on the outside surface of the preform while it is heated above the glass transition temperature. The top portion of Figure 3.15 shows the process. In the squeezing process, the preform length slightly increases. Because the flow of polymer is not uniform during squeezing, the core develops waves and kinks as shown schematically in the bottom of Figure 3.15.

A meandering core, however, is not necessarily problematic. Recall that when making a single-mode fiber, a $10cm$-long preform can draw into a $1km$ fiber. If the wavelength of spatial oscillations is on the order of millimeters to centimeters, the final fiber will have oscillations over 10 and 100 meters, respectively. A 10cm section of fiber, then, will appear straight. Furthermore, the long-scale oscillations

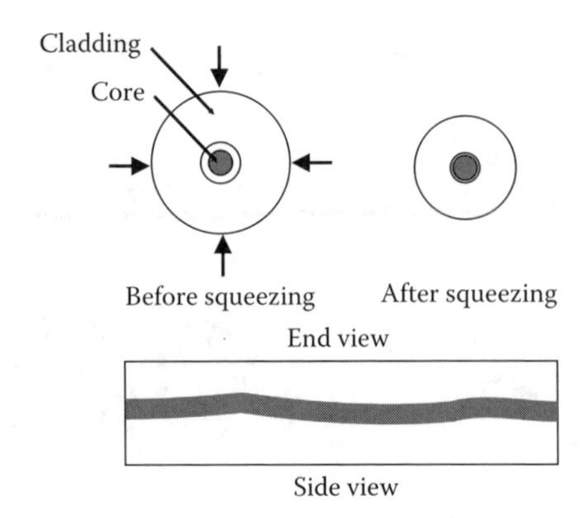

FIGURE 3.15 A core fiber is placed in the hole of a tube and radially squeezed until the core and cladding form intimate contact. The core, though, develops ripples in the process.

may also not be a problem because they are so much longer than a wavelength that the light can follow the path of the core adiabatically.

For telecom applications, however, core concentricity in single-mode fibers is crucial because two arbitrary pieces of fiber need to be connected to each other or spliced. Tolerances required in concentricity are typically in the $1\mu m$ range; so, fibers made using Dirk's method would most likely not meet such standards. The method is, however, useful in research applications, or for making specialized devices where connectors do not require a high degree of concentricity.

3.2.4.2.2 Liquid Monomer Cladding

Another wet method for making a preform starts with a liquid monomer cladding in a cylindrical vessel. A solid core fiber is placed into the vessel, and the liquid monomer polymerized. Such a process may work well if the core fiber is not readily dissolved by the cladding monomer. If the core and cladding material are similar, the core is often found to dissolve before the cladding polymerizes. To get around this problem, the cladding can be pre-polymerized until its consistency is similar to thick syrup. When the core fiber is placed into the pre-polymerized polymer, the core remains intact during polymerization.

Using this process, it is difficult to make a single-mode fiber using a dye-doped core fiber because even in a highly pre-polymerized cladding, the dye diffuses over long distances before the cladding is fully polymerized. The elevated temperature required to polymerize the cladding accelerates the diffusion process, but a lower temperature gives the cladding material more time to dissolve the core. To optimize the process, then, the competition between diffusion and solubility of the dye in the monomer must be balanced. This is difficult to experimentally control because the polymerization process tends to be chaotic. Furthermore, when a core fiber is placed into the pre-polymerized cladding, it is difficult to keep it straight. Nevertheless, while this process cannot be easily adapted to make a single-mode step-index fiber, it can be used to make a gradient-index fiber.

3.2.4.2.3 Ultraviolet Curing

In applications where many short sections of fiber are attached to each other with connectors (as one finds in local area networks and fiber to the home), the difficulty and cost of splicing single-mode fibers together makes them not suitable for such applications. While multimode fibers are more tolerant of bad splices (i.e., with cores that are not well aligned), their bandwidth is much lower. Graded-index fibers, on the other hand, have a higher bandwidth, have larger core regions, and are easier to splice.

Koike and coworkers pioneered two methods for making graded-index preforms based on ultraviolet curing and the interfacial gel technique. We begin with the UV curing method.

The desired refractive index profile has a parabolic radial dependence from the cylinder axis of the form

$$n(r) = n_0 \left(1 - \frac{1}{2}A' \left(\frac{r}{R_p}\right)^2\right), \tag{3.3}$$

where n_0 is the refractive index of the cladding material, R_p the radius of the plastic rod, and A' a constant. The idea is to make a polymer rod that is composed of a mixture of two materials of different refractive indices in a way that yields a radial dependence in the concentration profile.

In the mid-1970s to early 1980s, Ohtsuka and coworkers reported on making various gradient-index polymer rods that were used for making large numerical aperture imaging optics. The work in the early 1980s resulted in an acceleration of efforts in making polymer rods that could be heat drawn into fiber.[54] In 1981, Ohtsuka and coworkers made thermoplastic rods using photocopolymerization of methyl methacrylate (MMA) and vinyl benzoate (VB) with benzoyl peroxide (BPO) as the initiator.[54]

After removing the inhibitor with aqueous 0.5-N NaOH, the liquid monomer is washed three times with distilled water and distilled under reduced pressure. The BPO initiator is recrystallized from methanol and dried in vacuum. Since the central core region must have the highest refractive index, the two monomers selected (M1 and M2) must have the property that the lower refractive index one must have the higher reactivity. In the case of the polymers studied by Ohtsuka, the reactivity of MMA is $r_1 = 8.52$ and of VB is $r_2 = 0.07$. The refractive indices are $n_1 = 1.49$ and $n_2 = 1.5775$, respectively.

Figure 3.16 shows a schematic diagram of the apparatus used by Ohtsuka to make gradient-index rods by UV irradiation. The mixture of monomers and initiator is sealed in a glass tube and mounted on a rotating table that turns the tube around its axis at about 40rpm. The cylindrical UV light source (Ohtsuka used a high-pressure mercury vapor lamp) is apertured so that only a few centimeters of the sample length is exposed. The lamp is mounted on a vertical translation stage which moves the light source at a speed of 0.3 to 1.22 *mm/min*. As such, the region labeled C on Figure 3.16 has been exposed for a longer time than region A. Depending on the speed of the translation stage, the polymer is exposed from one to three hours.

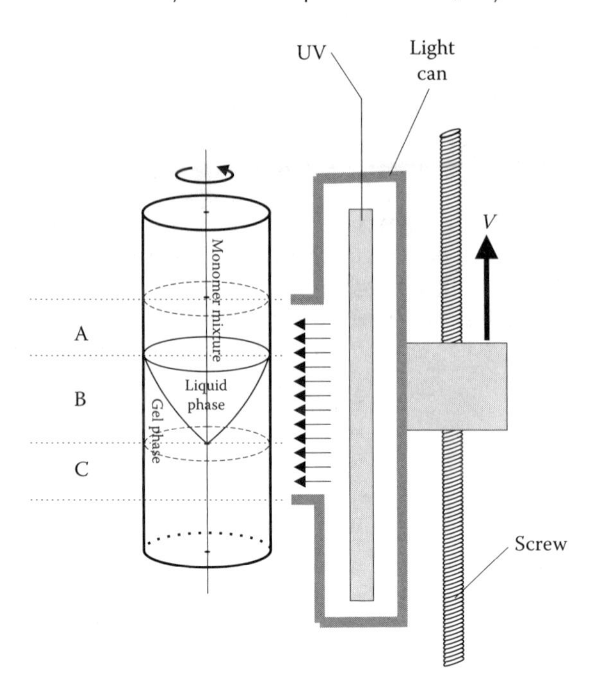

FIGURE 3.16 Schematic diagram of apparatus used to make gradient-index rods by UV irradiation. Adapted from Ref. [54] with permission of the Optical Society of America.

Because the lower refractive index polymer is made from a more reactive monomer, its polymerization rate is higher so its concentration is higher at the walls of the tube than the lower refractive index component. The polymerization process starts from the outer wall and moves inward; the concentration of the higher refractive index component grows toward the center of the sample. (Note that the sample is kept at 25°C through air cooling to prevent it from thermally polymerizing.) The vertical speed of the lamp is adjusted to ensure that the gel forms to the center of the rod before the light source is eclipsed. The whole rod is polymerized after the light source traverses its length.

After this process, the rod is in the form of a gel, which consists of partially polymerized monomer. To complete the process, the rod is heat-treated at 60°C for 40 hours, which fully polymerizes the remaining monomer. The glass tube is then shattered to remove the preform rod. The rod is heat drawn into a fiber. Instead of using the radiative heating process as described above, Ohtsuka used a hot-oil-bath oven in the mid-200°C range.

After the fiber is drawn, the refractive index profile is measured. (Methods used for measuring the refractive index are described in Chapter 5.) Figure 3.17 shows the refractive index profiles for fibers drawn at different draw ratios.[55] While the observed profiles are similar, the higher draw rates — which make thinner fibers — seem to yield a somewhat smaller refractive index. The straight line shows the theoretical parabolic profile. Clearly, the refractive index profile is

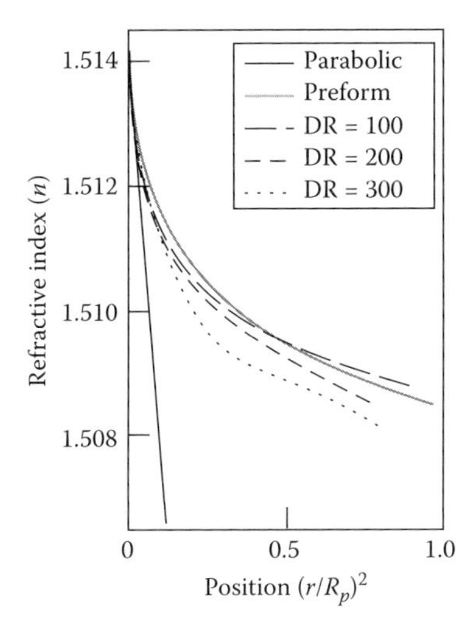

FIGURE 3.17 Refractive index profile of UV-cured preform and fibers for different draw rates — where MMA/VB = 4 wt/wt and T_d = 270°C. Straight line theory shows for a parabolic index profile. Adapted from Ref. [55] with permission of the Optical Society of America.

parabolic only at the center region of the fiber and falls off more slowly away from the center.

Figure 3.18 shows the dependence of the refractive index profile on draw temperature and composition.[55] Each figure also shows the refractive index of the preform used to make the fiber. The refractive index profiles of the fibers are only qualitatively similar. Over the last 20 years, the preform-making process has been improved and the refractive index profiles more nearly approximate a parabola.

3.2.4.2.4 Two-Step Polymerization

The two-step polymerization process can be used to produce gradient-index rods. The process is often called diffusion polymerization or the interfacial gel technique. In the diffusion polymerization method, a solid polymer rod of higher refractive index (called the mother rod) is placed in a monomer of lower refractive index. The monomer diffuses into the polymer and some polymer dissolves into the monomer during the polymerization process resulting in a compositional gradient. In the interfacial gel technique, the mother rod starts out partially polymerized in a gel state to more easily allow for the absorption and diffusion from the monomer bath. In both cases, the processing parameters and the time of polymerization determines the refractive index profile.

In the late 1970s, Iga and Yamamoto used diffusion polymerization to make gradient-index rods.[56] The higher-refractive index mother rod is polymerized at 80°C for about two hours inside a thick polypropylene tube so that it can be easily removed by peeling away the propylene. This is particularly important when the mother rod is soft. The mother rod is then dipped in a monomer liquid bath at

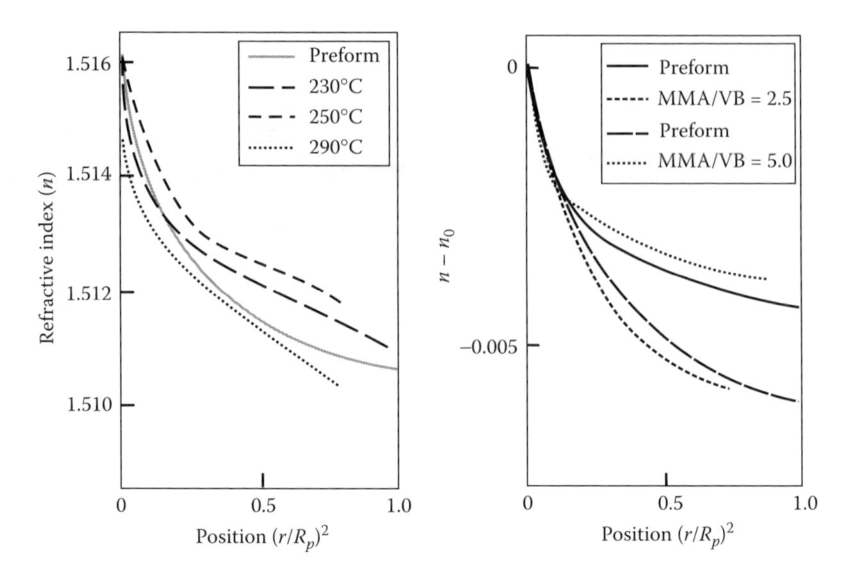

FIGURE 3.18 Refractive index profile as a function of draw temperature (left) and composition (right). On left plot, MMA/VB = 3.5 wt/wt and DR = 100. On right plot, $T_d = 250°C$. Adapted from Ref. [55] with permission of the Optical Society of America.

80°C for 5 to 25 minutes to allow the diffusion process to take place. After the rod is removed from the bath, it is heated for several hours at 80°C until the rod is completely polymerized.

Note that in cases where the bath monomer is highly reactive with the mother rod material, the mother rod can be dipped, removed, cooled, and re-dipped to effectively increase the diffusion time without dissolving the mother rod.

Some common monomers used for making GRIN optics and the refractive indices of the neat polymers made from these monomers are shown in Table 3.2. Table 3.3 shows the refractive index of PMMA polymer doped with common dopants used to elevate the refractive index.

Since the applications Iga and Yamamoto had in mind were fiber scopes for medical imaging, they were most concerned with optical properties in the visible. It is important to stress, however, that the refractive index varies with wavelength and with the molecular weight of the polymer so tabulated numbers need to be understood as approximate.

Clearly, there are many choices of monomers that can be used to tailor the resulting graded index rods. GRIN fibers have been made using DAI as the mother rod and MMA as the cladding or CR-39 as the mother rod and 4FMA as the cladding. In the DAI/MMA GRIN rods made with the diffusion method, the refractive index of the core is about $n(0) = 1.545$ and the refractive index in the cladding region is about $n(a) = 1.515$. When a VB-mother rod and MMA bath are used with benzoin methyl ether as the initiator, a much steeper refractive index profile results. For low loss (non-fluorinated) graded-index fiber applications, VB/MMA is common.[57,9]

TABLE 3.2
Refractive indices of polymers (in the visible) used to make GRIN rods[57,58,59,60]

Monomer Abbreviated Name	Common Name	Polymer Refractive Index
PS	styrene	1.58
VB	vinyl benzoate	1.578
DAI	diallyl isophthalate	1.57
VPAc	vinyl phenylacetate	1.567
CR-39	diethylene glycol bisallyl carbonate	1.51
PMMA	poly (methyl methacrylate)	1.49
4FMA	1,1,3-trihydroperfluoropropyl methacrylate	1.42
SAN	styrene-acrylonitrile copolymer	1.58
PC	polycarbonate	1.586
TPX	poly (4-methylpentene-1)	1.466

3.2.4.2.5 Cross-Linked Polymers

A glassy polymer is made of a tangle of chains, similar to a bowl of spaghetti, as shown in Figure 3.19a. When a glassy polymer is heated above its glass transition temperature, the chains can slide against each other, allowing the polymer to flow. Figure 3.19c shows the glassy polymer at the microscopic scale after it has been drawn into a fiber. When the fiber is drawn at temperatures just above the glass transition temperature, the draw tension is high and the chains tend to align, as shown in the figure. When drawn at higher temperatures, the chains remain more nearly randomly oriented.

Figure 3.19b shows a cross-linked polymer. The dots represent chains that are attached together with chemical bonds. A partially cross-linked polymer may

TABLE 3.3
Refractive indices of PMMA (in the visible) when doped with molecules typically used as dopants for making GRIN rods[61]

Dopant Abbreviated Name	Common Name	Polymer Refractive Index
BB	bromobenzene	1.58
BBP	benzyl *n*-butyl phthalate	1.578
DPS	diphenyl sulfate	1.57
TPP	triphenyl phosphate	1.567
BEN	benzyl benzoate	1.51

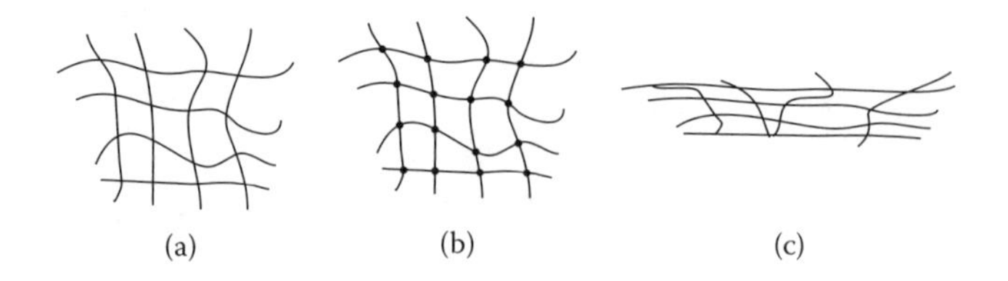

FIGURE 3.19 (a) A glassy polymer; (b) a cross-linked polymer; and (c) drawing a glassy polymer.

be able to be drawn under the right conditions, but a fully cross-linked polymer cannot. As such, a drawing process cannot be used to pull a preform if it is fully cross-linked.

There are several approaches for making a fiber with a cross-linked polymer. The simplest is to make a preform from a glassy polymer that is not cross-linked, draw it into a fiber, and cross-link it with subsequent heat treatment. Alternatively, the drawing process may supply enough heat to cross-link the fiber as it emerges from the tower. However, for this to work, the draw conditions need to be very carefully controlled to prevent cross-linking while the preform is in the furnace.

Another method to make a fiber with a cross-linked core material is to start with a capillary tube, fill it with monomer, and polymerize/cross-link the core. This method was used by Flipsen and coworkers to make a polymer network core with intermolecular hydrosilylation.[62] Figure 3.20 shows the cross-linking processes of the pre-polymer. Since the pre-polymer has a relatively low molecular weight and therefore a moderate viscosity, it is easy to process. The cross-linked material has a glass transition temperature of only $T_g \approx 14°C$; but retains high optical quality even when it is heated to well above the glass transition temperature. This is undoubtedly due to the very high density of cross-linking — the average molecular weight between cross-links is $M_C = 300g/mol$.

In addition to retaining good optical quality at elevated temperature, the cross-linked polycarbosiloxane polymer also has excellent mechanical properties. At $T = 20°C$, the material has an elasticity of modulus 350 MPa, stress at break of 16 MPa, and elongation at break of 22%.

The fiber is made by injecting the viscous core pre-polymer, along with a platinum catalyst (typically at 1 ppm catalyst), into a FEP (fluoroethylene-*co*-fluoropropylene) sheathing and cured for 24 hours at 150°C. The polycarbosiloxane polymer can be cured at a faster rate by adding more catalyst, but this increases the optical loss. Such increased optical loss is common when a monomer is polymerized too quickly, due to inhomogeneities that are introduced in the process. The FEP tub needs to be thoroughly cleaned before injecting the core. Flipsen and coworkers cleaned the inner walls by ultrasonic rinsing with 20% HNO_3 for 30 minutes followed by several rinse cycles using demineralized water, filtered ethanol, and filtered pentane.[62]

The fibers made with the process are necessarily large. The fibers made by Flipsen are typically 1.6mm diameter with a 0.95mm core diameter.

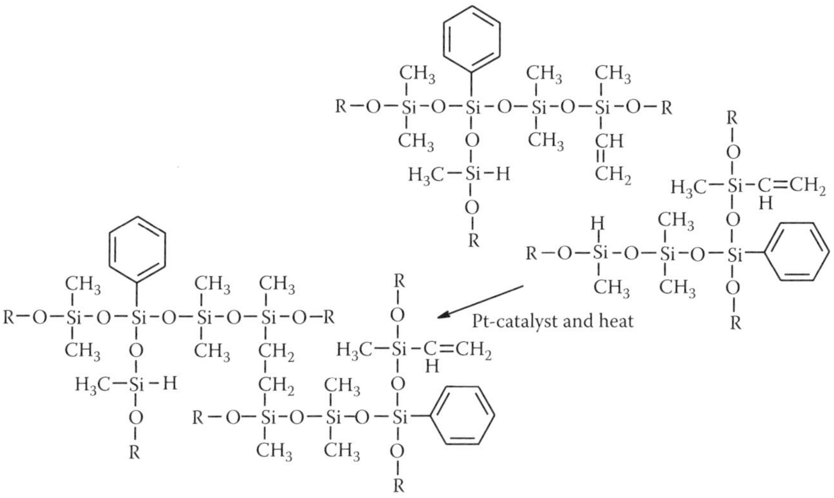

FIGURE 3.20 Making a polymer network by a hydrosilylation cross-linking reaction. *R* represents a monomer unit.

3.3 BIREFRINGENCE OF DRAWN FIBERS

The chains in a polymer align during drawing as illustrated in Figure 3.19. This mechanical anisotropy leads to a birefringence.[63] In addition to the chains being stretched due to the drawing process, a rapid rate of cooling leads to anisotropy due to differential cooling between the surface and the inside of the fiber. This is called quenching. Finally, if the fiber is made of two or more layers (for example, a core and a cladding made of two different polymers), the differential expansion rate can cause internal stresses that lead to stress-birefringence.

The draw-alignment mechanism in a monolithic polypropylene fiber was first studied by Hamza and Kabeel.[63] They used Fizeau fringes to measure the refractive index for light polarized along the fiber axis, n_p, and perpendicular to it, n_t. They found that the refractive index along the fiber axis is the larger one. This is expected from the fact that the refractive index is always larger along a chain since electrons can more easily move in that direction. Figure 3.21 shows a plot of the measured birefringence of the fiber surface, B, which is defined by

$$B = 2\frac{n_p - n_t}{n_p + n_t}, \tag{3.4}$$

as a function of the draw ratio and a fit to an exponential. The parameter B_0 is the birefringence for fully aligned chains while $A + B_0$ is the residual birefringence of the unstretched material.

Peng and Chu measured the birefringence of a fiber with a cladding and core using an Interphako microscope (Section 5.1.1 describes the principles underlying the Interphako microscope). The advantage of the Interphako microscope is that it

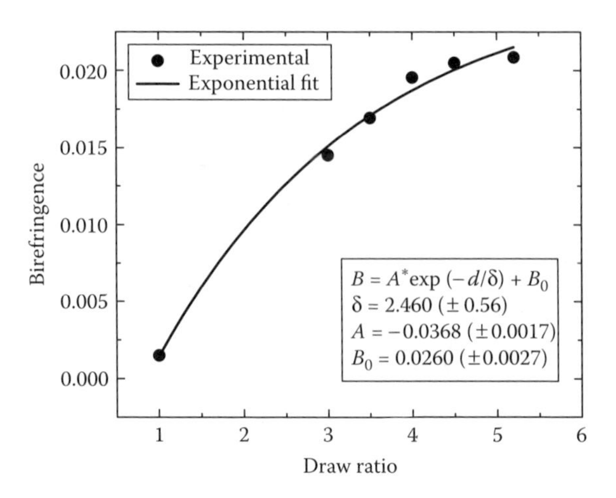

FIGURE 3.21 Birefringence at a wavelength of 546.1*nm* of a polypropylene fiber as a function of the draw ratio.

can determine the birefringence as a function of radial coordinate, so the average birefringence of the core and cladding can be determined separately.

Peng and Chu found that the birefringence of the core is smaller than for the cladding and that it increases with draw speed[64] — consistent with the results of Hamza and Kabeel. Subsequently, the fiber was repeatedly annealed in a cycle corresponding to a soak at 80°C for 2.5 hours followed by a slow cooling ramp to room temperature over an 8-hour period. This process was repeated until the birefringence no longer changed with additional cycling. The results are displayed in Table 3.4. The core birefringence changed dramatically while the cladding birefringence remained high. They concluded that differential thermal expansion between the core and cladding is the only mechanism that should persist after annealing. Clearly, even before annealing, stresses from differential thermal expansion is one of the dominant mechanisms in the cladding. As such, the refractive index in the cladding is thus lower for a propagating mode than it would be for the same isotropic polymer, leading to tighter confinement of the mode than would be predicted for the isotropic polymer.

TABLE 3.4
Birefringence of the core and cladding of a fiber after drawing[64]

Birefringence	Core	Cladding
After draw	2.585×10^{-4}	8.372×10^{-4}
After annealing	0.458×10^{-4}	4.047×10^{-4}

3.4 MECHANICAL PROPERTIES OF FIBERS

In addition to the drawing process's effect on the optical properties, there are also profound effects on the mechanical properties. In applications that require specialized mechanical properties, it is important to choose drawing conditions that optimize the appropriate properties of the polymer. In this section, we review how various mechanical properties of a fiber depend on the drawing history.

The drawing process causes polymer chains to align while thermal agitation causes the chains to randomize into an isotropic state. The net degree of alignment of a fiber is thus determined by these two competing processes. When a fiber is drawn at high temperature and pulled slowly so that the fiber is in its flowing state (well above the glass transition temperature) for a long time compared with the time it takes the chains to randomize, the resulting fiber will be nearly isotropic. If the fiber is pulled quickly at a low temperature, a high degree of chain anisotropy would be expected. Figure 3.22 shows the g-normalized measured force (F/g in units of grams) on a PMMA fiber for fixed preform feed speed and five different temperatures (bottom abscissa); and the Draw force measured for fixed temperature and variable feed space (top abscissa).[51] The behavior observed is as expected.

The final ordered state of the polymer is one of lower entropy than for the isotropic case. Therefore, entropic processes will cause a relaxation of alignment over time to maximize the entropy. Since room temperature is well below the glass transition temperature of most polymers used for making fibers, this relaxation process will be slow. However, the relaxation of anisotropy can be accelerated

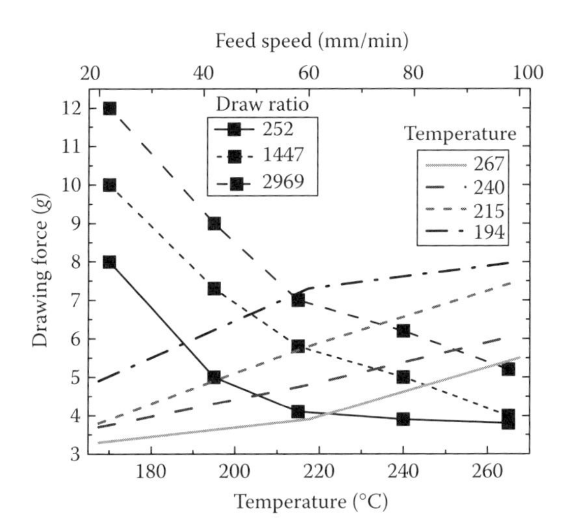

FIGURE 3.22 The draw force as a function of drawing temperature (bottom abscissa) and feed speed (top abscissa). Adapted from Ref. [51] with permission of the American Institute of Physics.

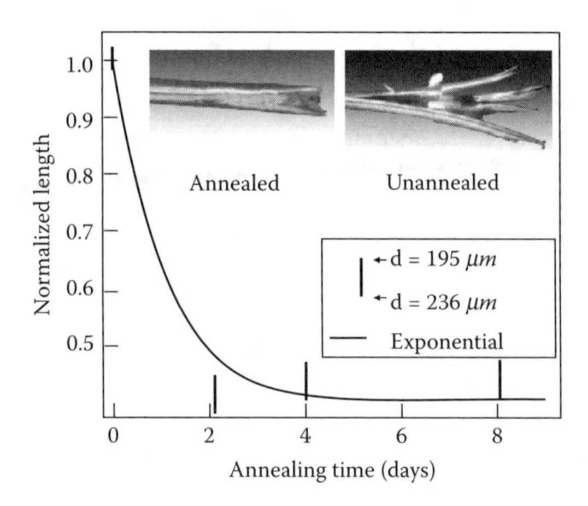

FIGURE 3.23 Normalized fiber length as a function of annealing time. Inset shows a fractured fiber under uniaxial strain before and after annealing. Adapted from Ref. [51] with permission of the American Institute of Physics.

at elevated temperature. Thus, if an isotropic fiber is heated, it is expected to shrink in length and expand in diameter.[65] Figure 3.23 shows the length as a function of annealing time for a PMMA polymer fiber doped with ISQ dye at a concentration of 0.25% by weight.[51] The solid curve is a fit to an exponential. Note that the fiber shrinks to about half its original length. Differential expansion due to inhomogeneities in the anisotropy leads to curling and bending in addition to shrinkage, so length measurements can have large uncertainties.

The bars showing the data in Figure 3.23 represent fibers with initial diameters that range from $195\mu m$ (top of bar) to $236\mu m$ (bottom of bar). All of these fibers were drawn at 160°C, well below the typical drawing temperature of 240°C. The degree of shrinking at the higher temperature (which is more typical in commercial drawing processes) is usually less than 6%. The inset in Figure 3.23 shows that when an annealed isotropic fiber is fractured through uniaxial strain, the break is relatively clean; but when an anisotropic polymer fiber is fractured, fraying at the fracture point results.

The mechanical properties of a polymer fiber are measured with a mechanical spectrometer. This instrument holds the fiber at each end and pulls it so that the elongation changes linearly with time. A gauge measures the force being applied to the fiber for a given elongation. The pulling force per unit area of the fiber end is called the stress and the change in fiber length divided by the initial length is called the strain. The gray curve in Figure 3.24 shows a typical stress versus strain curve for a DR1 dye-doped PMMA fiber. The stress is approximately a linear function until the yield point, where the fiber softens and small increments in the stress yield a large increase of the length. This effect is due to the polymer chains untangling and slipping against each other. Aptly, this is called strain softening.

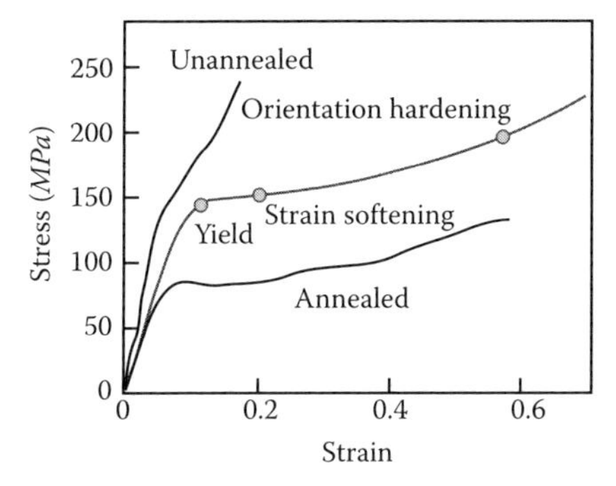

FIGURE 3.24 The gray curve shows a typical stress strain curve for optical fiber doped with DR1 dye with diameter $d = 0.1880mm$ and length $l_0 = 15.80mm$ drawn at a temperature of $240°C$. Also shown is the effect of annealing on the mechanical behavior of fibers drawn at low temperature (dye: ISQ; drawing temperature: $160°C$; strain rate: $0.2/min$; diameter: $0.2057mm$). Adapted from Ref. [51] with permission of the American Institute of Physics.

At higher strain, the chains become aligned and the fiber becomes stiffer. This is called orientational hardening. At higher strain the fiber breaks.

The two dark curves in Figure 3.24 show the stress-strain curve for ISQ dye-doped PMMA fiber that was fabricated at low temperature and high strain — before and after annealing. In the un-annealed fiber, the chains are already aligned so that the slope of the curve is everywhere higher than for the annealed fiber. So highly anisotropic fibers that are drawn at low temperature are stronger since it takes more stress to cause fracture. In this example, the un-annealed fiber breaks at about 250 MPa while the annealed one breaks at a stress of under 150 MPa. The stress at which the fiber fractures is called the tensile strength. Young's modulus, on the other hand, is related to the slope of the stress-strain curve in the limit of zero strain. The yield strength of a fiber is defined as the stress at the yield point. Table 3.5 shows the yield stress as a function of strain rate. The yield strength clearly increases as the strain rate increases.

TABLE 3.5
Effect of strain rate on yield strength[51]

Strain rate (min^{-1})	Yield strength (MPa)
0.1	75.7 ± 1.4
0.2	76.4 ± 2.9
0.3	80.5 ± 4.0
0.4	83.8 ± 5.6
0.5	85.0 ± 3.0

TABLE 3.6
Effect of annealing on the mechanical properties of fibers drawn at low temperature (dye: ISQ; drawing temperature: 160°C)[51]

	Un-annealed	Annealed
Young's modulus (GPa)	2.60 ± 0.14	1.55 ± 0.03
Yield point (MPa)	121 ± 10	78.5 ± 2.7
Tensile strength (MPa)	216 ± 31	127.2 ± 5.9
Ductility	20.4% ± 2.77%	78.0% ± 9.1%

Table 3.6 summarizes the mechanical properties of annealed and highly anisotropic un-annealed dye-doped PMMA fibers. The ductility is defined as the strain at failure, and is a measure of how much the material can be stretched. Note that the tensile strength and the ductility are inversely related to each other. As one increases the other decreases. Table 3.7 shows the mechanical properties of a dye-doped fiber that was drawn at high temperature. Note that the properties are very different from the fiber made at low temperature. Clearly, the mechanical properties of a polymer fiber is a sensitive and complicated function of the draw conditions. While there are qualitative similarities between many polymeric materials, the relationship between the processing conditions and the mechanical properties must be determined individually for each material.

Table 3.8 shows how dopant concentration affects the mechanical properties. Even small amounts of dopant can have a large effect. It is therefore no wonder that the draw conditions that work for making a fiber with one material often fail miserably for what appears to be a similar material. It must be stressed that the tabulations presented in this section are all for monolithic fiber. In layered structures that contain a core and cladding, more complications arise due to the

TABLE 3.7
Effect of annealing on the mechanical properties of fibers drawn at high temperature (dye: DR1; drawing temperature: 240°C; draw ratio: 21710)[51]

	Un-annealed	Annealed
Young's modulus (GPa)	1.77 ± 0.06	1.38 ± 0.19
Yield point (MPa)	61 ± 14	38.5 ± 3.6
Tensile strength (MPa)	70 ± 12	49 ± 11
Ductility	26.9% ± 6.6%	37.7% ± 9.3%

TABLE 3.8
Effect of DR1 concentration on the mechanical properties of fibers[51]

Dopant conc. (weight %)	Yield strength (*MPa*)	Tensile strength (*MPa*)	Young's modulus (*GPa*)	Ductility
0.15%	54.7 ± 1.8	53.7 ± 4.7	1.55 ± 0.19	4.74% ± 1.11%
0.075%	42.7 ± 3.6	45.1 ± 4.4	1.19 ± 0.08	5.66% ± 1.51%
0.0375%	65.1 ± 4.3	118.6 ± 16.5	1.63 ± 0.27	39.25% ± 1.21%
0.0094%	75.9 ± 2.1	143.8 ± 3.9	1.98 ± 0.09	42.60% ± 5.24%
0.0047%	64.8 ± 2.0	94.9 ± 4.3	1.58 ± 0.11	63.05% ± 1.9%
0.0023%	61.7 ± 2.1	83.2 ± 4.0	1.55 ± 0.05	64.54% ± 6.59%

dissimilarity of the component materials. Similarly, in photonic crystal materials such as holey fibers, the presence of an array of holes can also have a profound effect on the required draw conditions. As such, it is not a simple task to make a high-quality fiber. Most good fibers that are reported in the literature are the result of extensive trial and error.

4 Theory of Refractive Index and Loss

This chapter develops the theory for the relationships between molecular and bulk susceptibilities. The theory is applied to calculating the refractive index and loss of dye-doped polymers and mixtures of polymers. The chapter concludes with a practical example.

4.1 REFRACTIVE INDEX

The refractive index difference between the core and cladding must be precisely chosen to make a single-mode waveguide. In a fiber in which the cladding is a polymer and the core is a dye-doped polymer, the refractive index difference can be controlled by adjusting the concentration of the dye. We end this chapter by deriving the relationship between the concentration of a dye-doped polymer and its refractive index under several useful conditions.

4.1.1 ISOTROPIC POLYMER

For a homogeneous and isotropic material, the refractive index is independent of the plane of the electric field of the light beam. (Note that this is called the polarization of the light beam and should not be confused with the material polarization, which is defined as the dipole moment per unit volume.) Even if a molecule itself is highly anisotropic, the bulk refractive index will be isotropic if the molecules are randomly oriented.

We begin by considering a one-dimensional molecule, where the polarizability is nonzero along the long axis of the molecule and zero in the two orthogonal directions. Recall that the polarizability, α, is a measure of the induced electric dipole moment per unit of applied electric field, that is,

$$p_i = \alpha_{ij} E_j, \tag{4.1}$$

where summation notation over repeated indices is implied and p_i is the ith component of the electric dipole moment. *Note that we use Gaussian units throughout this chapter.* For the one-dimensional molecule, this leads to

$$p_z = \alpha_{zz} E_z, \tag{4.2}$$

where E_z is the component of the electric field along the long axis of the molecule.

The linear susceptibility is given by an orientational average over the polarizabilities of the collection of molecules. (For now, we neglect the polymer.)

If the number density of the molecules is N, the linear susceptibility is given by

$$\chi_N^{(1)} = \frac{N}{2} \int_{-1}^{+1} d(\cos\theta) a_{ii'}(\cos\theta) a_{jj'}(\cos\theta) \alpha^*_{i'j'} \tag{4.3}$$

where $a_{jj'}(\cos\theta)$ is the jj' element of the rotation matrix, which depends on the direction cosines. $\alpha^*_{i'j'}$ is the $i'j'$ component of the dressed polarizability tensor and the factor of two in the denominator is the normalization factor. For a one-dimensional molecule, Equation (4.3) yields

$$\chi_N^{(1)} = \frac{N}{3} \alpha^*_{zz}. \tag{4.4}$$

The dressed susceptibility takes into account that a molecule reacts not only to the applied external field, but also is affected by the induced dipole moments of the other molecules in the collection. The dressed susceptibility is related to the vacuum susceptibility through a local field model. We use the Lorentz local field, which gives the dressed polarizability:

$$\alpha^* = \left(\frac{n^2+2}{3}\right)^2 \alpha, \tag{4.5}$$

where we have dropped the z subscripts. n is the refractive index of the material, which in our case is made of the collection of molecules.

Equation (4.4) is the electronic susceptibility, and describes the dipole moment that is induced in the molecules as the electron cloud responds to the applied electric field. The polymer also contributes to the susceptibility. If the concentration of dopant molecules is low enough to not change the density of the polymer, the total susceptibility can be written as the sum of the bulk polymer susceptibility and the dopant contribution:

$$\chi^{(1)} = \chi^{(1)}_{poly} + \chi^{(1)}_N = \chi^{(1)}_{poly} + \frac{N}{3}\alpha^*_{zz}. \tag{4.6}$$

When making a single-mode polymer optical fiber, the refractive index difference between the core (dye-doped polymer) and the cladding (neat polymer) is about 0.005. The refractive index of the cladding is about 1.48, so the refractive index contribution of the dye molecules, while essential in making the waveguide, is necessarily small. As such,

$$\chi^{(1)}_{poly} \gg \frac{N}{3}\alpha^*_{zz}. \tag{4.7}$$

The refractive index of the dye doped polymer is given by

$$n = \left(1 + 4\pi\left(\chi^{(1)}_{poly} + \frac{N}{3}\alpha^*_{zz}\right)\right)^{1/2} = n_0\left(1 + \frac{4\pi N}{3n_0^2}\alpha^*_{zz}\right)^{1/2}, \tag{4.8}$$

where n_0 is the refractive index of the neat polymer,

$$n_0 = \left(1 + 4\pi\chi^{(1)}_{poly}\right)^{1/2}. \tag{4.9}$$

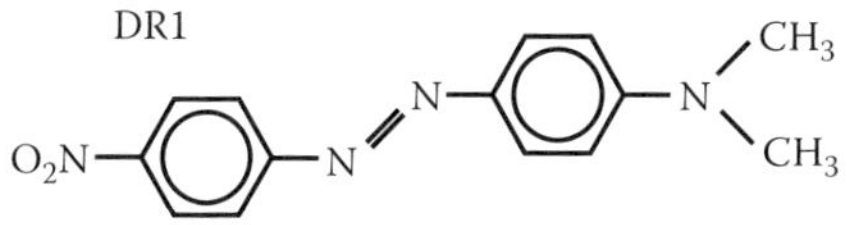

FIGURE 4.1 The DR1 molecule.

Now, we apply the approximation given by Equation (4.7) to expand Equation (4.8) to first order in the dye concentration,

$$\boxed{n = n_0 + \frac{2\pi N}{3n_0}\alpha^*_{zz}.} \tag{4.10}$$

Example: Disperse Red 1 Azo Dye

Since the result given by Equation (4.10) is in Gaussian units, the polarizability α^* has units cm^3 and the number density has units cm^{-3}. For the disperse red azo dye number 1 (DR1), shown in Figure 4.1, $\alpha^* = 9 \times 10^{-23} cm^3$ in the near IR part of the spectrum when $n_0 = 1.48$. To get a refractive index difference of $n - n_0 = 0.005$, we require a concentration of $N = 1.2 \times 10^{19} molecules/cm^3$. This corresponds to 0.6% dye by weight in the polymer. The relationship between number density and dye concentration will be discussed in Section 4.1.3.

We note that there are more sophisticated models for the refractive index.[66] As long as we are dealing with the off-resonant susceptibility, however, Equation (4.10) works well. Next we consider the case of an aligned system of molecules.

4.1.2 Birefringent Polymers

It is possible to make a polymer birefringent by stretching it. In doing so, the polymer chains and the dopant molecules align in the stretching direction. Axial stress can be applied to a fiber when pulling it at a lower temperature. This results in a refractive index that is larger on axis (direction of chain alignment) and smaller perpendicular to the axis than the neat polymer.

The birefringence of a material depends on the orientational order of its molecules. For simplicity, we treat the case of a collection of axially symmetric molecules that make an azimuthally symmetric sample ($n_{xx} = n_{yy} \neq n_{zz}$, so z is the unique axis). (The general treatment can be found in the literature.[66]) Let's define a function $G(\cos\theta)$, which gives the probability density of finding a molecule oriented between polar angles θ and $\theta + \Delta\theta$. The probability of finding a molecule in this range, $P(\theta, \Delta\theta)$, is given by

$$P(\theta, \Delta\theta) = G(\cos\theta)\Delta\cos\theta, \tag{4.11}$$

where we note that the result is independent of the axial angle ϕ because of symmetry.

The orientational distribution function can alternatively be represented as a series expansion of an orthonormal set of functions. The coefficients of this expansion are called the order parameters and the distribution function can be completely described by them. In optics applications, it is most convenient to represent this series in terms of the Legendre polynomials, because only the second-order order parameter, $\langle P_2 \rangle$, determines the refractive index of the material. In particular, $\langle P_2 \rangle$ is calculated from $G(\cos\theta)$ according to

$$\langle P_2 \rangle = \frac{2}{5}\int_{-1}^{+1} G(\cos\theta) P_2(\cos\theta) d\cos\theta, \tag{4.12}$$

where

$$P_2(\cos\theta) = \frac{1}{2}(3\cos^2\theta - 1). \tag{4.13}$$

If we know the order parameter $\langle P_2 \rangle$ for a given material, the linear susceptibility tensor for the molecular ensemble is given by

$$\chi_{zz}^{(1)} = \frac{N}{3}(1 + 2\langle P_2 \rangle)\alpha_{33}^*, \tag{4.14}$$

and

$$\chi_{xx}^{(1)} = \chi_{yy}^{(1)} = \frac{N}{3}(1 - \langle P_2 \rangle)\alpha_{33}^*, \tag{4.15}$$

where, to differentiate between the lab and molecular frame, we use the new notation that 3 is the long axis of the molecule and z the laboratory z-axis.

In the case where we have a polymer host and dopants, the susceptibilities are given by

$$\chi_{zz}^{(1)} = \chi_{zz}^{(1)poly}(1 + 2\langle P_2' \rangle) + \frac{N}{3}(1 + 2\langle P_2 \rangle)\alpha_{33}^*, \tag{4.16}$$

and

$$\chi_{xx}^{(1)} = \chi_{yy}^{(1)} = \chi_{xx}^{(1)poly}(1 - \langle P_2' \rangle) + \frac{N}{3}(1 - \langle P_2 \rangle)\alpha_{33}^*, \tag{4.17}$$

where $\langle P_2' \rangle$ is the order parameter of the polymer host. Applying the same procedure we used in getting Equation (4.10), the zz-tensor component of the refractive index of a guest-host polymer is

$$\boxed{n_{zz} = n_{zz}^0 + \frac{2\pi N}{3n_{zz}^0}(1 + 2\langle P_2 \rangle)\alpha_{33}^*,} \tag{4.18}$$

where n_{zz}^0 is the polymer's refractive index in the $\hat{z}$-direction:

$$n_{zz}^0 = \left[1 + 4\pi\chi_{zz}^{(1)poly}(1 + 2\langle P_2' \rangle)\right]^{1/2}. \tag{4.19}$$

The xx(and yy)-tensor component of the refractive index of a guest-host polymer is

$$\boxed{n_{xx} = n_{xx}^0 + \frac{2\pi N}{3n_{xx}^0}(1 - \langle P_2 \rangle)\alpha_{33}^*,} \tag{4.20}$$

where

$$n_{xx}^0 = \left[1 + 4\pi\chi_{xx}^{(1)poly}\left(1 - \langle P_2' \rangle\right)\right]^{1/2}. \tag{4.21}$$

Example: Poled Polymer

It is not always a trivial matter to determine the order parameter $\langle P_2 \rangle$. For illustration, we calculate the refractive index of a poled polymer, which is prepared by applying an electric field to align the molecules. Since the dopant molecules generally have a larger dipole moment and molecular birefringence than the polymer, the bulk birefringence from a poled polymer will be dominated by the dopants. As such, $n^0_{xx} = n^0_{yy} = n^0_{zz} \equiv n_0$. When the polymer does not contribute to the birefringence, the bulk birefringence is

$$n_{zz} - n_{xx} = \frac{2\pi N}{n_0} \langle P_2 \rangle \alpha^*_{33}. \tag{4.22}$$

For a poled polymer, when an external field, $\vec{E}$, aligns the dipole moment, $\vec{m}$, the induced second-order parameter is given by[66]

$$\langle P_2 \rangle = \frac{1}{15} \left(\frac{m^* E}{kT} \right)^2, \tag{4.23}$$

where T is the temperature. Note that Equation (4.23) is obeyed only when the molecules are free to reorient on time scales comparable to the time duration when the electric field is on. This will be obeyed when the material is above its glass transition temperature. When the material is cooled with the field applied, the birefringence is locked in place.

Substituting Equation (4.23) into Equation (4.22), we finally get

$$n_{zz} - n_{xx} = \frac{2\pi N}{15 n_0} \left(\frac{m^* E}{kT} \right)^2 \alpha^*_{33}. \tag{4.24}$$

4.1.3 Relationship Between Number and Weight Percent

The concentration of dopants in a polymer matrix are often quoted as a weight percent because this is experimentally the simplest to calculate. For example, if the dye and polymer are mixed in 1 to 9 proportions by weight, the dye concentration is simply 10 wt%. In all calculations of the susceptibility, the number density is required. In this section, we calculate the relationship between weight percent and number density. Note that there are several methods for deriving such a relationship depending on the approximations that are used. In each case, however, they yield approximately the same numerical results.

A dye-doped polymer is prepared by mixing a mass w_d of a dye in a mass w_p of polymer. The weight fraction of dye, w, is then

$$w = \frac{w_d}{w_d + w_p}. \tag{4.25}$$

We assume that the volume occupied by the polymer does not change when the dye is added. If the polymer's density is ρ, the volume occupied by the polymer, V, is

$$V = \frac{w_p}{\rho}. \tag{4.26}$$

The number of dye molecules added to the polymer, N_d is given by

$$N_d = \frac{w_d}{m_d} N_A, \tag{4.27}$$

where $N_A = 6.02 \times 10^{23}$ is Avagadro's Number and m_d is the atomic weight of the dye molecule.

Using Equations (4.26) and (4.27), we get the number density of dye, N,

$$\boxed{N = \frac{N_d}{V} = \frac{w_d}{w_p}\left(\frac{\rho N_A}{m_d}\right).} \tag{4.28}$$

We can invert Equation (4.25) to solve for the weight ratio w_d/w_p in terms of the weight fraction w and use this in Equation (4.28) to get

$$\boxed{N = \frac{w}{1-w}\left(\frac{\rho N_A}{m_d}\right).} \tag{4.29}$$

It is also convenient to convert from number density to weight fraction. Inverting Equation (4.29), we get

$$\boxed{w = \frac{N}{N + \left(\frac{\rho N_A}{m_d}\right)}.} \tag{4.30}$$

4.1.4 Mixing Polymers to Control the Refractive Index

In applications where a fiber is being used for nonlinear optical interactions, the dye concentration in the core needs to be maximized. Since the dye molecules elevate the refractive index of the core, the waveguide may become multimode at the wavelength of interest. To prevent this from happening, it is possible to use polymer mixtures in both the core and cladding to compensate for the dye's contribution to the refractive index.

In this section, we calculate the refractive index of a two-polymer blend using the Maxwell Garnet theory (which applies for all concentrations). Subsequently, we add the effects of the dye-dopants using the approach described in Sections 4.1.1 and 4.1.2.

The underlying basis for the Maxwell Garnet theory is to model a composite material as being made of macroscopic regions of each material — which we will call clumps. By macroscopic, we mean a piece of material that behaves as a continuous dielectric. In many local field models, a single molecule is approximated as a small dielectric ellipsoid and the resulting local field models are found to work well. As such, our "macroscopic" assumption will still work in the dilute regime. It is, however, important to point out that the size of the "macroscopic" regions need to be small compared to the wavelength of light to avoid scattering.

In order to simplify the calculation of the average electric fields within a material, it is most straightforward to assume a shape for the material clumps. Furthermore, we will assume that the higher concentration material is a continuous

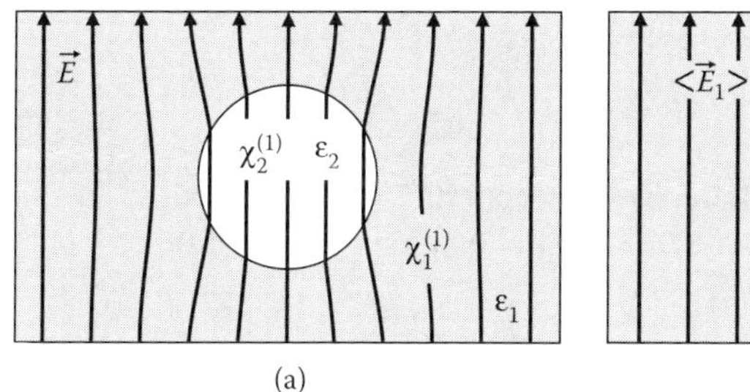

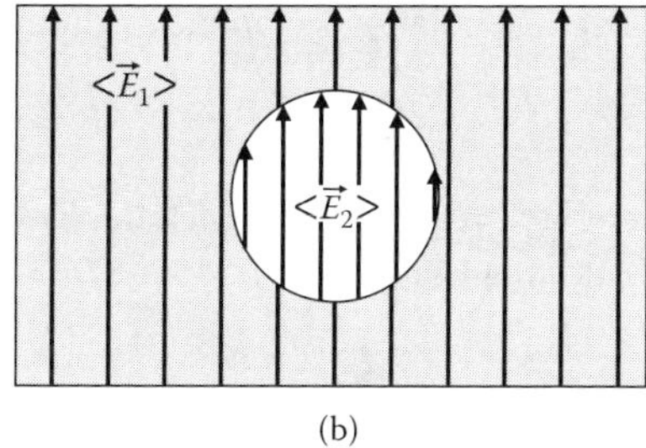

FIGURE 4.2 In the Maxwell Garnet treatment for a) a spherical inclusion, b) the average field in the sphere $\langle \vec{E}_2 \rangle$ and in the surrounding medium $\langle \vec{E}_1 \rangle$ are uniform.

(and contiguous) host while the clumps are randomly dispersed in the material. Many researchers have studied such models in great detail to understand the implications of picking a particular geometry for the composition of the material. In our treatment, we will assume that the clumps are spheres. The end result will turn out to be independent of the shape that is chosen.

Figure 4.2a shows a spherical inclusion with linear susceptibility $\chi_2^{(1)}$ in a medium of linear susceptibility $\chi_1^{(1)}$. A uniform applied electric field is distorted by the sphere.[66] The average electric field in the host material can be shown to be uniform, as depicted in Figure 4.2b.

For an inclusion of volume fraction f, and average electric fields in the inclusion and surrounding medium $\langle \vec{E}_2 \rangle$ and $\langle \vec{E}_1 \rangle$ respectively, the average field, $\langle \vec{E}_{av} \rangle$, in the medium is

$$\langle \vec{E}_{av} \rangle = (1 - f)\langle \vec{E}_1 \rangle + f \langle \vec{E}_2 \rangle. \tag{4.31}$$

The average electric field is uniform, so we will express the average fields as scalar quantities. Furthermore, because it is understood that we are dealing with average fields, we drop the brackets and the 'av' subscript so that Equation (4.31) becomes

$$E = (1 - f)E_1 + f E_2. \tag{4.32}$$

Maxwell Garnett starts with the relationship between the average electric field inside the sphere and outside the sphere, which for a sphere that is much smaller than the wavelength of light is treatable as a simple electrostatic boundary value problem. The magnitudes of the fields are related to each other according to

$$E_2 = \frac{3\epsilon_1}{2\epsilon_1 + \epsilon_2} E_1. \tag{4.33}$$

With the average polarizations given by

$$P_1 = \chi_1^{(1)} E_1, \tag{4.34}$$

and

$$P_2 = \chi_2^{(1)} E_2, \tag{4.35}$$

the average polarization in the medium is

$$P = \chi^{(1)} E = (1 - f)P_1 + f P_2 \tag{4.36}$$

where $\chi^{(1)}$ is the average or effective susceptibility of the composite medium. Substituting Equations (4.34) and (4.35) into Equation (4.36) yields

$$\chi^{(1)} E = (1 - f)\left(\frac{\epsilon_1 - 1}{4\pi}\right) E_1 + f\left(\frac{\epsilon_2 - 1}{4\pi}\right) E_2. \tag{4.37}$$

We can eliminate E_1 and E_2 from Equation (4.37) with the help of Equations (4.32) and (4.33):

$$\chi^{(1)} = \frac{(1 - f)\left(\frac{\epsilon_1 - 1}{4\pi}\right) + f\left(\frac{\epsilon_2 - 1}{4\pi}\right)\left(\frac{3\epsilon_1}{2\epsilon_1 + \epsilon_2}\right)}{1 - f + f\left(\frac{3\epsilon_1}{2\epsilon_1 + \epsilon_2}\right)}. \tag{4.38}$$

After the algebraic dust settles, Equation (4.38) can be used to express the effective dielectric function $\epsilon = 1 + 4\pi\chi^{(1)}$:

$$\epsilon = \epsilon_1 \left[\frac{\epsilon_2 + 2\epsilon_1 + 2f(\epsilon_2 - \epsilon_1)}{\epsilon_2 + 2\epsilon_1 - f(\epsilon_2 - \epsilon_1)}\right]. \tag{4.39}$$

This is the Maxwell Garnet result. For a small volume fraction, the Maxwell Garnet effective dielectric function reduces to

$$\epsilon \approx \epsilon_1 \left(1 + 3f\left(\frac{\epsilon_2 - \epsilon_1}{\epsilon_2 + 2\epsilon_1}\right)\right). \tag{4.40}$$

There are many different methods for calculating dielectric models. While we will not delve into the subject here, the Bruggeman model warrants mention. The dielectric constant is given by solving.[156]

$$f\frac{\epsilon_2 - \epsilon}{\epsilon_2 + 2\epsilon} + (1 - f)\frac{\epsilon_1 - \epsilon}{\epsilon_1 + 2\epsilon} = 0. \tag{4.41}$$

Solving for ϵ, we get

$$\begin{aligned}\epsilon = {} & \frac{3}{4}f(\epsilon_2 - \epsilon_1) - \frac{1}{4}(\epsilon_2 - 2\epsilon_1) \\ & + \frac{1}{4}\left(9f^2(\epsilon_2 - \epsilon_1)^2 - 6f(\epsilon_2 - 2\epsilon_1)(\epsilon_2 - \epsilon_1) + (\epsilon_2 + 2\epsilon_1)^2\right)^{1/2}.\end{aligned} \tag{4.42}$$

While the Maxwell Garnet theory works well in the low concentration limit of species number 2 when $\epsilon_2 > \epsilon_1$, the Bruggeman model is better when both species are of comparable volume fraction.

We can test the models using symmetry arguments. For example, if we interchange ϵ_1 and ϵ_2 and simultaneously interchange f with $1 - f$, the model should give the same results. The Bruggeman model given by Equation (4.41) is

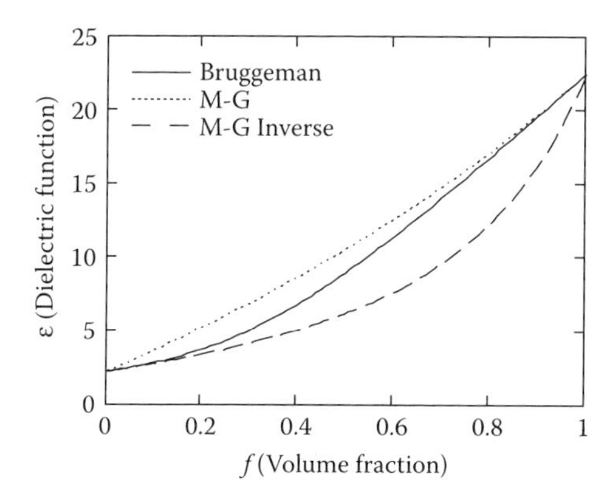

FIGURE 4.3 The dependence of the dielectric function on the volume fraction with $\epsilon_1 = 2.2$ and $\epsilon_2 = 22.5$

clearly symmetric. The Maxwell Garnet model, however, is not. Figure 4.3 shows a plot of the Bruggeman model using Equation (4.42), the Maxwell Garnet theory using Equation (4.39), and the Maxwell Garnet theory with the parameters interchanged. Note that the Bruggeman model is both consistent with the low concentration limit of species 2 in the Maxwell Garnet theory (corresponding to $f = 0$) and at $f = 1$ is consistent with the interchanged Maxwell Garnet model. The Bruggeman model is therefore more useful for polymer blends over a broad range of volume fraction.

In the case of polymer fibers, the refractive index differences are small. A polymer blend made with two materials of refractive index 1.48 and 1.56 yields a range of refractive indices which exceeds the needs of most applications. In this case, the dielectric constants are 2.19 and 2.43, respectively. This range corresponds to a fractional difference of 0.1. We can therefore expand the dielectric models to first order in this difference. Both the Bruggeman and the Maxwell Garnet models give the same result:

$$\epsilon = \epsilon_1 + (\epsilon_2 - \epsilon_1) f. \tag{4.43}$$

Using Equation (4.43), the refractive index of the polymer is

$$\boxed{n = \left[n_1^2 + \left(n_2^2 - n_1^2\right) f\right]^{1/2} \approx n_1 \left[\left(1 - \frac{1}{2} f\right) + \frac{1}{2} f \frac{n_2^2}{n_1^2}\right],} \tag{4.44}$$

where we have made the approximation that the refractive index difference between the two polymers is small. When calculating the refractive index of a dye-doped blend, we use Equation (4.44) as the polymer's refractive index in Equations (4.18) and (4.20).

To make Equation (4.44) useful in determining the refractive index of a blend, we need to convert the volume fraction to a weight percent. For a polymer density of species i as ρ_i, volume V_i will be made by using a mass of polymer w_i. The volume fraction of polymer 2 is then

$$f = \frac{V_2}{V_1 + V_2} = \frac{\frac{w_2}{\rho_2}}{\frac{w_1}{\rho_1} + \frac{w_2}{\rho_2}}. \tag{4.45}$$

For a polymer 2 weight fraction of

$$w' = \frac{w_2}{w_1 + w_2}, \tag{4.46}$$

the volume fraction given by Equation (4.45), upon using Equation (4.46) to eliminate w_2 becomes

$$\boxed{f = \frac{1}{1 + \left(\frac{1-w'}{w'}\right)\frac{\rho_2}{\rho_1}}.} \tag{4.47}$$

We can invert Equation (4.47) to express the weight fraction in terms of the volume fraction:

$$\boxed{w' = \frac{1}{1 + \left(\frac{1-f}{f}\right)\frac{\rho_1}{\rho_2}}.} \tag{4.48}$$

4.2 OPTICAL LOSS

The optical loss is separated into two broad categories: intrinsic loss, which originates from the material and is process-independent; and extrinsic loss, which arises from impurities or materials processing. For example, bubbles that form during the drawing process lead to loss, but bubbles are not an intrinsic property of the material and can be eliminated by baking the material to reduce trapped volatiles and can be prevented from forming by drawing the fiber at a low enough temperature. All materials absorb light at wavelengths corresponding to electronic or nuclear resonances in the molecules. As such, aside from meticulously choosing molecules with the desired windows of transparency when used in making a material, these losses cannot be affected by processing.

There are many sources of extrinsic and intrinsic loss. As an example of intrinsic loss, the hydrogen atoms in a polymer act like masses on a spring and therefore absorb light at the characteristic frequency of the "spring" and its harmonics. As such, this type of loss can be lowered by replacing the hydrogen atoms with heavier ones such as deuterons (a hydrogen atom with an extra neutron in the nucleus), since this pushes the absorption resonance further into the infrared. One disadvantage of deuterated polymers is that the deuterons diffuse to the surface of the material by exchanges between sites where they are replaced with hydrogen nuclei, which are abundant in our surroundings. In turn, these hydrogen nuclei diffuse into the material via the surface through exchanges with deuteron nuclei, thus converting the material to a non-deuterated form. It is also possible to fluorinate a polymer, where all hydrogens are replaced with fluorine atoms. This type of material is less susceptible to diffusion because of the greater mass of the fluorine atom as well as the fact that fluorine is not an isotope of hydrogen, making it impossible for the two nuclei to exchange while conserving energy. However, fluorinated polymers often have undesirable material properties such as being more brittle than the hydrogenated version of the polymer.

Similarly, if dopant molecules are incorporated into a polymer, the characteristic wavelength of absorption depends on the structure of the molecule. To avoid loss due to the dopant, it can also be deuterated or fluorinated. In addition, the position of the electronic absorption peak can be shifted by engineering molecules of varying size or shape.

The number of possible extrinsic sources of loss are huge and include scattering from rough surfaces, impurities, inhomogeneities of the material, and imperfections in the polymer. In addition to scattering, a bend or kink in a fiber can cause light to refract out of the core. The mechanisms for such losses can be quite complex. For example, when a fiber is bent, the geometry changes and in addition, the stress induces a birefringence.

In the remainder of this chapter, we treat several of these mechanisms. Some are calculated in detail while the more complicated cases are described qualitatively.

4.2.1 Absorbance

We begin by considering the intrinsic mechanism of absorption. Since the optical loss is related to the imaginary part of the refractive index, many of the calculations presented above can be reused to model absorption loss. In Section 4.4, a simple harmonic oscillator model of the refractive index is presented. This leads to a description of how the refractive index and loss depend on the wavelength.

For a plane wave propagating along $\hat{z}$, the electric field is given by

$$\vec{E} = Re[E_0 \exp(i(nkz - \omega t))]\hat{\rho}, \tag{4.49}$$

where E_0 is the amplitude of the electric field, $k = 2\pi/\lambda$, ω the angular frequency, $\hat{\rho}$ the electric polarization direction, and n the refractive index. With a complex refractive index,

$$n = n_R + in_I, \tag{4.50}$$

the intensity, which is proportional to the time average of the square of the electric field (Equation (4.49)) becomes

$$I = I_0 \exp\left[-4\pi \frac{n_I z}{\lambda}\right]. \tag{4.51}$$

The absorption coefficient, α (not to be confused with the polarizability!), is defined by the equation

$$I = I_0 \exp[-\alpha z], \tag{4.52}$$

so the absorption coefficient is related to the imaginary part of the refractive index,

$$\boxed{\alpha = 4\pi \frac{n_I}{\lambda}.} \tag{4.53}$$

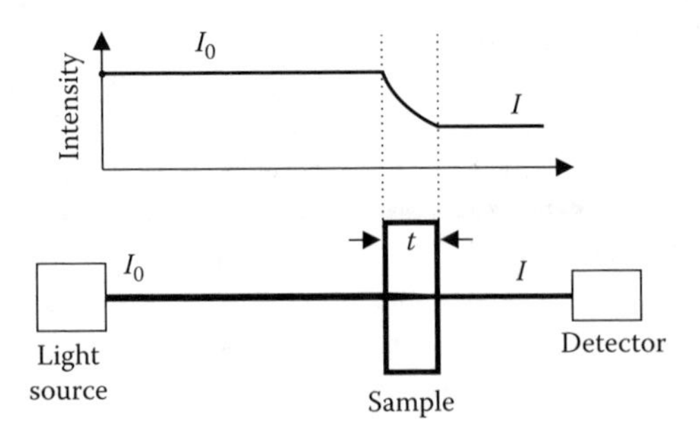

FIGURE 4.4 The optical absorption measurement.

In a spectrometer, a sample of thickness t is placed in a beam of intensity I_0, and the transmitted intensity, I is measured with a photodetector. Figure 4.4 shows a diagram of the experiment. The quantity measured is called the absorbance, A, which is given by

$$\boxed{A = -\log[I/I_0].} \tag{4.54}$$

Using Equations (4.51) and (4.52) the absorbance given by Equation (4.54) is related to the absorption coefficient and the imaginary part of the refractive index according to

$$\boxed{A = \alpha t \log[e] = 4\pi \frac{t n_I}{\lambda} \log[e].} \tag{4.55}$$

When the optical loss of a fiber is quoted in *dB/m*, it is calculated according to

$$\boxed{Loss = -10\frac{1}{t}\log[I/I_0],} \tag{4.56}$$

where the units depend on the unit of length of t. For low-loss fiber, the units used are usually *dB/km* and for polymer fiber are given in dB/m. The loss is clearly related to the other quantities:

$$\boxed{Loss = \frac{10}{t}A = 10\alpha\log[e] = 40\pi\frac{n_I}{\lambda}\log[e].} \tag{4.57}$$

As given by Equation (4.57), the loss can be expressed in terms of α by a constant numerical factor $10\log e$. This yields

$$\boxed{Loss = 4.343\alpha \qquad \text{and} \qquad \alpha = 0.230 Loss.} \tag{4.58}$$

4.2.2 Linear Susceptibility

The linear dielectric function and the linear susceptibility are related to the complex refractive index. If we start with the constitutive equation,

$$\vec{D} = \vec{E} + 4\pi\vec{P}, \tag{4.59}$$

where $\vec{D}$ is the electric displacement and $\vec{P}$ the polarization, and use the relationship between $\vec{D}$ and $\vec{E}$,

$$\vec{D} = \epsilon \vec{E}, \tag{4.60}$$

where ϵ is the dielectric function, and the relationship between $\vec{P}$ and $\vec{E}$,

$$\vec{P} = \chi^{(1)} \vec{E}, \tag{4.61}$$

we can solve Equation (4.59) through (4.61) to get

$$\epsilon = 1 + 4\pi \chi^{(1)}. \tag{4.62}$$

Note that this calculation assumes that the material is linear, which means that the polarization is a linear function of the electric field; and, for simplicity, we have suppressed the tensor nature of the linear susceptibility $\chi^{(1)}$ and the dielectric function ϵ.

The complex refractive index derives from Equation (4.62)

$$n = \sqrt{\epsilon} = (1 + 4\pi \chi^{(1)})^{1/2} = (1 + 4\pi \chi_R + i 4\pi \chi_I)^{1/2}, \tag{4.63}$$

where we have written the first order susceptibility, $\chi^{(1)}$, in terms of the real and imaginary parts, χ_R and χ_I. It is useful to write the real and imaginary parts of the refractive index in terms of the real and imaginary parts of $\chi^{(1)}$. As such, we must rework Equation (4.63).

For simplicity, we define the real quantities A and B as

$$A + Bi = (1 + 4\pi \chi_R) + i 4\pi \chi_R, \tag{4.64}$$

which then allows us to rewrite Equation (4.63) in terms of its magnitude and phase:

$$n = ((A^2 + B^2)^{1/2} \exp[i \tan^{-1}(B/A)])^{1/2}, \tag{4.65}$$

which reduces to

$$n = (A^2 + B^2)^{1/4} \left[\cos\left(\frac{1}{2} \tan^{-1}(B/A)\right) + i \sin\left(\frac{1}{2} \tan^{-1}(B/A)\right) \right]. \tag{4.66}$$

Equation (4.66) can be simplified using the trigonometric identities,

$$\cos\left(\frac{1}{2} \tan^{-1}(B/A)\right) = \left(\frac{1 + \frac{A}{\sqrt{A^2+B^2}}}{2}\right)^{1/2}, \tag{4.67}$$

and

$$\sin\left(\frac{1}{2} \tan^{-1}(B/A)\right) = \left(\frac{1 - \frac{A}{\sqrt{A^2+B^2}}}{2}\right)^{1/2}. \tag{4.68}$$

Combining Equations (4.66), (4.67), and (4.68) and combining terms yields

$$n = \left(\frac{\sqrt{A^2 + B^2} + A}{2}\right)^{1/2} + i \left(\frac{\sqrt{A^2 + B^2} - A}{2}\right)^{1/2}. \tag{4.69}$$

Finally, Equations (4.69) and (4.64) yield the relationships between the complex refractive index, n, and the susceptibility $\chi^{(1)}$,

$$\boxed{n_R = \frac{1}{\sqrt{2}}\left(\sqrt{(1 + 4\pi\chi_R)^2 + (4\pi\chi_I)^2} + 1 + 4\pi\chi_R\right)^{1/2},} \tag{4.70}$$

and

$$\boxed{n_I = \frac{1}{\sqrt{2}}\left(\sqrt{(1 + 4\pi\chi_R)^2 + (4\pi\chi_I)^2} - [1 + 4\pi\chi_R]\right)^{1/2}.} \tag{4.71}$$

Clearly, Equations (4.70) and (4.71) show that the real part (and imaginary part) of the refractive index is related to both the real and imaginary parts of the linear susceptibility.

An important limiting case of Equations (4.70) and (4.71) is in the off-resonant regime where the imaginary part of the refractive index is small compared to the real part. For an optical fiber that is to be used in transmission, high transparency is required so this limit is obeyed. The lowest-order term for the real and imaginary parts are

$$n_R = (1 + 4\pi\chi_R)^{1/2}, \tag{4.72}$$

and

$$n_I = \frac{2\pi\chi_I}{\sqrt{1 + 4\pi\chi_R}} = \frac{2\pi\chi_I}{n_R}. \tag{4.73}$$

Note that these limiting cases follow trivially from Equation (4.63).

With the relationships derived in this section, the optical loss of a material can be modeled in analogy to the refractive index. When the loss is included, however, the refractive index and dielectric function are complex so the expressions must be generalized with caution. For example, the refractive index of a composite will depend on both the refractive index and loss of the substituent materials. Similarly, the optical loss will also depend on both the refractive index and loss. While the loss of a homogeneous mixture is not often calculated, it is useful to understand how the intrinsic material loss affects the derivations that previously assumed a real refractive index. For illustration, we treat the case of a mixture of two polymers.

The simplest case is the two-component system, such as a mixture of polymers when the concentration of one of the polymers is small. Since the dielectric function is the basic material property, we start by considering the effective medium theory given by the real and imaginary parts of Equation (4.43),

$$\epsilon_R = \epsilon_{R1} + (\epsilon_{R2} - \epsilon_{R1}) f, \tag{4.74}$$

$$\epsilon_I = \epsilon_{I1} + (\epsilon_{I2} - \epsilon_{I1}) f. \tag{4.75}$$

To cast Equations (4.74) and (4.75) in terms of the complex refractive index, we use the relationship

$$\epsilon = \epsilon_R + i\epsilon_I = n^2 = n_R^2 - n_I^2 + 2in_R n_I. \tag{4.76}$$

Writing each dielectric function in Equations (4.74) and (4.75) in the form of Equation (4.76), we get

$$n_R n_I = n_{R1} n_{I1} + (n_{R2} n_{I2} - n_{R1} n_{I1}) f \equiv a, \tag{4.77}$$

and

$$n_R^2 - n_I^2 = n_{R1}^2 - n_{I1}^2 + \left[\left(n_{R2}^2 - n_{I2}^2\right) - \left(n_{R1}^2 - n_{I1}^2\right)\right] f \equiv b. \tag{4.78}$$

We can then solve Equations (4.77) and (4.78) for the real and imaginary parts of the refractive index of the composite, yielding

$$n_I^2 = \frac{(b^2 + 4a^2)^{1/2} - b}{2}, \tag{4.79}$$

and

$$n_R^2 = \frac{2a^2}{(b^2 + 4a^2)^{1/2} - b}. \tag{4.80}$$

The machinery derived in the previous sections of this chapter can then be used to generalize the expressions for the refractive index of a composite to calculate the loss. The generalized expressions, however, are algebraically messy and do not add any insights, so they are not presented here.

4.3 BENDING LOSS

The problem of bending loss is too complex to treat here in detail. We therefore begin by presenting a qualitative description of macro-bending loss in the multimode fiber using ray tracing, and then present the results for a single-mode fiber.

4.3.1 MULTIMODE FIBER

When a fiber is bent, there are two effects that can cause light to couple out. First, just from the geometrical changes, some of the rays that were confined in the straight fiber will be beyond the critical angle around the bend. Second, at the bend, the resulting stress in the polymer induces a birefringence.

Figure 4.5a shows five representative rays that illustrate the purely geometrical contribution when there is no refractive index change. The gray region shows the range of ray angles that end up coupling out of the fiber. Note that all of the rays pictured are in a plane that contains the central axis of the fiber. The loss in the limit of no refractive index change can be calculated numerically by following all rays that start within a solid angle less than the critical angle and emanating from a plane perpendicular to the fiber axis far from the bend.

Figure 4.5b shows the effect of birefringence. The solid ray represents the geometrical limit where there is no refractive gradient. In the presence of a refractive gradient, the ray will bend in the direction of the larger refractive index. Furthermore, rays of differing polarization will refract by differing amounts because the refractive index change is usually birefringent. The two dashed rays show that for a positive gradient (in the direction represented by the arrow), the ray will couple

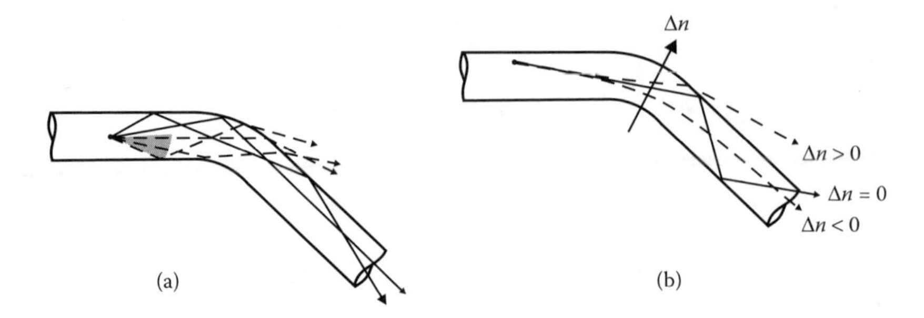

FIGURE 4.5 (a) Tracing rays from a point source that are within the critical angle for the straight fiber. (b) Ray tracing in a bent fiber for a refractive index gradient at the bend that is positive, negative, and zero. Note that all the rays considered are in a plane that contains the fiber axis.

out the fiber. For a negative refractive index change, the ray refracts toward the axis and is therefore not lost.

Once again, it is possible to numerically determine the loss by tracing rays provided that the refractive index ellipsoid is known at all points in the material. Bløtekjær calculated the birefringence of a section of fiber of constant radius of curvature under the approximation that the fiber radius, a, is small compared with the bending radius of curvature, R.[67] It is a useful exercise to discuss some of the highlights of his calculations as well as analytical results.

Figure 4.6a shows the coordinates used for the bent fiber and Figure 4.6b shows a close-up of an infinitesimal piece of length Δz. The arrows represent the forces applied to the infinitesimal piece. The forces on top and bottom labeled F and shown as solid arrows are due to the material above and below the fiber and lead

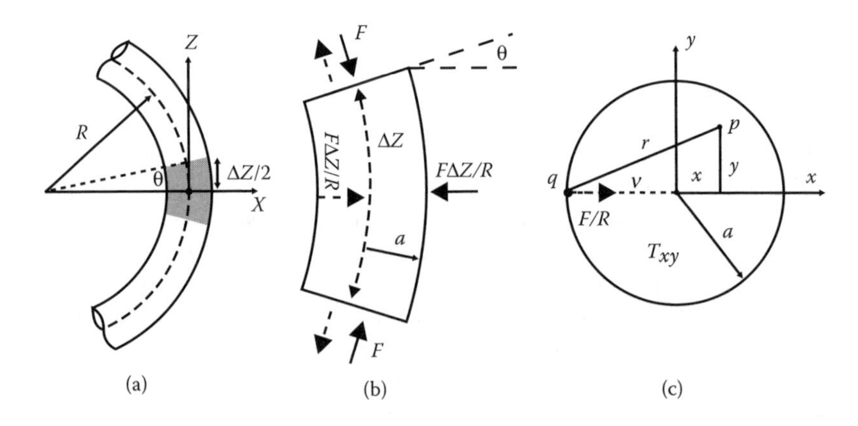

FIGURE 4.6 (a) The coordinate system of the bent fiber; (b) an infinitesimal part of the fiber of length δz with all forces labeled; and (c) the coordinates of the cross-section of the fiber in the $z = 0$ plane. Adapted from Ref. [67] with permission of the Optical Society of America.

mostly to a compression of the fiber in the $\hat{z}$-direction. The dotted arrows show the reaction force of the infinitesimal piece on the part of the fiber above and below. The vector sum of these two forces cancels along the $\hat{z}$-axis, so only an $\hat{x}$-component remains. If the fiber is to remain in equilibrium (i.e., if it's stationary), the total force acting on it must vanish. As such, there must be a force along -x of magnitude $F_x = 2F\sin\theta$. Referring to the shaded part of Figure 4.6a and Figure 4.6b for small angle θ, $\sin\theta \approx \theta = (\Delta z/2)/R$. So finally, we get

$$F_x = F\Delta z/R. \tag{4.81}$$

The dotted arrow shows the force that balances the reaction forces.

For small angles, the tensional force F is mostly along $\hat{z}$, so there is a net compression along the fiber axis. The strain along $\hat{z}$, ϵ, is related to the stress T_{zz} through Young's modulus E according to $T_{zz} = E\epsilon$. (Recall that the strain is the fractional increase in length and the stress is the force per unit area at the point of application of the force.) Since the fiber cross-sectional area is given by πa^2, the strain is given by

$$\epsilon = \frac{T_{zz}}{E} = \frac{F}{E\pi a^2}. \tag{4.82}$$

Note that Young's modulus may be expressed in terms of the Lamé constants λ and μ,

$$E = \frac{\mu(3\lambda + 2\mu)}{\lambda + \mu}. \tag{4.83}$$

For the case of a pure elongation or compression (no $\hat{x}$-component of the applied stress), the strain in the $\hat{x}$- and $\hat{y}$-directions is given by

$$\epsilon_{xx} = \epsilon_{xx} = -\sigma\epsilon = -\frac{\lambda}{2(\lambda + \mu)}\epsilon, \tag{4.84}$$

where σ is the Poisson ratio. Note that the Poisson ratio, Young's modulus, and the Lamé coefficients are all empirically determined mechanical properties of the polymer fiber.

Figure 4.6c shows a slice of the infinitesimal piece defined by the plane $z = 0$. Recall that the force in the $\hat{x}$ direction is given by $F_x = \int T_{xz} dA_z$, where the integral spans the xy-plane. As such, using Equation (4.81) and noting that the stress is constant in this plane (there are no body forces, i.e., no forces applied inside the fiber), we get

$$F_x = F\Delta z/R = \int_{z=0} T_{xz} dA_z = T_{xz}\pi a^2. \tag{4.85}$$

Solving Equation (4.85) for the stress, we get

$$T_{xz} = \frac{F\Delta z}{R\pi a^2}. \tag{4.86}$$

If the stress T_{xz} is uniform in the infinitesimal section of fiber, then the force per unit volume, f, is given by

$$\vec{f} = -\frac{T_{xz}}{\Delta z}\hat{x} = -\frac{F}{R\pi a^2}\hat{x}. \tag{4.87}$$

In equilibrium, Bløtekjær assumes that this force must be balanced by a force in the opposite direction that can be assumed to be applied along a line over the length of the infinitesimal piece of fiber, Δz, that passes through point q on Figure 4.6c. As such, this force on the surface has a linear density (i.e., force per unit length) of F/R. Physically, we can therefore treat this problem as a section of fiber with a uniform volume force that acts to the left, which is balanced by a force to the right on the surface of the fiber that is applied along a line of length Δz at $x = -a$.

Bløtekjær used the more general results from Muskhelishvili[68] on the application of two line forces to a cylinder, to calculate the displacement of the material as a function of the x and y coordinates. The result for the component of the displacement in the x direction, u, is given by[67]

$$u = -\frac{\lambda}{2(\lambda+\mu)} x\epsilon - \frac{F}{4\pi\mu R}\left[\frac{2(\lambda+2\mu)}{\lambda+\mu}\ln\frac{r}{a} + 1 - \cos 2\nu - \frac{\mu(x/a)}{\lambda+\mu} - \frac{1}{2}\left(\left(\frac{x}{a}\right)^2 + 3\left(\frac{y}{a}\right)^2\right)\right], \tag{4.88}$$

where all of the variables are defined in Figure 4.6c. The y-component of the displacement is given by[67]

$$v = -\frac{\lambda}{2(\lambda+\mu)} y\epsilon - \frac{F}{4\pi\mu R}\left[\frac{2\mu\nu}{\lambda+\mu} - \sin 2\nu - \frac{\mu(y/a)}{\lambda+\mu} + \frac{x}{a}\cdot\frac{y}{a}\right]. \tag{4.89}$$

Note that the first term in Equations (4.89) and (4.89) is from Poisson contraction due to the axial force, given by Equation (4.84) noting that $u = \epsilon_{xx} \cdot x$ and $v = \epsilon_{yy} \cdot y$.

The strain tensor is calculated from Equations (4.88) and (4.89) by differentiation: $\epsilon_{xx} = \partial u/\partial x$, $\epsilon_{xy} = \partial u/\partial y$, $\epsilon_{yy} = \partial v/\partial y$, and $\epsilon_{yx} = \partial v/\partial x$. The refractive index is related to the strain. In the limit of small strain, the magnitude of the birefringence is

$$\Delta n = \frac{p_{11} - p_{12}}{2n^3}\left[(\epsilon_{xx} - \epsilon_{yy})^2 + 4\epsilon_{xy}^2\right]^{1/2}, \tag{4.90}$$

where the angle of the fast axis with respect to the x axis, ϕ, is given by

$$\cot\phi = \frac{\epsilon_{xx} - \epsilon_{yy} - \left[(\epsilon_{xx} - \epsilon_{yy})^2 + 4\epsilon_{xy}^2\right]^{1/2}}{2\epsilon_{xy}}, \tag{4.91}$$

and where p_{11} and p_{12} are the elastooptic constants, and n is the stress-free refractive index.

Figure 4.7 shows a plot of the magnitude of the birefringence as a function of position and the angle the fast axis makes with respect to the $\hat{x}$. At the center of the fiber ($x = 0$ and $y = 0$), the magnitude of the birefringence is given by[67]

$$\Delta n_0 = -\frac{p_{11} - p_{12}}{2n^3} \cdot \frac{3\lambda + 2\mu}{\lambda+\mu} \cdot \frac{a}{R}\epsilon = -\frac{p_{11} - p_{12}}{n^3} \cdot (1+\sigma) \cdot \frac{a}{R}\epsilon. \tag{4.92}$$

By using a combination of the geometrical effect and birefringence, it is possible to calculate the bending loss in a multimode fiber using ray tracing.

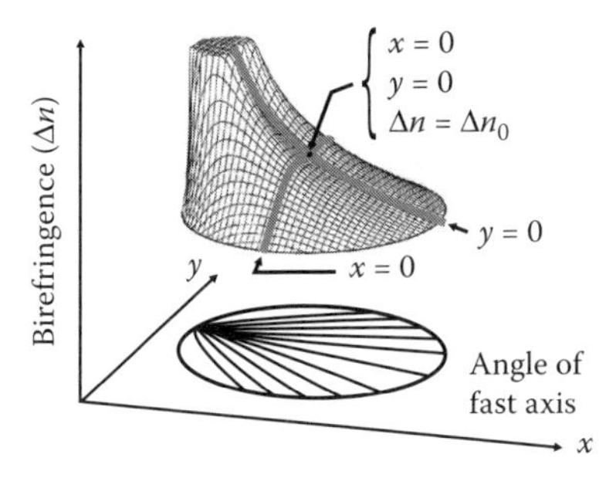

FIGURE 4.7 The birefringence of a bent fiber. The gray curves show the intersection between the surface and the $x = 0$ and $y = 0$. Adapted from Ref. [67] with permission of the Optical Society of America.

4.3.2 Single-Mode Fiber

The approach to determining the bending loss in a single-mode fiber is more complex. It is always possible to solve for the modes of an arbitrary geometry and refractive index profile using numerical means; however, it is more enlightening to at least solve for an approximate analytical solution that gives some physical insight into the problem. Unfortunately, given the complexity of the calculation, the result for the fiber must be stated without proof.

Using the results of Marcuse,[69] the bending loss for the fundamental mode in the weakly guiding limit is given by

$$\alpha = \frac{\sqrt{\pi}\kappa^2 \exp\left[-\frac{2}{3}\left(\gamma^3/\beta^2\right) R\right]}{2\gamma^{3/2} V^2 \sqrt{R} \ln^2(\gamma a)}, \tag{4.93}$$

where R is the radius of curvature of the bend, a the core radius, β is the wavevector of the mode,

$$\gamma = \left(\beta^2 - n_2^2 k^2\right)^{1/2}, \tag{4.94}$$

$$\kappa = \left(n_1^2 k^2 - \beta^2\right)^{1/2}, \tag{4.95}$$

$$V = ka\left(n_1^2 - n_2^2\right)^{1/2}, \tag{4.96}$$

k is the free space wavevector, n_1 is the refractive index of the core, and n_2 is the refractive index of the cladding. For mode ν, the loss in the weakly guiding limit is[69]

$$\alpha = \frac{\sqrt{\pi}\kappa^2 (\gamma a)^{2\nu-3/2} a^{3/2} \exp\left[-\frac{2}{3}\left(\gamma^3/\beta^2\right) R\right]}{2^{2(\nu-1)}(\nu-1)!(\nu+1)! V^2 \sqrt{R} \ln^2(\gamma a)}, \tag{4.97}$$

while in the tightly bound mode limit,[69]

$$\alpha = \frac{2a\kappa^2 \exp(2\gamma a) \exp\left[-\frac{2}{3}\left(\gamma^3/\beta^2\right) R\right]}{e_\nu V^2 \sqrt{\pi\gamma R}}, \tag{4.98}$$

where $e_\nu = 2$ for $\nu = 0$ and $e_\nu = 1$ otherwise.

4.3.2.1 Rayleigh Scattering

When the wavelength of light is long compared with the size of the scatterer, the scattering process is called Rayleigh scattering. Rayleigh scattering is thus relevant when visible light scatters from atoms, molecules, small aggregates of molecules, and inhomogeneities in a material that is too small to see with a visible-light microscope. The scattering process can be treated quantum mechanically[70] or using classical electrodynamics.[71] Here, we chose the classical approach due to its simplicity.

We assume that the incident light beam is in the form of a plane wave that propagates in the direction of the unit vector $\hat{n}_0$ with an electric field of the form $\vec{E}_{inc}$:

$$\vec{E}_{inc} = \vec{\epsilon}_0 E_0 \cdot e^{ik\hat{n}_0 \cdot x} \cdot e^{-i\omega t}, \tag{4.99}$$

where $\vec{\epsilon}_0$ is the complex polarization unit vector, E_0 is the magnitude of the electric field, and k is the wavevector $k = 2\pi/\lambda$. For a plane wave, the magnetic field $\vec{B}_{inc}$ is related to the electric field according to

$$\vec{B}_{inc} = \hat{n}_0 \times \vec{E}_{inc}. \tag{4.100}$$

For a scattered wave polarized along $\vec{\epsilon}$ and of field amplified $\vec{E}_{sc}$, the differential cross-section is related to the ratio of scattered intensity to the incident intensity:

$$\frac{d\sigma}{d\Omega}(\vec{n}, \vec{\epsilon}; \vec{n}_0, \vec{\epsilon}_0) = \frac{r^2 \frac{c}{8\pi} \left| \vec{\epsilon}^* \cdot \vec{E}_{sc} \right|^2}{\frac{c}{8\pi} \left| \vec{\epsilon}_0^* \cdot \vec{E}_{inc} \right|^2}, \tag{4.101}$$

where the complex conjugate of the polarization vector is particularly important when dealing with elliptically polarized light.

To solve for the differential cross-sections, we need to express the scattered field in terms of the incident field. At the site of an individual scatterer, the electric field is harmonic in time (at frequency ω) so the induced electric dipole moment $\vec{p}(t)$ is also harmonic with amplitude $\vec{p}_0$:

$$\vec{p}(t) = \vec{p}_0 e^{-i\omega t}. \tag{4.102}$$

It is straightforward to show that a harmonically oscillating electric dipole moment leads to an electric field in the radiation zone (i.e., far from the radiating dipole) of the form

$$\vec{E}_{sc} = k^2 \left(\frac{e^{ikr}}{r} \right) \left[(\hat{n} \times \vec{p}_0) \times \hat{n} - \hat{n} \times \vec{m}_0 \right], \tag{4.103}$$

where r is the distance from the scatterer to the observation point. The last term accounts for the presence of a harmonically oscillating magnetic dipole moment of magnitude $\vec{m}_0$.

Figure 4.8 shows a diagram of the scattering geometry. Let's consider the special case of scattering from a dielectric sphere with dielectric constant ϵ_2 inside a material of dielectric constant ϵ_1. The scatterer could be an inhomogeneity in

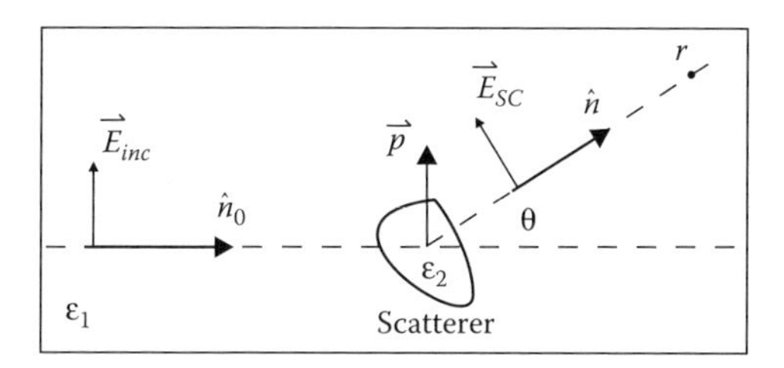

FIGURE 4.8 Rayleigh scattering.

the polymer, an impurity, or a molecule. Since the wavelength of the light λ is large compared with the radius a of the scatterer, the incident field can be assumed uniform. It is easy to show that the induced dipole moment of a sphere in a uniform applied electric field is given by[72]

$$\vec{p}_0 = \left(\frac{\epsilon_2 - \epsilon_1}{2\epsilon_1 + \epsilon_2} \right) a^3 \vec{E}_{inc}. \tag{4.104}$$

For p — polarized light (the electric field of the incident and scattered light are both in the plane of incidence — i.e., in the plane of the page in Figure 4.8), substituting Equation (4.104) into Equation (4.103) and ignoring the magnetic dipole term, Equation (4.101) yields

$$\frac{d\sigma}{d\Omega}(\theta) = k^4 a^6 \left| \frac{\epsilon_2 - \epsilon_1}{2\epsilon_1 + \epsilon_2} \right|^2 \cos^2\theta. \tag{4.105}$$

Equation (4.105) shows that the amount of scattering is inversely proportional to the fourth power of the wavelength. As such, blue light scatters more strongly than red light. As discussed in Chapter 3, when two monomers such as MMA and styrene are co-polymerized, the material becomes foggy, as shown in Figure 3.14. When the concentration of scatterers is small, the polymer shows a violet tint. As the number of scattering centers increases, multiple scattering becomes dominant and all colors scatter a multiple number of times. In this limit, the material appears white. Also note that the strength of the differential scattering cross-section is proportional to the sixth power of the size of the scatterer, so Rayleigh scattering is a sensitive function of the size.

It is a simple matter to estimate the fraction of scattered photons f_{sc} that either leave the waveguide or scatter backward by calculating the fraction of photons that lie outside the critical angle, θ_{cr}. Assuming that the light travels along the axis of the fiber, we get

$$f_{sc} = \frac{\int_{\theta_{cr}}^{\pi} \cos^2\theta \sin\theta d\theta}{\int_0^{\pi} \cos^2\theta \sin\theta d\theta} = \frac{1 + \cos^3\theta_{cr}}{2}. \tag{4.106}$$

Since the total cross-section is given by

$$\sigma = \int_0^{\pi} \int_0^{2\pi} d\Omega \frac{d\sigma}{d\Omega}(\theta) = \frac{4\pi}{3} k^4 a^6 \left| \frac{\epsilon_2 - \epsilon_1}{2\epsilon_1 + \epsilon_2} \right|^2, \tag{4.107}$$

the fraction of the light that is scattered out of a fiber I_{sc}/I_{inc} of length L and density of impurities $N = 3/4\pi a^3$, assuming that N is small enough to avoid multiple scattering and assuming that the process is incoherent, is

$$I_{sc}/I_{inc} = N f_{sc} \sigma L = \frac{4\pi}{3} N L k^4 a^6 \left(\frac{1 + \cos^3 \theta_{cr}}{2} \right) \left| \frac{\epsilon_2 - \epsilon_1}{2\epsilon_1 + \epsilon_2} \right|^2 \equiv \alpha L, \tag{4.108}$$

where α is the absorption coefficient defined in Equation (4.52).

It is sometimes more convenient to express the loss in terms of the refractive index. Recalling that $n = \sqrt{\epsilon}$, defining $\delta n = n_2 - n_1$ and $n = n_1$, and eliminating a by writing it in terms of N, Equation (4.108) becomes

$$\boxed{\alpha = \frac{1}{6\pi} \frac{k^4}{N} (1 + \cos^3 \theta_{cr}) \left| \frac{\Delta n}{n} \right|^2 .} \tag{4.109}$$

It is interesting to evaluate Equation (4.109) for typical numbers. First, we note that the long wavelength limit for Rayleigh scattering implies that $a \ll \lambda$. In the visible to the near IR, this implies that $a < 10^{-5} cm$ so that $N > (3 \times 10^{15}/4\pi) cm^{-3}$. Taking some typical values for polymers ($n = 1.5$, $\Delta n = 10^{-5}$, $\theta_{cr} = 30°$) and assuming a wavelength of $\lambda = 0.5 \times 10^{-4} cm$, we get $\alpha = 2.4 \times 10^{-6} cm^{-1}$. Using Equation (4.58), the loss is $1 \times 10^{-5} dB/cm = 1 dB/km$. These are reasonable values considering that the Rayleigh scattering contribution to PMMA is about 50dB/km and for Polystyrene is about $200 dB/cm$ (see Figure 1.8).

It is important to stress that several approximations were used in deriving Equation (4.109). As such, the results are most useful in building a qualitative understanding of the Rayleigh scattering process rather than for getting precise quantitative results. However, the quantitative predictions given by Equation (4.109) are a good order-of-magnitude estimate.

4.3.3 Microbending Loss

So far, we have treated the topic of what is generally called macro-behind loss. In this case, the radius of curvature is large compared with the smallest dimension of the waveguiding region. In contrast, microbends are on the scale of refractive index variations that are comparable to or smaller than the size of the waveguiding region. Indeed, microbending loss can be appreciable when such variations are on the order of a wavelength of the guiding light.

The topic of microbending losses, while important, is too complex to treat because the microgeometry is not always easy to quantify. For example, a plastic jacket that is placed on a fiber induces stress, which can result in microsized variations in the refractive index or radius of the fiber. Light can then scatter from these fluctuations. Similarly, larger fluctuation can be imprinted at the interface between the core and cladding due to impurities, imperfections in the material, or fluctuations due to processing, such as differential cooling.

When the fluctuations in the refractive index or radius are smaller than the wavelength of light, the amount of scattering can be estimated using scattering theory. For example, in the case of refractive index fluctuations, the Rayleigh scattering formula (Equation (4.109)) as derived in Section 4.3.2 can be used. Such specialized treatment is possible only when the microgeometry of the fiber is either measured for a particular fiber or calculated under the appropriate conditions.

4.4 DISPERSION

The linear susceptibility depends on the wavelength. The functional form of this wavelength dependence is called dispersion. In this section, we look at a simple spring model of the susceptibility and apply it to the refractive index and optical loss.

4.4.1 THE HARMONIC OSCILLATOR

Consider a charge q of mass m that is attached to a spring with a natural angular frequency of oscillation ω_0 and equilibrium length 0. In the presence of an electric field of the form

$$E = E_\omega \cos \omega t = \frac{1}{2}\left(E_\omega \exp[-i\omega t] + E_\omega \exp[+i\omega t]\right), \tag{4.110}$$

the equation of motion of the charge in one dimension is

$$m\left[\frac{d^2x}{dt^2} - 2\Gamma\frac{dx}{dt} + \omega_0^2 x\right] = \frac{1}{2}q\left(E_\omega \exp[-i\omega t] + E_\omega \exp[+i\omega t]\right), \tag{4.111}$$

where Γ is a damping term that accounts for the loss of energy from the oscillator.

The right-hand side of Equation (4.111) is the driving term due to the external electric field. When the oscillating field is first turned on, the charge undergoes transient motion that eventually decays until the charge oscillates at the frequency of the applied electric field. In the steady state, we assume that the position of the charge is of the form

$$x = \frac{1}{2}\left(A \exp[-i\omega t] + A^* \exp[+i\omega t]\right), \tag{4.112}$$

where A is a complex amplitude. Substituting Equation (4.112) into Equation (4.111), we solve for the amplitude:

$$A = \left(\frac{qE_\omega}{m}\right)\frac{1}{\omega_0^2 - 2i\Gamma\omega - \omega^2}. \tag{4.113}$$

To illustrate how we can use the above results, we apply them to a non-interacting collection of oscillators of number density N. The polarization, P, for an electron ($q = -e$) is given by

$$P = -Nex. \tag{4.114}$$

But the polarization at frequency ω can be written as

$$P = \frac{1}{2}P_{\omega}\exp[-i\omega t] + \frac{1}{2}P_{-\omega}\exp[+i\omega t]. \tag{4.115}$$

Using Equations (4.112) to (4.115) and $P_{\omega} = \chi^{(1)}(\omega)E_{\omega}$ we get

$$\chi^{(1)}(\omega) = \left(\frac{Ne^2}{m}\right)\frac{1}{\omega_0^2 - 2i\Gamma\omega - \omega^2}. \tag{4.116}$$

Finally, we can get the real and imaginary parts of $\chi^{(1)}(\omega)$ from Equation (4.116):

$$Re\left[\chi^{(1)}(\omega)\right] = \left(\frac{Ne^2}{m}\right)\frac{\omega_0^2 - \omega^2}{(\omega_0^2 - \omega^2)^2 + 4\Gamma^2\omega^2}, \tag{4.117}$$

and

$$Im\left[\chi^{(1)}(\omega)\right] = \left(\frac{Ne^2}{m}\right)\frac{2\Gamma\omega}{(\omega_0^2 - \omega^2)^2 + 4\Gamma^2\omega^2}. \tag{4.118}$$

Figure 4.9 shows the dispersion of the linear susceptibility as calculated using Equations (4.117) and (4.118). The dispersion of the refractive index and loss can be calculated from the complex susceptibility as discussed in Sections 4.2.1 and 4.2.2.

It is interesting to look at the dispersion in the real and imaginary parts of the refractive index (Equations (4.70) and (4.71)) using the harmonic oscillator model results given by Equations (4.117) and (4.118) and comparing these with the off-resonant model given by Equations (4.72) and (4.73). Figure 4.10 shows the results. Clearly, both models agree except for the on-resonant limit.

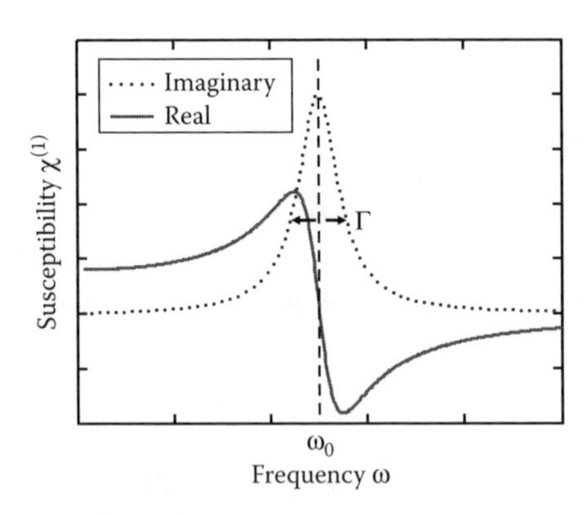

FIGURE 4.9 The dispersion of the real and imaginary parts of the linear susceptibility as calculated from the spring model.

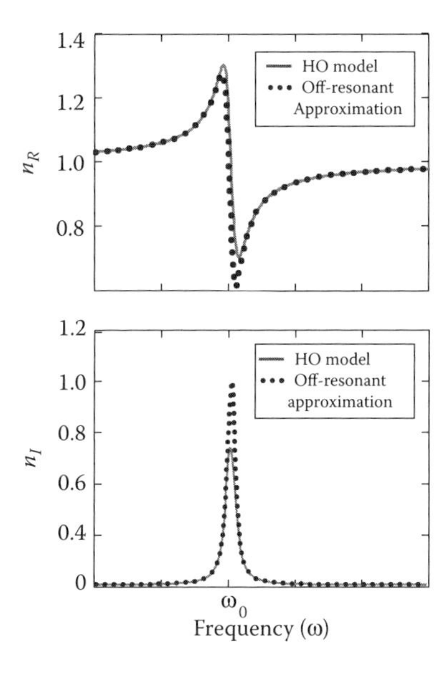

FIGURE 4.10 The harmonic oscillator model for the real and imaginary parts of the refractive index. The dotted curve shows the off-resonance approximation.

4.5 A PRACTICAL EXAMPLE

In this section, we give an example of how the information in this chapter can be applied to a practical problem.

4.5.1 PROBLEM

A fiber with a large electrooptic susceptibility needs to be prepared for a switching system. DR1 dye-doped in PMMA is used in the core and a polystyrene/PMMA cladding surrounds the core. The maximum poling electric field in the material is $1MV/cm$, above which dielectric breakdown damages the material irreversibly. An electric field is applied while the fiber is being drawn. The fiber is required to operate as a single-mode guide at $\lambda = 1064nm$ for light that is polarized along the poling field but it must not guide light when it is polarized perpendicular to the poling field. If the core material is prepared to be DR1 $12wt\%$ in PMMA, calculate the composition of the cladding if the core is required to be $9\mu m$ in diameter. Assume that the polarizability of the molecule at the operational wavelength is given by $\alpha^*_{zz} = 9 \times 10^{-23} cm^3$, the dipole dressed moment of DR1 is $m^* = 14 \times 10^{-18} statcoul \cdot cm$, and the refractive indices of PMMA and polystyrene are 1.48 and 1.59, respectively. Also assume that the glass transition temperature of the 12 $wt\%$ DR1 in PMMA is $T_g = 85°C$.

4.5.2 Solution

The electrooptic susceptibility scales with the concentration of dyes, so a core made of 12% DR1 dye will have a large electrooptic coefficient, yet will retain the good optical quality of the polymer host. The refractive index of the core is then given by Equations (4.18) and (4.20) where the order parameter of the dye molecules will be given by Equation (4.23). When a polymer fiber is drawn, its temperature is well above the glass transition temperature. As the fiber cools below the glass transition temperature, the orientational order will lock in place. The order parameter is thus evaluated at the glass transition temperature. Note that if the fiber is pulled at low tension, there is no birefringence due to the polymer, so $\langle P_2' \rangle = 0$.

The refractive index along the poling direction we define to be n_{zz} while the perpendicular one is n_{xx}. The single-mode waveguiding condition holds for any refractive index when $V < 2.405$. When $V < 2.405$, as the refractive index difference becomes smaller, the tail of the waveguiding mode will penetrate further into the cladding. In a perfect core and an infinite cladding, the single-mode guiding condition is met until the core and cladding index are the same. In real applications with finite cores and imperfect waveguides, a non-zero difference between core and cladding is required. To ensure that there is no single-mode waveguiding for x-polarized modes, we use the condition that the cladding refractive index, n_2 matches the refractive index $n_2 = n_{xx}$. Subsequently, the single-mode condition is calculated for n_{zz} in term of n_2.

To solve for the refractive index, we need to know the number density, N, of molecules. Using the density of PMMA $\rho = 1.16g/cm^3$ and the molecular weight of DR1 $m_d = 297$, Equation (4.29) gives the number density for the 12 wt% DR1,

$$N = \frac{w}{1-w}\left(\frac{\rho N_A}{m_d}\right) = \frac{0.12}{1-0.12}\left(\frac{1.16 \cdot 6.02 \times 10^{23}}{297}\right) = 3.21 \times 10^{20}. \tag{4.119}$$

Since we assume that the host polymer is isotropic, $n_{xx} = n_{zz} = n_0$, and (4.18) and (4.20) become

$$n_{zz} = n_0 + \frac{2\pi N}{3n_0}(1 + 2\langle P_2 \rangle)\alpha_{33}^* \tag{4.120}$$

and

$$n_{xx} = n_0 + \frac{2\pi N}{3n_0}(1 - \langle P_2 \rangle)\alpha_{33}^*. \tag{4.121}$$

The condition for single-mode guiding of the z-polarized light is given by Equation (2.50). With $n_2 = n_{xx}$, this yields

$$n_{zz}^2 - n_{xx}^2 < 5.784\frac{\lambda_0^2}{\pi^2 d^2}. \tag{4.122}$$

Substituting Equations (4.120) and (4.121) into Equation (4.122), and keeping terms only to first-order in $\langle P_2 \rangle$ yields

$$\langle P_2 \rangle < 5.784 \cdot \frac{1}{(2n_0 + b/3)b}\left(\frac{\lambda_0}{\pi d}\right)^2, \tag{4.123}$$

where

$$b = \frac{2\pi N\alpha^*}{n_0} = \frac{2\pi \cdot \left(3.21 \times 10^{20}\right) \cdot \left(9 \times 10^{-23}\right)}{1.48} = 0.1226. \tag{4.124}$$

Equation (4.123) with the help of Equation (4.124) yields

$$\langle P_2 \rangle < 5.784 \cdot \frac{1}{(2 \cdot 1.48 + 0.1226/3)\, 0.1226} \left(\frac{1.064}{9\pi}\right)^2 = 2.2226 \times 10^{-2}. \tag{4.125}$$

Before proceeding with the calculation of the core and cladding refractive index, we first need to check if the electric field required to get this order parameter is within the dielectric breakdown voltage of the polymer. Inverting Equation (4.23), we get

$$\begin{aligned} E &= \frac{kT}{m^*}\sqrt{15\langle P_2\rangle} \\ &= 300 \cdot \frac{1.38 \times 10^{-16} \cdot 358}{14 \times 10^{-18}} \sqrt{15 \cdot 2.2226 \times 10^{-2}} \\ &= 0.6118 \times 10^6, \end{aligned} \tag{4.126}$$

where the factor of 300 is used to convert the result from statvolt/cm (Gaussian units) to volt/cm. This field is less than the breakdown voltage of the polymer.

We can now calculate the core and cladding refractive index. Using Equations (4.120) and (4.121), we get

$$\begin{aligned} n_1 = n_{zz} &= n_0 + \frac{2\pi N}{3n_0}\left(1 + 2\langle P_2\rangle\right)\alpha^*_{33} \\ &= 1.48 + \frac{2\pi \cdot 3.21 \times 10^{20}}{3 \cdot 1.48}(1 + 2 \cdot 2.2226 \times 10^{-2}) \cdot 9 \times 10^{-23} \\ &= 1.5227 \end{aligned} \tag{4.127}$$

and

$$\begin{aligned} n_2 = n_{xx} &= n_0 + \frac{2\pi N}{3n_0}\left(1 - \langle P_2\rangle\right)\alpha^*_{33} \\ &= 1.48 + \frac{2\pi \cdot 3.21 \times 10^{20}}{3 \cdot 1.48}(1 - 2.2226 \times 10^{-2}) \cdot 9 \times 10^{-23} \\ &= 1.5200. \end{aligned} \tag{4.128}$$

Given the refractive index of the cladding, we can now calculate the composition by inverting Equation (4.44), which we solve for the volume fraction of polystyrene,

$$\begin{aligned} f_{PS} &= \frac{n_2^2 - n_{PMMA}^2}{n_{PS}^2 - n_{PMMA}^2} \\ &= \frac{(1.5200)^2 - (1.48)^2}{(1.59)^2 - (1.48)^2} = 0.3553^+, \end{aligned} \tag{4.129}$$

where the "+" indicates that the weight fraction of polystyrene needs to be a tad above that value to make $V < 2.405$. Since it is often easier to prepare a ratio of weights, we convert f_{PS} to w'_{PS} using Equation (4.48):

$$w'_{PS} = \frac{1}{1 + \left(\frac{1-f_{PS}}{f_{PS}}\right)\frac{\rho_{PMMA}}{\rho_{PS}}} = \frac{1}{1 + \left(\frac{1-0.3553}{0.3553}\right)\frac{1.16}{1.05}} = 0.333, \tag{4.130}$$

where we have used a density of $1.05 g/cm^3$ for polystyrene. To make the cladding, then one would use 1 part of polystyrene to 2 parts PMMA by weight.

4.6 POLARIZATION

The polarization of a ray or mode can have a strong influence on the waveguiding properties. For example, if the waveguide is not circular in geometry and/or refractive index, one polarization may support a mode while the orthogonal polarization may correspond to a radiation mode. Polarization-maintaining fibers are routinely made out of glass by inserting a metal rod on either side of the core in the preform. During drawing, the differential mechanical properties between the polymer and metal yields stress birefringence in the core. As a result, a mode polarized along the direction of higher refractive index will be bound while the orthogonal mode will radiate.

The topic of polarization is an important one. For example, light can scatter from one polarization to another, and radiate — leading to loss. In a birefringent fiber, the polarization can change as the light propagates, or a polarized mode can become depolarized. While we do not treat such processes in this book, below we discuss two methods for characterizing the polarization of light. (More details about polarization can be found in books devoted to the topic.[73])

Stokes Parameters

The most general state of polarization can be experimentally prepared by launching two collinear linearly polarized beams of arbitrary polarization and phase. Since we are interested solely in the polarization, we consider only plane waves. Each beam, then, is of the form

$$\vec{E}_j = \hat{\epsilon}_j a_j \exp[i(\vec{k} \cdot \vec{x} - \omega t + \delta_j)], \tag{4.131}$$

where $\hat{\epsilon}_j$ with $j = 1$ and $j = 2$ is the polarization along the $\hat{\epsilon}_1$ and $\hat{\epsilon}_2$ directions, respectively. δ_j is the phase of beam j and a_j the amplitude. Note that the polarization vectors are orthogonal so that $\hat{\epsilon}_2 \cdot \hat{\epsilon}_2 = 0$.

As such, we can characterize the most general state of polarization of a beam using the four real quantities a_1, a_2, δ_1, and δ_2. However, these quantities are not always experimentally easy to determine for a beam of unknown polarization. Stokes recognized that if a beam is expressed as

$$\vec{E} = \hat{\epsilon}_1 a_1 \exp[i(\vec{k} \cdot \vec{x} - \omega t + \delta_1)] + \hat{\epsilon}_2 a_2 \exp[i(\vec{k} \cdot \vec{x} - \omega t + \delta_2)], \tag{4.132}$$

then certain combinations of these constants could be easily determined. These combinations are called the Stokes parameters, which are of the form

$$\begin{aligned}
s_0 &= |\hat{\epsilon}_1 \cdot \vec{E}|^2 + |\hat{\epsilon}_2 \cdot \vec{E}|^2 = a_1^2 + a_2^2, \\
s_1 &= |\hat{\epsilon}_1 \cdot \vec{E}|^2 - |\hat{\epsilon}_2 \cdot \vec{E}|^2 = a_1^2 - a_2^2, \\
s_2 &= 2\mathrm{Re}[(\hat{\epsilon}_1 \cdot \vec{E})^*(\hat{\epsilon}_2 \cdot \vec{E})] = 2a_1 a_2 \cos(\delta_2 - \delta_1), \\
&\text{and} \\
s_3 &= 2\mathrm{Im}[(\hat{\epsilon}_1 \cdot \vec{E})^*(\hat{\epsilon}_2 \cdot \vec{E})] = 2a_1 a_2 \sin(\delta_2 - \delta_1).
\end{aligned} \tag{4.133}$$

Each Stokes parameter is determined by placing a "filter" in the beam and measuring the transmitted intensity with a detector. To get s_0, no filter is used (i.e., the identity filter). Thus, the detector measures the intensity, which is clearly proportional to s_0. Next, a horizontal filter is placed before the detector, so that $|\hat{\epsilon}_2 \cdot \vec{E}|^2$ is measured. Twice the resulting value is subtracted from s_0 to get s_1; that is, $s_1 = s_0 - 2|\hat{\epsilon}_2 \cdot \vec{E}|^2$.

Next, a polarizer is set at 45° to the horizontal, yielding a measure of

$$\begin{aligned}
\left|\left[\frac{\hat{\epsilon}_1 + \hat{\epsilon}_2}{\sqrt{2}}\right] \cdot \vec{E}\right|^2 &= \frac{1}{2}|(\hat{\epsilon}_1 \cdot \vec{E}) + (\hat{\epsilon}_2 \cdot \vec{E})|^2 \\
&= \frac{1}{2}[|\hat{\epsilon}_1 \cdot \vec{E}|^2 + |\hat{\epsilon}_2 \cdot \vec{E}|^2 + 2\mathrm{Re}[(\hat{\epsilon}_1 \cdot \vec{E})^*(\hat{\epsilon}_2 \cdot \vec{E})]].
\end{aligned} \tag{4.134}$$

Thus, we can solve Equation (4.134) for s_2 with the help of the result obtained for s_0:

$$s_2 = 2\mathrm{Re}[(\hat{\epsilon}_1 \cdot \vec{E})^*(\hat{\epsilon}_2 \cdot \vec{E})] = 2\left|\left[\frac{\hat{\epsilon}_1 + \hat{\epsilon}_2}{\sqrt{2}}\right] \cdot \vec{E}\right|^2 - s_0. \tag{4.135}$$

Finally, to get s_3, a quarter-wave plate and polarizer combination is placed into the beam, which projects out the right circularly polarized component of the light. Recall that the right-hand circularly polarized unit vector is given by $(\hat{\epsilon}_1 + i\hat{\epsilon}_2)/\sqrt{2}$, so the transmitted intensity is proportional to

$$\begin{aligned}
\left|\left[\frac{\hat{\epsilon}_1 + i\hat{\epsilon}_2}{\sqrt{2}}\right] \cdot \vec{E}\right|^2 &= \frac{1}{2}|(\hat{\epsilon}_1 \cdot \vec{E}) + i(\hat{\epsilon}_2 \cdot \vec{E})|^2 \\
&= \frac{1}{2}[|\hat{\epsilon}_1 \cdot \vec{E}|^2 + |\hat{\epsilon}_2 \cdot \vec{E}|^2 + 2\mathrm{Im}[(\hat{\epsilon}_1 \cdot \vec{E})^*(\hat{\epsilon}_2 \cdot \vec{E})]].
\end{aligned} \tag{4.136}$$

Solving for s_3, we get

$$s_3 = 2\mathrm{Im}[(\hat{\epsilon}_1 \cdot \vec{E})^*(\hat{\epsilon}_2 \cdot \vec{E})] = 2\left|\left[\frac{\hat{\epsilon}_1 + i\hat{\epsilon}_2}{\sqrt{2}}\right] \cdot \vec{E}\right|^2 - s_0. \tag{4.137}$$

To review, the process of getting the Stokes parameters is as follows. First, s_0 is determined from a measure of the beam intensity. Then, a horizontal polarizer is placed in the beam. s_1 is calculated by subtracting twice the intensity passing the horizontal polarizer from s_0. Next, the intensity passing a polarizer set at 45° is measured, from which twice s_0 is subtracted to get s_2. Finally, the intensity passing a half-wave plate set to transmit right-handed circularly-polarized light is measured, and s_2 subtracted from the result to get s_3. Given these four parameters, the polarization state is completely specified. However, only three of the parameters are independent since they are related according to $s_0^2 = s_1^2 + s_2^2 + s_3^2$. This makes sense since the absolute phase of a beam cannot be measured; so, the three independent Stokes parameters can be used to solve for the two intensities a_1 and a_2 — and the phase difference $\delta = \delta_1 - \delta_2$.

We note that the choice of Stokes parameters is not unique, and many other conventions are possible. Instead of using the basis set of two orthogonal linearly polarized states, for example, we can choose the two orthogonal states of circular polarization. As long as the two states used are orthogonal, four Stokes parameters can be determined using a process analogous to the one described above.

4.6.1 The Poincaré Sphere

Recall that the Stokes parameter s_0 is associated with the intensity of the light. As such, for a beam with fixed intensity, the remaining three Stokes parameters are related to each other through

$$s_0^2 = s_1^2 + s_2^2 + s_3^2, \tag{4.138}$$

so, only two of these parameters are independent. Note that Equation (4.138) represents a sphere of radius s_0 in the Stokes parameter space (i.e., s_1, s_2, s_3, and s_0 are like x, y, z and r in regular space). As such, the two independent parameters can be viewed as the polar and azimuthal angles that define a point on a sphere of radius s_0. This is called the Poincaré sphere and the angles that represent a point on the sphere can be related to the Stokes parameters according to

$$s_1 = s_0 \sin\theta \cos\phi \tag{4.139}$$

$$s_2 = s_0 \sin\theta \sin\phi \tag{4.140}$$

and (4.141)

$$s_3 = \cos\theta.$$

Note that the polar angle spans $0 \leq \theta \leq \pi$ and the azimuthal angle spans $0 \leq \phi \leq 2\pi$.

Thus, we have two equivalent ways of expressing the state of polarization; the Stokes parameters or a point on the surface of a sphere.

4.6.2 Some Examples of Polarization States

It is useful to consider specific examples of a state of polarization to build an intuition of the meaning of these representations of polarization. Let's start with linearly polarized light of unit intensity so that $s_0 = 1$ and $\delta = 0$. This yields

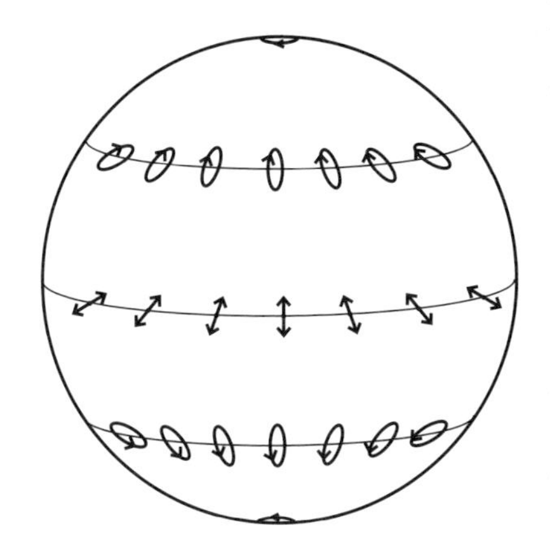

FIGURE 4.11 The polarization states on the Poincaré sphere.

$(s_0, s_1, s_2, s_3) = (1, 2a_1^2 - 1, 2a_1\sqrt{1 - a_1^2}, 0)$. Note that a_1 is associated with the vertical component of the electric field, whose magnitude can vary between $0a_1 \leq +1$. Therefore, for vertically polarized light, $(s_0, s_1, s_2, s_3) = (1, 1, 0, 0)$ and for horizontal polarization, we have $(s_0, s_1, s_2, s_3) = (1, 0, 1, 0)$. For any linear polarization, $s_3 = 0$, so linear polarization can be represented as points on the equator of the Poincaré sphere.

For circularly polarized light, $a_1 = a_2$ and $\delta = \pm\pi/2$. As such, $(s_0, s_1, s_2, s_3) = (1, 0, 0, \pm 1)$, which falls on the north and south poles of the Poincaré sphere. Elliptical polarization states vary smoothly from being circular at the poles to linear at the equator. For a constant polar angle, the major axis of the ellipse remains fixed in length but rotates its orientation in proportion to the azimuthal angle ϕ. Figure 4.11 shows a sketch of the polarization states at various representative points on the Poincaré sphere.

5 Characterization Techniques and Properties

This chapter focuses on experimental methods used to measure the refractive index and optical absorbance — undoubtedly the two most important properties of a fiber that define its waveguiding properties and how much light is lost. Those measurements that are amenable to measuring fibers, polymers, and preforms will be highlighted.

5.1 REFRACTIVE INDEX

The refractive index is a measure of the speed of light in a material. Most methods used to measure the refractive index rely on the wave properties of light, such as interference, diffraction, and refractive bending. The most useful measurements are those that give the refractive index as a function of the position. So, we concentrate on position-dependent techniques. At the end of the chapter, methods for measuring the optical bandwidth are reviewed. This is an important property, which determines the ultimate speed of a fiber-optic transmission system.

5.1.1 INTERPHAKO

The interphako microscope is commonly used to measure the refractive index profile of gradient index rods, polymer fibers, and preforms and is based on interference of two wavefronts. This method was first used by Ohtsuka and Shimizu to measure the refractive index profiles of gradient-index rods and fibers.[58] In these studies, the refractive index of each rod was assumed to vary only radially. This assumption was reasonable based on the radial diffusion process that was used to make the rod. Furthermore, the radial dependence was assumed to be a truncated polynomial, which simplified the analysis. In the following section, we begin by illustrating the principle behind the interphako technique by applying the analysis to a thin slab of material whose refractive index changes along the surface in only one direction. The following section shows how the technique can be applied to a rod-shaped sample.

5.1.1.1 One-Dimensional Slab

Figure 5.1 shows a diagram of the apparatus. For the sake of illustration, we assume that the refractive index varies only along the x-direction. Furthermore, we assume that the sample is planar and that the refractive index variation is small

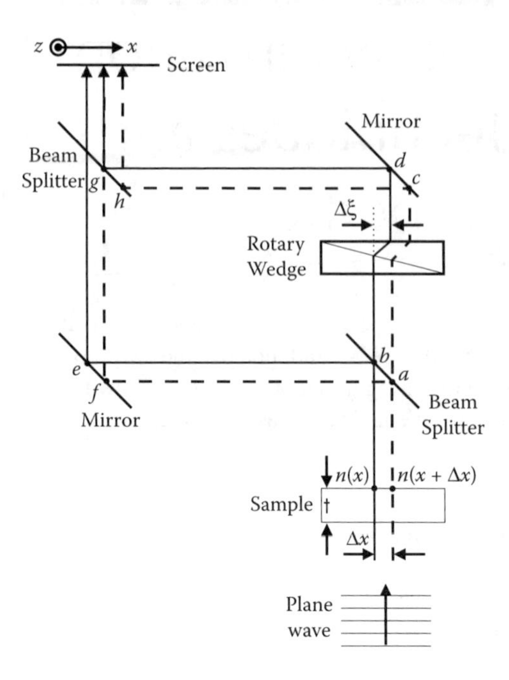

FIGURE 5.1 The interphako microscope shown with two representative ray trajectories.

so that deflection of the beam is minimal. (The general two-dimensional case is similar — but of course, much more complicated — so will not be treated here.)

A plane wave is incident on the sample. The solid and dashed lines trace the paths of two points on a phase front that are originally separated by a distance Δx. A rotary wedge displaces the upward-traveling beam by an amount $\Delta\xi$ so when the two beams recombine at the second beam splitter, different parts of the wavefront interfere. For example, the part of the wavefront that follows the path $a \rightarrow f \rightarrow g$ (dashed line) ends up at the same place on the screen as the part of the beam that follows $b \rightarrow d \rightarrow g$. Furthermore, two points that are originally separated by Δx end up separated by an amount $\Delta\xi$ at the screen.

Let's now calculate the intensity at point g. Assuming that the beam splitters, mirrors, and rotary wedge are balanced so that the intensity of each ray is the same, the electric field at the screen is

$$E = E_0 \frac{1}{2} \left(\exp\left[ikn(x)t + i\phi_1\right] + \exp\left[ikn(x + \Delta x)t + i\phi_2\right] \right), \tag{5.1}$$

where $n(x)$ is the refractive index at x, k is the wave vector, ϕ_1 is the optical path length along $b \rightarrow d \rightarrow g$, and ϕ_2 is the optical path length along $a \rightarrow f \rightarrow g$. Squaring the magnitude of Equation (5.1) yields the intensity:

$$I = I_0 \cos^2\left[2k\left\{n(x + \Delta x) - n(x)\right\}t + 2\left(\phi_2 - \phi_1\right)\right]. \tag{5.2}$$

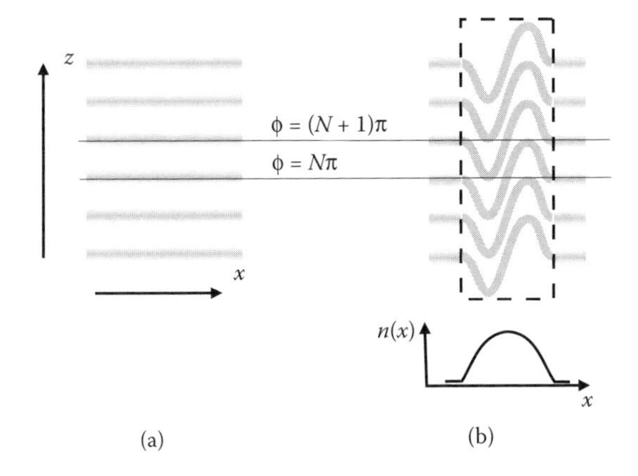

FIGURE 5.2 The view of the fringes on the screen a) without a sample and b) with a sample.

Noting that for small Δx,

$$n(x + \Delta x) - n(x) \approx \frac{\partial n(x)}{\partial x} \Delta x, \tag{5.3}$$

so

$$I = I_0 \cos^2 \left[2k \left\{ \frac{\partial n(x)}{\partial x} \Delta x \right\} t + \phi' \right], \tag{5.4}$$

where ϕ' is twice the phase difference between the two paths in the interferometer.

First, let's consider the case with no sample. If the beam splitter is adjusted so that the two beams hitting the screen are skewed while still remaining parallel in the plane of the page (as in Figure 5.1) but have slightly different k-vectors perpendicular to the page, a set of fringes will appear on the screen as shown in Figure 5.2a. The z-direction in Figure 5.1 is perpendicular to the plane of the page in Figure 5.1.

Figure 5.2b shows the fringe pattern on the screen for a sample with a parabolic refractive index profile. The dashed box represents the projection of the sample on the screen (i.e., the idealized diffractionless shadow cast by the sample). The shape of the fringes can be calculated from Equation (5.4); or, using Equation (5.4) and the measured fringes, the refractive index profile can be determined, as follows.

A fringe corresponds to a bright spot, or constructive interference between the two beams. The Nth fringe corresponds to $I = I_0$ in Equation (5.4), or

$$2k \left(\frac{\partial n(x)}{\partial x} \Delta x \right) t + \phi' = N\pi. \tag{5.5}$$

If, as in Figure 5.3, we focus on two consecutive fringes (for example, fringe N and $N + 1$) from Figure 5.2, the phase at some point between the fringes a distance z from fringe N is

$$\phi(z) = N\pi + \frac{\pi}{\Delta z} z, \tag{5.6}$$

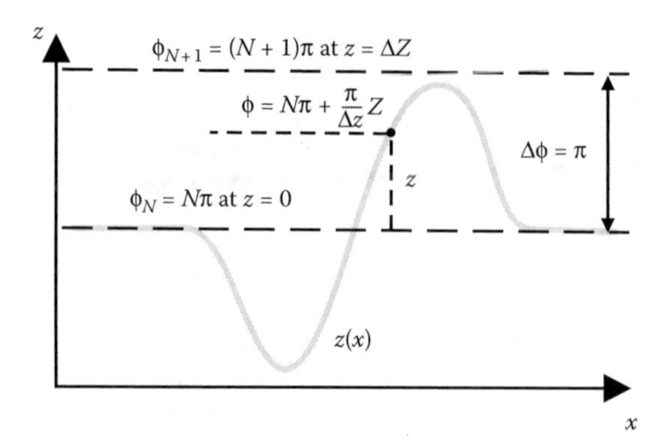

FIGURE 5.3 Calculating phase from fringe shape.

where Δz is the distance between two consecutive fringes when no sample is present.

Equation (5.5) gives the phase on a fringe maximum while Equation (5.6) gives the additional phase shift between fringes. As such, to generalize Equation (5.5) to the region between fringes, we get

$$2k\left(\frac{\partial n(x)}{\partial x}\Delta x\right)t + \phi' = \phi(z) = N\pi + \frac{\pi}{\Delta z}z. \tag{5.7}$$

But, since z is a function of x, we can solve for the refractive index by integrating Equation (5.7). Without loss of generality, we define $z = 0$ as the point on a bright fringe in the absence of a sample. As such, we get $\phi' = N\pi$. We can therefore now solve for the refractive index at x:

$$\boxed{n(x) - n(x_0) = \frac{\pi}{2kt\Delta z\Delta x}\int_{x_0}^{x} z(x)dx.} \tag{5.8}$$

If the refractive index is known at $x = x_0$, then Equation (5.8) can be used to determine the refractive index at any other point in the sample simply by integrating the function $z(x)$.

5.1.1.2 Cylindrical Sample

The focusing properties of a cylindrical rod in air results in severe beam deflection, which makes it impossible to use the interphako technique. To fix this problem, the rod is submerged in index matching fluid whose refractive index is matched to the refractive index of the outer shell of the rod. Figure 5.4 shows the sample and a typical ray that passes perpendicular to the rod's axis.

The thickness of the sample was constant in the calculation given in Section 5.1.1, so the optical path length depends only on the refractive index.

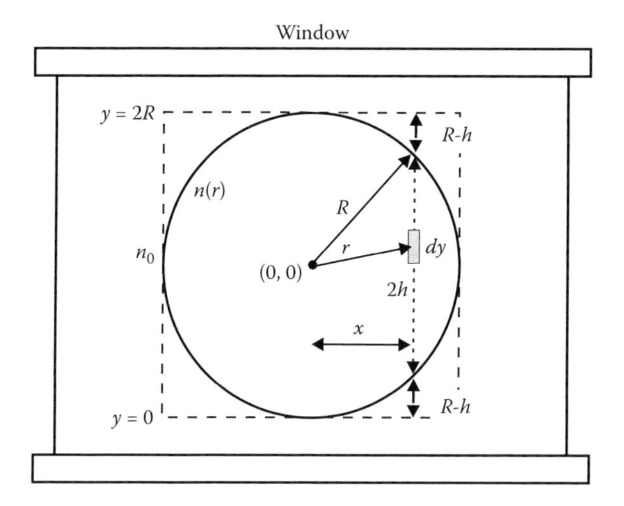

FIGURE 5.4 The optical path through a rod in index matching liquid.

For the cylinder, the optical path length depends on both the refractive index and the thickness of the rod traversed by the light beam. In particular, we need to calculate the optical path length through the sample cell as a function of the position, x, that the ray enters the rod. The only part of the sample whose optical path length depends on x is inside the dashed box. The optical path length is thus given by

$$\ell(x) = \int_0^{2R} n(x, y)dy = 2(R - h)n_0 + \int_{R-h}^{R+h} n(x, y)dy, \tag{5.9}$$

where we have split the integral in two parts: the path through the rod and the path through the index matching liquid of refractive index n_0.

Since we are interested in the radial dependence of the refractive index, we rewrite Equation (5.9) in terms of the radius (see Figure 5.4) using $x^2 + y^2 = r^2$, $ydy = rdr$ for fixed x, and $x^2 + h^2 = R^2$ to get the optical path length

$$\ell(x) = 2R\left(1 - \sqrt{1 - \frac{x^2}{R^2}}\right) n_0 + \int_{R\left(1-\sqrt{1-\frac{x^2}{R^2}}\right)}^{R\left(1+\sqrt{1-\frac{x^2}{R^2}}\right)} n(r)\frac{1}{\sqrt{1 - x^2/r^2}}dr. \tag{5.10}$$

Using Equation (5.10) to replace $n(x)t$ in Equation (5.1) with $\ell(x)$; and following the same procedure as we used from Equation (5.1) through Equation (5.8), we get

$$\ell(x) - \ell(x_0) = \frac{\pi}{2k\Delta z\Delta x}\int_{x_0}^{x} z(x)dx. \tag{5.11}$$

To get the refractive index profile, Equation (5.10) is substituted into Equation (5.11). Clearly, this integral equation is very complex; and, in general, needs to be solved numerically.

The simplest approach to using Equations (5.10) and (5.11) for estimating the refractive index profile is to assume a functional form for the refractive index and solve for the free parameters in the function. For example, if a gradient-index rod is made by a diffusion process, the radial dependence of the refractive index can be assumed to be a symmetric function of r. Furthermore, if the refractive index peaks in the center and is approximately parabolic, the function often used is

$$n(r) = n_0 \left(1 - ar^2 + br^4 + cr^6 + dr^8\right). \tag{5.12}$$

The second method for dealing with gradient-index rods is to cut a slice of the rod parallel to its axis. The resulting slab can then be analyzed according to Equation 5.8. For a small-diameter fiber, it is difficult to get a thin slice through the center of the fiber; and such a slice is difficult to hold. Figure 5.5 shows how a thin fiber can be polished and held in place. In this process, the fiber is first placed inside an epoxy, which is allowed to cure around the fiber. The resulting solid is then glued to a substrate and sanded until almost half the fiber diameter is removed. Subsequently, the surface is polished smooth. The epoxy slab is removed from the substrate and the polished surface is glued to a clean substrate. The exposed side is sanded and polished until a thin sliver of fiber remains. With this structure,

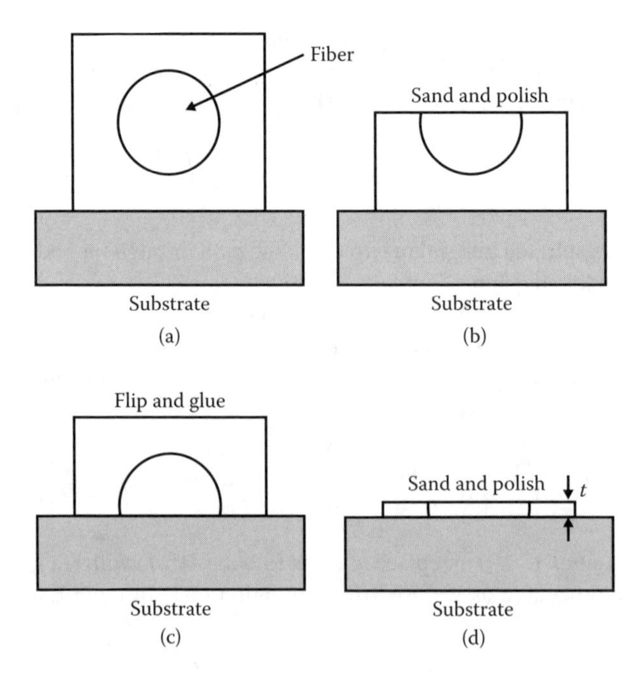

FIGURE 5.5 Making a thin slice of fiber. a) The fiber is embedded in an epoxy and glued to a substrate. b) The structure is polished until half the fiber is removed. c) The fiber/epoxy is removed from the substrate, flipped, and glued so that the fiber is attached to the substrate. d) The structure is polished until only a thin slice of the fiber remains.

the refractive index can be measured using the much simpler flat slab analysis as described in Section 5.1.1.

5.1.2 Beam Deflection Technique

The Direct Deflection Method (DDM) was developed by Canfield and coworkers.[74] When a beam of light passes through a material with a transverse refractive index gradient, it deflects in the direction of the gradient. A measure of beam deflection as a function of position in a material can be used to map out the refractive index profile. The method is simple to use, is compact, and has good resolution. As with any refractive index measurement, corrections for topographical deviations of the surface must be addressed.

5.1.2.1 DDM Theory

The direct deflection method works best when the sample is a thin slice of material. For the case of a gradient index rod, the slice is taken parallel or perpendicular to the rod's axis. (Making such a slice for a large rod is straightforward. For a fiber, the epoxy technique can be used as described in Section 5.1.1.) Figure 5.6 illustrates the deflection due to a sample and how it is measured with a CCD camera array.

The radius of curvature of a ray in a refractive index gradient is governed by the refractive bending equation,[75]

$$\rho = \frac{1}{|\nabla(Log\, n)|} = \frac{n}{|\nabla n|}, \tag{5.13}$$

where ρ is the radius of curvature of a light ray in the slice. Note that this equation holds for a small gradient, which implies that the radius of curvature is large. In step-index single mode fibers, the refractive index difference between the core and

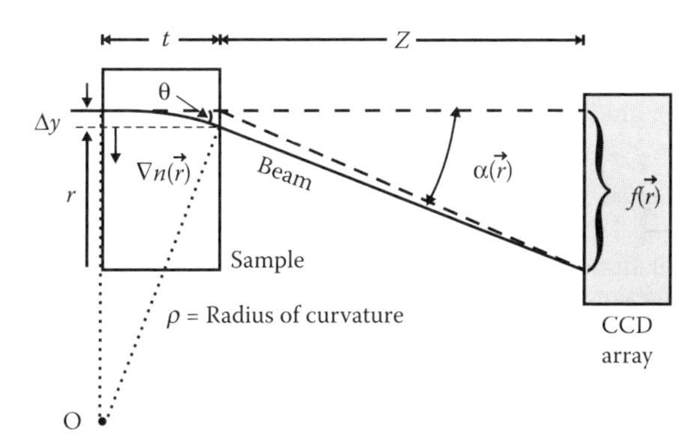

FIGURE 5.6 Schematic of deflection experiment and parameters. Adapted from Ref. [74] with permission of the Optical Society of America.

cladding is $\sim 10^{-3}$ or less, so in most gradient index samples, the deflections will be small, yielding a large radius of curvature and therefore the small-angle approximation will apply. From simple geometry, the deflection angle θ, is related to the radius of curvature ρ, the thickness of the slice, t, and the vertical displacement Δy through the slice according to

$$\rho^2 = t^2 + (\rho - \Delta y)^2, \tag{5.14}$$

where we assume that ∇n is constant perpendicular to the slice.

For small deflection, $(\Delta y)/\rho \ll 1$, Equation (5.14) yields

$$\Delta y = \frac{t^2}{2\rho}. \tag{5.15}$$

Using the small angle approximation in Equation (5.15) ($\Delta y/t = \tan\theta \approx \theta$) and solving for θ, we get

$$\theta = \frac{t}{2\rho}. \tag{5.16}$$

When the ray exits the sample, it refracts due to Snell's Law, so the exit angle of the ray, α (again for small deflection) is given by

$$\theta = \frac{\alpha}{n}. \tag{5.17}$$

Combining Equations (5.13), (5.16), and (5.17), we have

$$|\nabla n| = \frac{2\alpha}{t}. \tag{5.18}$$

Defining $f(\vec{r})$ as the spot position on the CCD camera ($f(\vec{r}) \equiv 0$ with no deflection) when the beam in the preform slice is at $\vec{r}$, and defining z as the distance from the slice to the camera, then $\alpha(\vec{r}) \approx f(\vec{r})/z$. Note that in general, the amount of beam deflection must be expressed in terms of two independent angles, which we define to be $\vec{\alpha}(\vec{r}) = \alpha(\vec{r})\hat{\alpha}$. The unit vector $\hat{\alpha}$ lies in the plane of the CCD array and describes the direction of deflection while $\alpha(\vec{r})$ is the magnitude of the deflection angle for a beam at position $\vec{r}$ in the slice. The direction of deflection is along the refractive index gradient, so Equation (5.18) can be written in vector form

$$\nabla n = \frac{2\vec{\alpha}(\vec{r})}{t} = \frac{2f(\vec{r})\hat{\alpha}(\vec{r})}{zt}. \tag{5.19}$$

Integrating Equation (5.19) along $d\vec{r}$, the refractive index difference between two points inside the sample is related to the amount of deflection on the CCD camera according to

$$n(\vec{r}_2) - n(\vec{r}_1) = \frac{2}{zt}\int_{\vec{r}_1}^{\vec{r}_2} f(\vec{r})\,\hat{\alpha}\cdot d\vec{r}. \tag{5.20}$$

Note that the coordinate system in the sample is arbitrary so $d\vec{r}$ represents any arbitrary infinitesimal displacement in the sample.

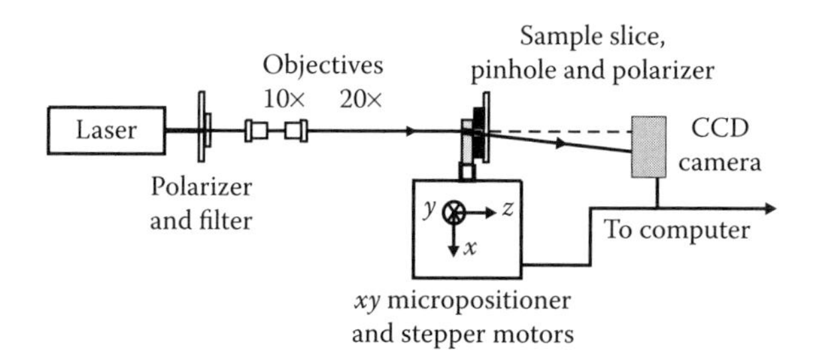

FIGURE 5.7 Diagram of deflection experiment. Adapted from Ref. [74] with permission of the Optical Society of America.

Equation (5.20) is used to relate the refractive index difference between two arbitrary points in the sample to the sum of measured deflection angles between those two points. Experimentally, a finite number of data points is taken, so the integral is approximated by a sum, where the step size dr is the transverse distance the sample is moved between deflection maltreatments. To determine the absolute index profile in a sample, the refractive index at one point in the sample must be known.

5.1.2.2 DDM Experiment

Figure 5.7 shows a diagram of the experimental setup used by Canfield and coworkers, who applied the direct deflection measurement to determine the refractive index profile for a thin slice of a fiber preform. A preform is made according to the methods outlined in Chapter 3. A $1mm$ thin disk is sliced from the preform, and is polished to make it optically flat and smooth. Fine sandpaper, $5\mu m$ and $3\mu m$ lapping film is used in sequence followed by polishing with $3\mu m$, $1\mu m$, and then $0.1\mu m$ liquid alumina polishing suspensions. The slice is flipped frequently during the polishing stages to maintain parallel, plane faces and to randomize any remaining scratch patterns.

Any light source can be used in a DDM run. Canfield used a HeNe laser ($\lambda = 632.8\ nm$). Because the sample can be birefringent, the degree of deflection as a function of polarization is a useful measurement. As such, the light source should be polarized. As seen in Figure 5.7, the light passes through a polarizer and is focused and recollimated to a smaller diameter spot using microscope objectives. In the experiment shown, the diameter of a typical HeNe lase is reduced to a $1/e$ value of $21\mu m$ with 10x and 20x microscope objectives. The spot diameter can be determined with either a knife-edge experiment or using a commercial beam profile measuring apparatus.

We briefly discuss the knife edge experiment for measuring the beam profile. While chopping the beam, a razor blade is translated across the beam at the sample location and the beam intensity is measured using a photodetector and lock-in amplifier. *Assuming that the laser beam is Gaussian*, the intensity passing the knife edge as a function of the position is given by the complementary error function

$$erfc(N) = k \int_{-\infty}^{N} \exp(-ax^2)\,dx, \tag{5.21}$$

where k is a numerical constant, N is the position of the knife edge in units of the smallest step provided by the stepper motor that is used to position the sample, and a is the inverse square of the $1/e$ width of the beam. A fit of the data to this function is used to get the beam width.

It is important to properly collimate the beam in the DDM technique so that the spot size is as small as possible while still retaining good collimation. The small spot size ensures that the beam probes a single index gradient value while collimation ensures that the spot size is as small as possible at the camera, making an accurate measurement of the deflection angle possible. In principle, the spot size at the CCD should be smaller than the amount of deflection measured by the CCD.

To increase sensitivity, the camera needs to be far from the sample so that very small deflections can be resolved. Lenses after the sample are avoided because small deflections (paraxial rays) would be refocused onto the optic axis, negating deflection information. The beam is incident perpendicular to the plane of the sample, which is held in an x-y translation stage driven by stepper motors. A pinhole is placed immediately behind the sample to act as a spatial filter that blocks scattered light and multiple reflections from inside the sample. Furthermore, the pinhole aids in centering the spot. The angle subtended by the pinhole and incident laser beam should be large enough not to affect the beam through its full deflection range. Canfield and coworkers found that a $200\mu m$ diameter pinhole met these needs. The beam then passes a second polarizer and moves onto the CCD array. The second polarizer is used in conjunction with the first as an analyzer to determine the sample's birefringence. The entire setup (excluding the computer) can fit on a small breadboard, making it compact and portable.

A measure of the deflection angle requires that the position of the center of the beam at the CCD plane be accurately determined. The central point can be determined by measuring its intensity profile with the CCD array, and determining the position of the peak intensity. Given variations in the beam profile and small sample inhomogeneities that lead to beam distortion, the brightest pixel may not be a good measure of the position of the beam. The best method for dealing with beam variations, as determined by Canfield and coworkers,[74] is to calculate the spot center by spatially averaging the brightest pixels in the frame and recording this intensity-averaged pixel location at the center of the spot. To calibrate the experiment, deflection data is taken without the sample present and the average x and y coordinates are used as the undeflected beam position. This undeflected coordinate is then subtracted from every deflection data point taken with a sample. The deflection data should be saved in an x-y column format to preserve the vector information of the deflection angle.

The deflection data is obtained for a sample that is translated as an x-y raster scan, running through all x-positions in the programmed range for a given

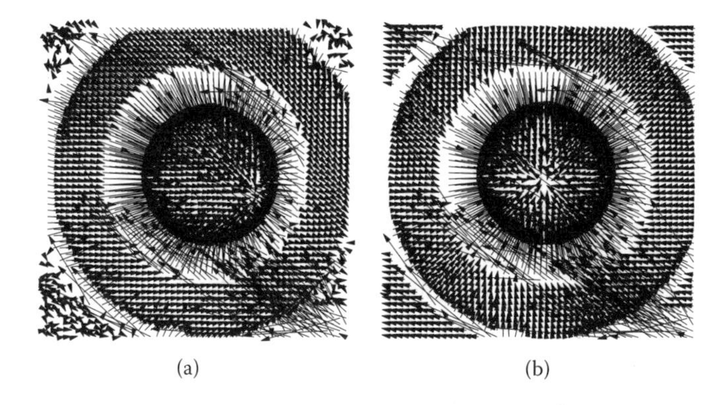

FIGURE 5.8 Deflection angle vector plot. a) Raw data. b) Data corrected for nonparallel faces. Adapted from Ref. [74] with permission of the Optical Society of America.

y-position, then returning to the initial x-position before moving on to the next y-position. (Canfield found a step size of about $0.85\mu m$ works best.) This protocol prevents backlash in the mechanical stages from biasing the data. Clearly, the run-time of such an experiment depends on the size of the sample and the stepper motor increment.

5.1.2.3 Examples of Measured Refractive Index Profiles

In the first step of data acquisition, the deflection angles are measured as a function of position inside a thin sample. At each point, the deflection angle is represented by a vector $\vec{\alpha}(\vec{r})$. A vector field plot of the deflection angle is therefore the most convenient way to display the raw data. Figure 5.8a shows such a plot for a slice taken from a GRIN rod that was prepared by the interfacial gel technique. This full two-dimensional data scan includes the regions outside the preform, which are found at the corners of the figures. Note that the air-cladding boundary gives rise to large discontinuities in the magnitude of the deflection angle due to the large refractive index gradient; i.e., due to the discontinuity in refractive index at the boundary ($n = 1$ to $n \approx 1.5$). Those vectors at the air-cladding interface which extend far beyond the boundary of the scan area are removed to reduce clutter in the plots.

One must be reminded that most refractive index measurements, in reality, measure the optical path length. Since the optical path length is a product of refractive index and distance of propagation, variations in a sample's thickness also result in deflections of the beam. As such, samples with non-parallel faces or curved surfaces may yield deflection data that is dominated by contributions from surface variations rather than the refractive index. As such, samples must be prepared very carefully. Furthermore, the sample's physical shape must also be determined. It is important to note that any experiment that measures the refractive index of a sample is sensitive to variations of the surface topology. As such, this is not a problem unique to DDM and needs to be taken into account in all experimental designs.

One example of thickness variations is illustrated in Figure 5.8a. A net linear bias of the vectors toward the right is clearly observed. This is due to non-parallel faces of the sample. For the sample presented in the figure, measurements of the thickness from one edge of the sample to the other varies linearly from 2.671*mm* to 2.759*mm*; a wedge angle of 0.0002°. Canfield found that the measured bias in the deflection data is consistent with this wedge angle. If the preform is cylindrically symmetric, all vector deflections must sum to zero. The simplest method for correcting the data, provided that the sample's refractive index profile is radially and azimuthally symmetric, is to calculate the average of all the deflection vectors, and then to subtract this average from each deflection data point. Figure 5.8b shows such a correction when applied to Figure 5.8a. It is important to note that regions with large deflections (such as the interfaces) should be omitted from the averaging process. (Due to the large gradients, both the magnitude of deflection and uncertainty are large, yielding a large uncertainty to the correction factor). The core region is usually chosen for the averaging procedure both because it comprises a comparatively large portion of the sample, and is the region of interest.

Once the data is corrected, $f(\vec{r})$ is calculated by taking the square root of the sum of the squares of the x- and y-coordinates for each deflection data point in the matrix from which the absolute deflection angle is calculated (recall that $\alpha(\vec{r}) \approx f(\vec{r})/z$). This angle is used in Equation (5.20), where the dot product in the integrand reduces to $\cos\phi\cos\alpha$ if the numerical integration is performed in the $\hat{x}$ direction, or $\sin\phi\cos\alpha$ if it is performed in the y-direction. Figure 5.9 illustrates the coordinate system used. Angle ϕ corresponds to the usual angle in the x-y plane. The value $\cos\phi$ (or $\sin\phi$) is found by dividing the x-component (or y-component) of the deflection vector by its magnitude $f(\vec{r})$, with the condition that $\cos\phi = 1$ (or $\sin\phi = 1$) if $f(\vec{r}) = 0$ to avoid singularities.

With the data corrected and the deflection angles determined from the experiment, the refractive index difference between any two points is determined by numerical integration, and the full index profile is obtained by integrating between

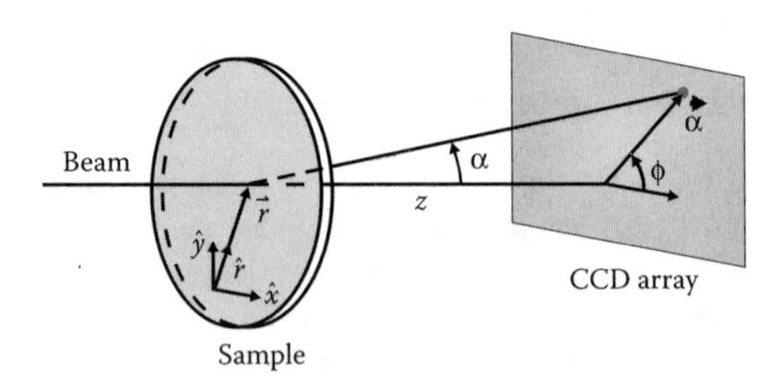

FIGURE 5.9 Experimental diagram showing beam deflection coordinates in the sample and in the CCD array. Adapted from Ref. [74] with permission of the Optical Society of America.

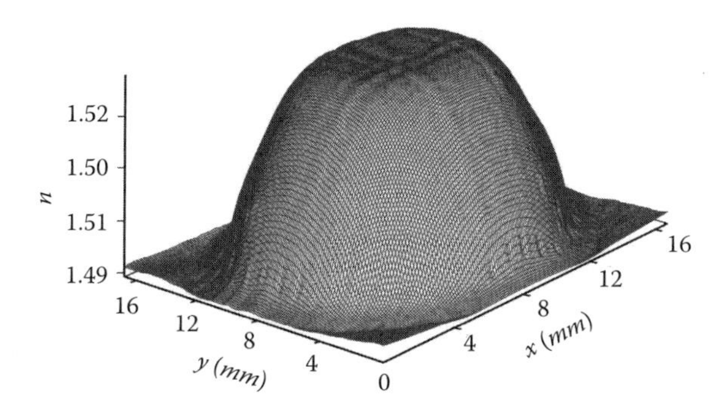

FIGURE 5.10 Refractive index profile determined from a GRIN rod measured with DDM. Adapted from Ref. [74] with permission of the Optical Society of America.

a fixed reference point and each point in the sample. Equation (5.20) (for the x-direction of integration) is more conveniently expressed in difference form as

$$\Delta n[R, C] \equiv n[R, C] - n[R, 1] = \frac{2\,dr}{z\,t} \sum_{j=1}^{C} f[R, j] \cos\phi[R, j] \cos\alpha[R, j], \tag{5.22}$$

where C and R represent the column and row numbers of the matrix and $[R, j]$ denotes the R, j matrix element. So, $\phi[R, j]$ is the angle ϕ when the light source enters the sample at a coordinate $\vec{r}$ which is represented by the coordinate pair R, j. For integration along the y-axis, $\sin\phi$ replaces $\cos\phi$. Figure 5.10 shows the resulting 3D refractive index profile for a 2D GRIN preform scan with the absolute refractive index (the cladding index value is 1.491) along the vertical axis. The corners are in the cladding and the central peak spans the core. The small-amplitude ridges visible are the result of minor discontinuities in individual deflection data points which, once entered into the summation, propagate through the rest of that particular row. If the effect of these discontinuities leads to a large-amplitude ridge, the amplitude can often be greatly reduced by repeating the numerical integration in the reverse direction (i.e., summing from C to 1 instead of 1 to C) and averaging the two results.

One-dimensional scans are faster to obtain, and the bias correction process is much simpler. Furthermore, the effects of thickness variations are easier to see. Figure 5.11 shows an index profile along the disk's diameter as determined by Canfield from a one-dimensional scan (in this case, along the y-axis) for a sample with a slight wedge shape. Figure 5.11a shows the raw data where the wedge effect is large and obvious. The dotted line represents the result of a linear fit to the data bias. When subtracted, the profile appears much improved, as shown in the middle graph.

Another common bias that is often observed is due to a small curvature in the surface of the disk that can arise in the polishing process, or from polymer relaxation over time. (Note that surface reflections from such a sample can be used

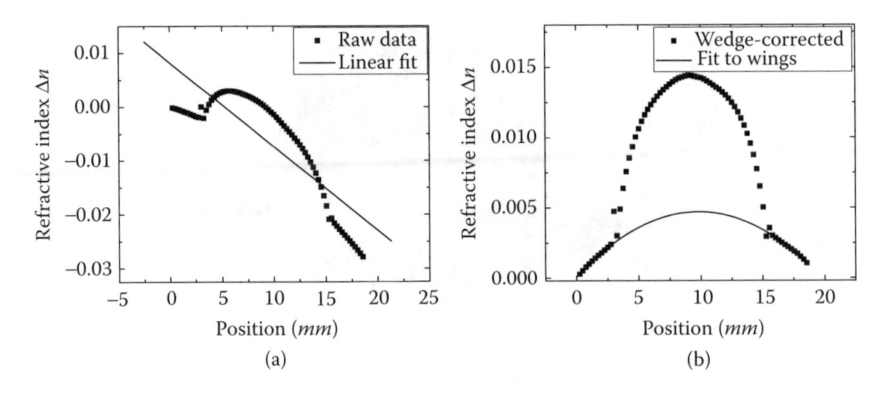

FIGURE 5.11 a) Measured refractive index profile and linear fit. b) Wedge-corrected data and fit to parabola in cladding. Adapted from Ref. [74] with permission of the Optical Society of America.

to independently confirm that the surface is indeed curved.) A very small parabolic deviation in the surface of the preform sample slice is evident in the cladding region of Figure 5.11b, which should otherwise be flat because the index profile within the cladding should be constant. This bias can be corrected by fitting a parabola to the cladding sections of the profile and subtracting the fit from the data. The resulting profile appears in Figure 5.12. Although the magnitude of these biases may seem insignificant, they can have an enormous effect on the refractive index profile measured. The sample investigated by Canfield was fabricated to have a parabolic refractive index profile. A parabolic fit to the corrected data shows that the profile is not precisely parabolic.

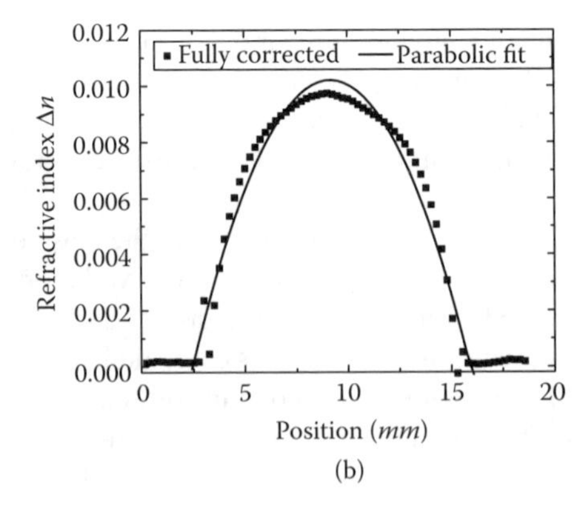

FIGURE 5.12 Refractive index data that is fully corrected for wedge and surface curvature with a fit to a parabolic refractive index profile. Adapted from Ref. [74] with permission of the Optical Society of America.

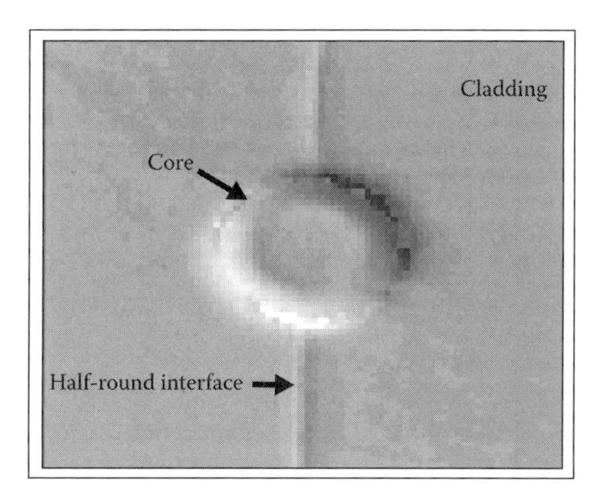

FIGURE 5.13 Magnitude of deflection in a two-dimensional scan of a step-index preform. Intensity represents magnitude of deflection. Adapted from Ref. [74] with permission of the Optical Society of America.

The DDM method also works well for step-index profiles. A step-index polymer optical fiber preform that was fabricated by Dennis Garvey and coworkers at Washington State University [49] was characterized with DDM by Canfield and coworkers. To understand the results, it is worthwhile to review how the preform is fabricated. The core is made of dye-doped PMMA from a core preform of $10mg$ DR1 dye dissolved in $10ml$ MMA (.07% by weight), and has a diameter of $723\mu m$ at the time of fabrication. The preform is squeezed at 120 °C for 96 hours before the sample slice is obtained and the profile measured. The magnitude of the raw deflection data obtained by the data-acquisition program for a 2D scan is shown in Figure 5.13, where the intensity is a linear function of the deflection angle.

The core and the region where the two half-rounds are fused together are clearly outlined in the plot. The process of fusing the cladding during squeezing results in a seam that is a noticeable discontinuity in the DDM experiment. (This seam is not easily detectable visually when inspecting the preform.) The cause is most likely due to stresses and incomplete fusing at the interface.

Figure 5.14 shows a high-resolution ($1.7\mu m$/point) one-dimensional scan along the diameter of the disk. The core-cladding interface is shown by the $723\mu m$ step. The rounding of the profile from a pure step, and the peaks and dip in the core region, result mostly from the finite size of the laser beam spot, which is shown to scale as the shaded circle (this scan was obtained using 4x and 5x objectives in place of the 10x and 20x objectives, resulting in a larger beam spot diameter). Since the spot has a finite width, the beam will be partially deflected to varying degrees as its cross-section encounters the interfacial region, resulting in the rounded step profile. The data shown in Figure 5.14 was corrected by Canfield for a slight parabolic bias. It is important to note that when a finite-sized beam straddles an

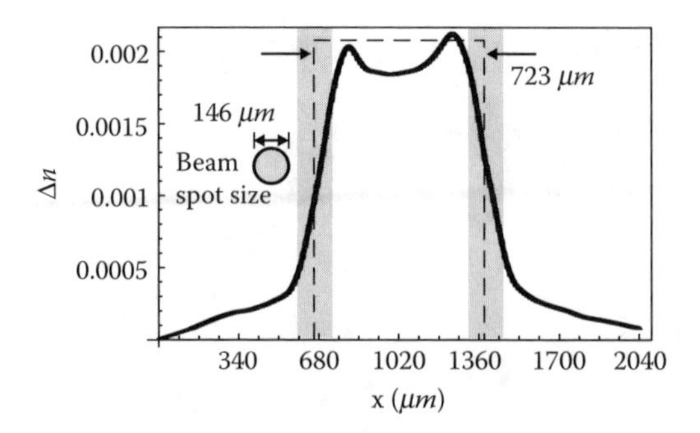

FIGURE 5.14 One-dimensional scan of a step-index preform. Adapted from Ref. [74] with permission of the Optical Society of America. The refractive index is corrected for wedge and surface curvature.

interface, the beams on each side of the interface may deflect by differing amounts leading to two bright spots on the CCD array. The data analysis in these cases may require caution.

5.1.3 Refracted Near-Field Technique

5.1.3.1 Introduction

The refracted near-field (RNF) technique was applied to measuring the refractive index profile of an optical fiber by White.[76] The principle of this technique is that a beam spot of light is focused into the core of a fiber with a numerical aperture that is larger than the numerical aperture of the fiber waveguide. As such, the rays that exceed the critical angle will refract out of the fiber. Figure 5.15 shows a diagram of the rays. Since the angle of a light ray entering the fiber depends on the refractive index at the point of entry, the fraction of the light refracting out of the fiber core will depend on the refractive index difference between the core and the cladding. As such, if the guiding modes of the fiber are blocked, then a measure of the total intensity leaving the fiber is a function of the refractive index of the entry point. The entry point is scanned and the light cone leaving the waveguide is measured to determine the refractive index profile. Note that this technique applies only when the refractive index profile has only a radial dependence and when the waveguide supports many modes. A single-mode fiber will thus not be amenable to this technique.

5.1.3.2 Theory

We begin by treating the waveguide using geometrical optics so that the reader can gain an insight into how and why the technique works. Figure 5.16 shows a schematic diagram of the sample cell. The fiber is embedded in a liquid cell where

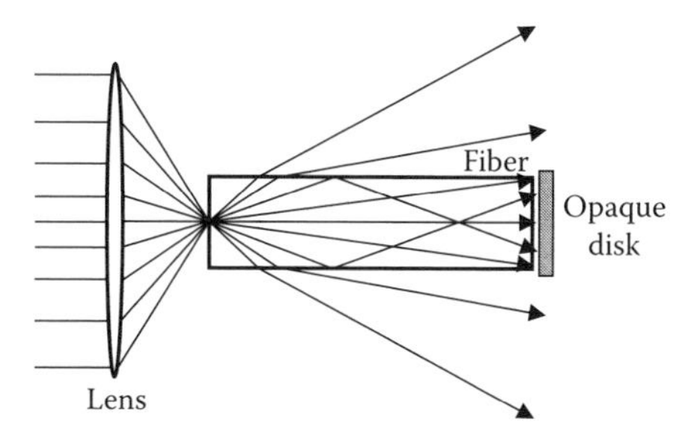

FIGURE 5.15 In the refracted near-field technique, light is focused into the core of a fiber and the integrated power of the rays that leave the waveguide are measured.

the refractive index of the liquid, n_L, is approximately equal to the refractive index of the cladding. One ray in the cone of rays entering the fiber is shown in the drawing.

The intensity of light exiting the fiber is related to the range in angle θ_1 of the light in the core that exceeds the critical angle for wave guiding. As such, we need to calculate the relationship between the angle θ_1 and θ_2. While we know that the dotted ray inside the core will follow a curved trajectory, we will first do the calculation assuming that the trajectory is a straight line and will later show that this yields the correct result. All the angles used in our calculations that follow are labeled on the diagram.

At the first interface, Snell's Law yields

$$\sin\theta_1 = n(r)\sin\theta_4. \tag{5.23}$$

The transmitted ray then hits the fiber/liquid interface at an angle of $90 - \theta_4$, so Snell's Law gives,

$$n(r)\cos\theta_4 = n_L\cos\theta_3. \tag{5.24}$$

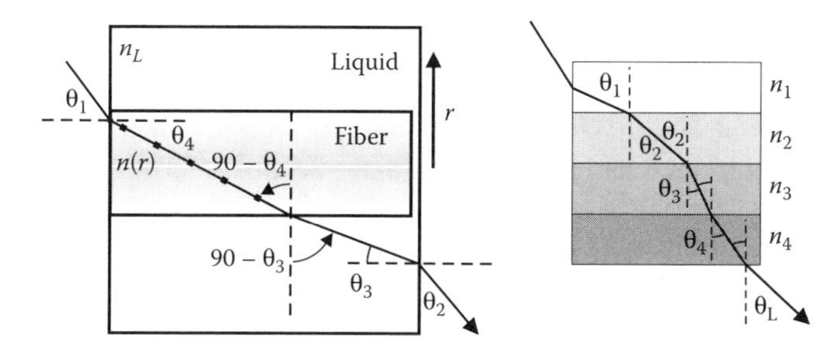

FIGURE 5.16 Ray tracing of a refracted ray that leaves the waveguide (left) and refraction of the ray as it travels through the graded index core as represented by a series of infinitesimal slabs (right).

Finally, the liquid air interface yields

$$n_L \sin\theta_3 = \sin\theta_2. \tag{5.25}$$

Combining Equations (5.23), (5.24) and (5.25) to eliminate θ_3 and θ_4 yields

$$n^2(r) - n_L^2 = \sin^2\theta_1 - \sin^2\theta_2. \tag{5.26}$$

Before proceeding, we digress to explain why the curved ray trajectory does not affect our results. The right part of Figure 5.16 shows the graded index region as a series of slabs. We find that Snell's Law yields $n_1 \sin\theta_1 = n_2 \sin\theta_2 = n_3 \sin\theta_3 \ldots = n_L \sin\theta_L$. Clearly, $n_1 \sin\theta_1 = n_L \sin\theta_L$, so we can ignore all of the rays in the intermediate-layers. Note that while the result for the exit angle is unchanged by ignoring these intervening slabs, the offset of the beam will depend on the geometry of the layers. The refractive near-field technique only depends on the angle, so the offset is not an issue.

A final point of caution is to note that some of the light, even if the incident angle is beyond the critical angle, will reflect back into the waveguide due to Fresnel reflections. Since the amount of reflection is small, and after several bounces the light all leaves the fiber, this effect can be ignored provided that the length of the fiber is long enough to allow the non-guided modes to leave the waveguide.

Now we are prepared to get the refractive index. We note that at the critical angle, the refracted ray exits the fiber along the surface, so $\theta_3 = \theta_2 = 0$. The cone of light in the incident beam that ends up leaving the fiber is thus given by Equation 5.26 with $\theta_2 > 0$,

$$\sin^2\theta_1 > n^2(r) - n_L^2. \tag{5.27}$$

The situation, however, is complicated by leaky modes, which are guided modes that leak out of the waveguide. The leaky and guided modes both travel down the core while the leaky and refracted rays radiate away from the core. The strategy, then, is to block the leaky rays, leaky modes, and guided modes; and only detect refracted rays.

Figure 5.17 shows the experiment. The fiber is bent away from the detection region so that the guided leaky modes and guided modes are eliminated. A disk-shaped beam block is placed in the radiation zone of the fiber and the size is made big enough to block the leaky rays. In the process, some of the refracted rays are also blocked. However, only the refracted rays are imaged onto the detector.

As the fiber is translated perpendicular to the beam, the refractive index that is probed changes, leading to a change in the divergence angle of the refracted light cone. As such, the power of the light that passes around the beam block changes. It is this change in intensity that is used to determine the refractive index. In the following discussion, we assume that the disk is large enough to block all of the leaky modes. At the end of the section, the required disk size will be discussed.

For a Lambertian light source, the radiance profile is of the form

$$L(\theta) = L_0 \cos\theta. \tag{5.28}$$

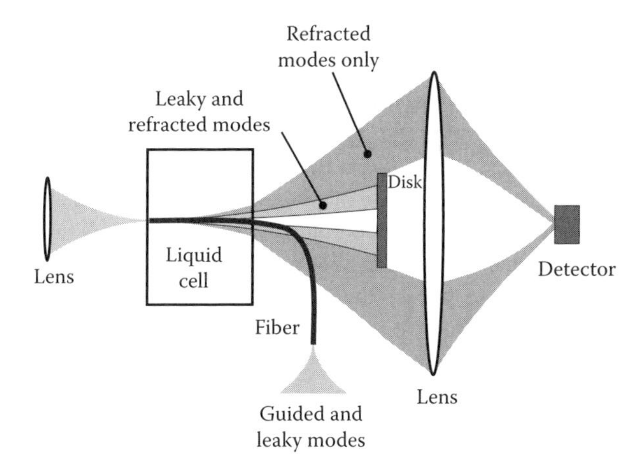

FIGURE 5.17 The refracted near-field experiment. The fiber leaves the liquid cell through a hole at the back end.

If the disk at the output subtends an angle $\theta_2 = \theta_2^0$ and the total light cone of the refracted rays is defined by $\theta_2 = \theta_2^{max}$, this will correspond to a range of input angles between $\theta_1 = \theta_1^0$ and $\theta_1 = \theta_1^{max}$, where each θ_1 is related to θ_2 through Equation (5.26). The power detected at the output is thus given by an integral over the input solid angle of the radiance (Equation 5.28) according to

$$P(\theta_1^0) = \int_0^{2\pi} d\phi \int_{\theta_1^0}^{\theta_1^{max}} L_0 \cos\theta \cdot \sin\theta d\theta. \tag{5.29}$$

The total input power is given by $P(0)$, so we can express Equation (5.29) as

$$P(\theta_1^0) = P(0) \left[\frac{\sin^2\theta_1^0 - \sin^2\theta_1^{max}}{-\sin^2\theta_1^{max}} \right] = P(0) \left[1 - \frac{\sin^2\theta_1^0}{\sin^2\theta_1^{max}} \right]. \tag{5.30}$$

We note that in the experiment, the input beam numerical aperture (θ_1^{max}) is fixed and the position of the disk, and therefore the angle subtended by it (θ_2^0), is fixed. As such, we can use Equation (5.26) to express Equation (5.30) as

$$P(\theta_1^0) = P(0) \left[1 - \frac{n^2(r) - n_L^2 + \sin^2\theta_2^0}{\sin^2\theta_1^{max}} \right]. \tag{5.31}$$

If the refractive index difference between the fiber and the index matching liquid is small (which is usually true), then

$$n^2(r) - n_L^2 \approx 2n_L \Delta n(r). \tag{5.32}$$

Substituting Equation (5.32) into Equation (5.31) and solving for the refractive index difference yields

$$\boxed{\Delta n(r) = \frac{1}{2n_L} \left[\left(1 - \frac{P(\theta_1^0)}{P(0)} \right) \sin^2\theta_1^{max} - \sin^2\theta_2^0 \right].} \tag{5.33}$$

Note that the measured power, $P(\theta_1^0)$, is a function of the position r but is not explicitly stated. So, the value of $\Delta n(r)$ is a linear function of the measured power. The other terms are constants that are determined from the experimental geometry.

Finally, we must determine the position of the disk that is required to block the leaky radiated modes. We note that the numerical aperture of a graded-index fiber depends on the radial coordinate of where the light is launched. For example, if the refractive index profile is a monotonically decreasing function of the radial coordinate, a beam launched into the fiber near the cladding will be more weakly guiding than one launched in the center. As such, the size of the disk required depends on the range of refractive index measured.

Saunders simplified White's result for the condition of the minimum required beam block angle θ_2^0,[77]

$$\sin^2\theta_2^0 = \left[n_L^2 - n^2(a)\right] + \frac{\alpha}{2}N^2, \tag{5.34}$$

where N is the local numerical aperture, α is the beam shape parameter (where the refractive index profile relative to the background refractive index is approximated by $\Delta n(r) = n_0(1 - (r/a)^\alpha)$), and $n(a)$ is the refractive index at the edge of the waveguide. To estimate the size of disk required for a particular fiber therefore requires some knowledge about the fiber's refractive index properties.

5.1.3.3 Experimental Results

Figure 5.18 shows a profile of a step-index fiber as measured by White.[76] The dashed curve shows the expected result for a perfect step. The eccentricity of the core is clearly seen with this measurement. Figure 5.19 shows the profile of a graded-index fiber that was made with chemical vapor deposition. The central dip is a true property of the fiber. Note that in both cases, the cladding and the surrounding index matching liquid is clearly observed.

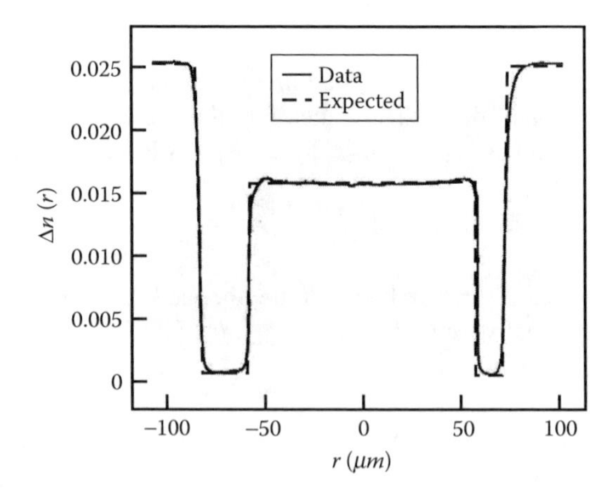

FIGURE 5.18 A refractive index profile of a step index fiber measured with the refracted near-field technique. Adapted from Ref. [76] with permission of Chapman and Hall Ltd.

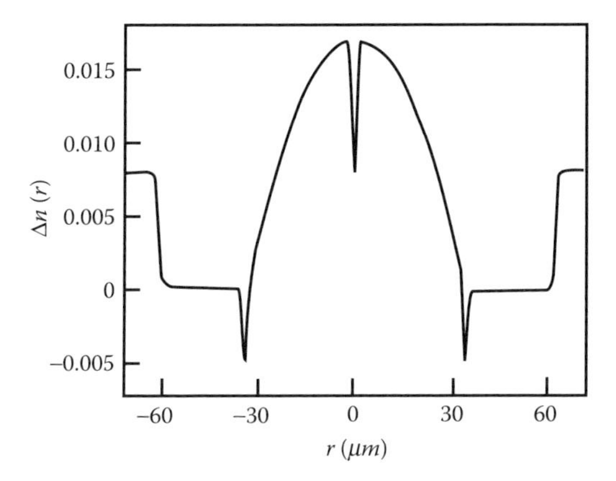

FIGURE 5.19 A refractive index profile of a graded-index fiber measured with the refracted near-field technique. Adapted from Ref. [76] with permission of Chapman and Hall Ltd.

The RNF technique, while straightforward to perform, needs to be analyzed with care. For example, most light sources are not Lambertian so the system needs to be calibrated. Such a calibration can be done by removing the fiber and measuring the intensity that gets past the disk as a function of the position of the disk as it is moved along the axis of the light source. Such a calibration procedure effectively gives the beam profile. There are also many subtle issues with regard to how the fiber is held in place in the cell, how it is pulled through the cell wall, and how the disk is held in place without the support blocking some of the refracted light. Since we do not treat these issues here, the reader is referred to the extensive literature on the topic.[78]

5.1.4 Fresnel Reflection Technique

The Fresnel Reflection Technique (FRT) is perhaps the most straightforward method. It uses the fact that the reflectance from a smooth interface depends on the indices of refraction of the two materials. Defining the refractive index in the material carrying the incident beam as n_L and the refractive index of the transmitted beam's medium as $n(r)$, the reflectance at normal incidence is given by

$$R = \left(\frac{n(r) - n_L}{n(r) + n_L} \right)^2 . \tag{5.35}$$

We can invert this equation to solve for the refractive index $n(r)$,

$$n(r) = n_L \left(\frac{1 + \sqrt{R}}{1 - \sqrt{R}} \right) . \tag{5.36}$$

Since the FRT method is simple to understand and straightforward to implement, the treatment here provides the experimental details of detector calibration

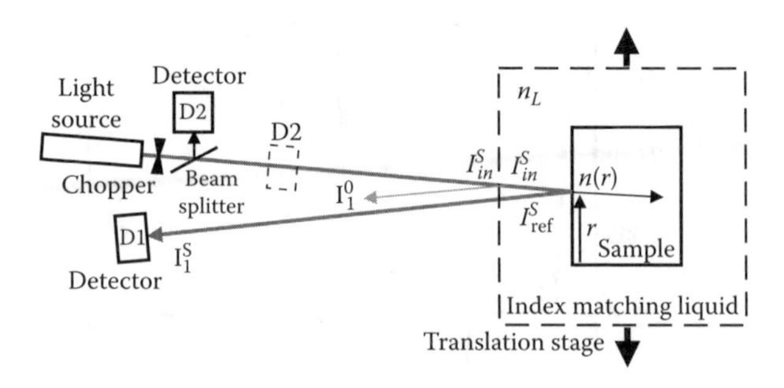

FIGURE 5.20 The Fresnel reflection experiment.

and experimental procedures so that the reader can become familiar with how intensity measurements are performed. The procedure takes advantage of the fact that all required parameters are ratios of intensities, so it is not important to measure an absolute intensity. As such, the voltage measured by a detector acts as a proxy for the intensity and we just call the measured quantity the intensity. Clearly, the detector voltage must be linearly proportional to the intensity for this proxy to hold.

There are many experimental configurations that can be used to measure the reflectance. Figure 5.20 shows one of these, where the incident beam is close enough to normal incidence that Equation (5.35) holds. As such, the laser source is necessarily far away from the sample. Though the sample is shown inside a cell of index matching liquid of refractive index n_L, the sample can also be measured in air (the role of the cell is described later). The sample sits on a translation stage so that the point probed by the laser can be varied across the surface of the sample.

The experimental procedure is as follows: Detector D_2 is placed in the position shown in Figure 5.20 by the dashed rectangle and the intensity it reads, I_2', is recorded. (It is advisable to use a chopper to modulate the beam and a lock-in amplifier to reduce background noise.) The detector is then placed in its permanent position, shown as the solid box $D2$ and the intensity it reads, I_2^0, is recorded. The calibration factor k is then given by

$$k = \frac{I_2'}{I_2^0}, \tag{5.37}$$

so that a reading of I_2 in detector $D2$ corresponds to an intensity incident on the sample cell, I_{in}, given by

$$I_{in} = kI_2. \tag{5.38}$$

The role of detector $D2$, called the reference, is to provide a measure of the incident intensity, which can change due to drift in the light source during an experiment. Since this experiment measures small changes of the reflected intensity as the incident beam is scanned across the surface of the sample, it is important that any small changes in the incident light beam be carefully taken into account.

To get the actual intensity incident on the sample, we must consider the reflectance due to the interface between the air and the index matching liquid. We can do this simply by measuring the reflectance due to the front surface of the sample cell. Assuming that the response of detector $D1$ is the same as $D2$ (if not, they can be calibrated to each other), then the reflectance from the sample cell, R_C, is

$$R_C = \frac{I_1^0}{I_{in}} = \frac{I_1^0}{kI_2}, \tag{5.39}$$

where I_1^0 is the intensity reflected from the sample cell surface and measured using detector D1. The intensity incident on the sample, I_{in}^S, is

$$I_{in}^S = f(1 - R_C)I_{in}, \tag{5.40}$$

where f is the fraction of light absorbed by the index matching liquid, which needs to be measured separately (see Section 4.2 on optical loss).

If the light reflected from the sample is I_{ref}^S, the intensity reaching the detector $D1$ is

$$I_1^S = f(1 - R_C)I_{ref}^S, \tag{5.41}$$

where again we take into account absorption and reflection by the index matching liquid. The sample reflectance is given by

$$R = \frac{I_{ref}^S}{I_{in}^S}. \tag{5.42}$$

Using Equations (5.40) and (5.41) in Equation (5.42), we get

$$R = \frac{I_1^S/f\,(1 - R_C)}{f\,(1 - R_C)\,I_{in}} = \frac{1}{kf^2\,(1 - R_C)^2} \cdot \frac{I_1^S}{I_2}, \tag{5.43}$$

where the last equality was obtained with the help of Equation (5.38). According to Equation (5.43), the reflectance is determined simply from the ratio of the voltages read by the two detectors (all the other parameters are previously determined and held constant). To get the refractive index profile, the reflectance, R, is measured as a function of position on the surface of the sample, and Equation (5.36) is used to get the refractive index.

One of the strengths of this technique is that a highly optically absorbing sample can be measured. Furthermore, the birefringence of the sample can be determined by performing separate reflectance measurements for two perpendicular polarizations of the incident beam. The two axes that diagonalize the refractive index tensor are called the principle axes, which are the most convenient to use in most applications. A principle axis can be found by rotating the polarization of the incident beam until the reflected beam has the same polarization as the incident beam.

Finally, we comment on the need for the index matching liquid. In typical measurements, the variations in the refractive index are small compared to the mean refractive index. As such, the measurement yields small changes in intensity of the reflected beam with a large background intensity. This background intensity can be decreased by making the refractive index difference between the material surrounding the sample and the sample small; and this is done by submersing the sample in index matching liquid. Note, however, that it is most convenient if the refractive index of the index matching liquid is smaller than the smallest refractive index on the profile of the material. Equation (5.35) shows that the reflectance is independent of the sign of the refractive index difference between the sample and the surrounding medium, so the analysis of the data becomes ambiguous if the refractive index difference measured changes sign over the surface.

To understand the accuracy of a particular measurement, the experiential uncertainties must be determined and propagated through all the equations. It is implicitly implied that such an analysis is performed for any measured quantity. The Fresnel reflection technique serves as a simple example of how error is propagated.

In general, consider the function, $f(x_1, x_2, \ldots)$, that depends on the measured quantities $x_1, x_2, \ldots$ The uncertainty in f, Δf, is simply given by the differential of f:

$$\Delta f(x_1, x_2, \ldots) = \frac{\partial f}{\partial x_1}\Delta x_1 + \frac{\partial f}{\partial x_2}\Delta x_2 + \frac{\partial f}{\partial x_3}\Delta x_3 + \cdots, \tag{5.44}$$

where Δx_i is the uncertainty of x_i. Applying Equation (5.44) to Equation (5.36), we get

$$\Delta n = n_L \frac{1}{\sqrt{R}} \frac{1}{(1-\sqrt{R})^2} \cdot \Delta R. \tag{5.45}$$

Even small uncertainties in the reflectance, ΔR, lead to a large uncertainty in the refractive index, Δn. Since the reflectance depends on the calibration factor as well as several measured intensities, the uncertainty in the reflectance will be given by a differential in terms of all of the measured quantities. We will not go through these details here. We note, however, that all uncertainties are eventually related to the experimentalist's estimate of how well a particular instrument measures a particular quantity. Estimating experimental uncertainties is sometimes an art that is best learned through practice.

5.1.5 Other Refractive Index Measurement Techniques

There are many other refractive index profile measurement techniques[78] that we will not present in detail, though they are worthwhile to mention. First is the Near Field Measurement (NFM), which applies to waveguides. The underlying principle is that the refractive index profile of a waveguide determines the intensity profile of the light in the waveguide. As such, the near-field intensity profile measured at the end of a waveguide determines the refractive index profile from a model of the expected profile. This technique, while simple, is not amendable to measuring an arbitrary refractive index profile.

Another is the transverse interferometric measurement (TIM), which was recently adapted to measuring the refractive index profiles of large-core fibers.[79]

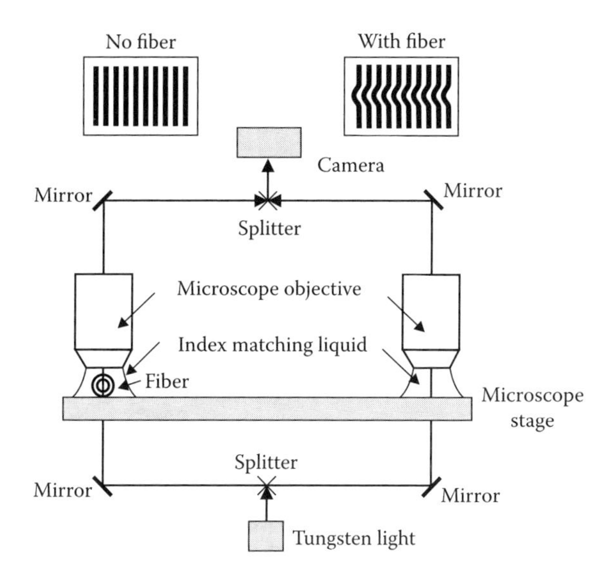

FIGURE 5.21 The Leitz transmitted-light interference microscope can be used to measure the refractive index profile of a fiber.

The fiber is placed on a transparent substrate and emersed in a puddle of index matching liquid so that the light does not refract at the cladding-air interface. A 20x microscope objective is placed above the sample and the end of the objective makes contact with the index matching liquid to make a liquid bridge. The reference arm of the interferometer is identical, but without the fiber. A tungsten white light source is used for illumination. Figure 5.21 shows a highly schematic representation of the setup, which is called a Leitz transmitted-light interference microscope. The two interfering light spots make a series of fringes. When no fiber is present, the fringes are parallel lines — as shown in the upper-left part of Figure 5.21. With the fiber in the sample arm, the fringes are curved. As with the interphako spectrometer, the fringe pattern can be analyzed to determine the refractive index profile.

5.1.6 Measurements on Gradient-Index Fibers

The process for making gradient-index (GI) polymer optical fibers (POF) is designed to make a parabolic refractive index profile. The typical process for making such GI POFs in commercial processes is similar to what is described in Chapter 3. Shi and coworkers measured the refractive index profile of GI POFs made of PMA polymer and benzyl benzoate at the dopant — which diffuses and makes the index gradient. They used both the refracted near-field technique and the transverse interferometric measurement and found similar results.[80]

Figure 5.22 shows the measured refractive index profile and two theoretical models. Shi and coworkers found that the refractive index profile could not be

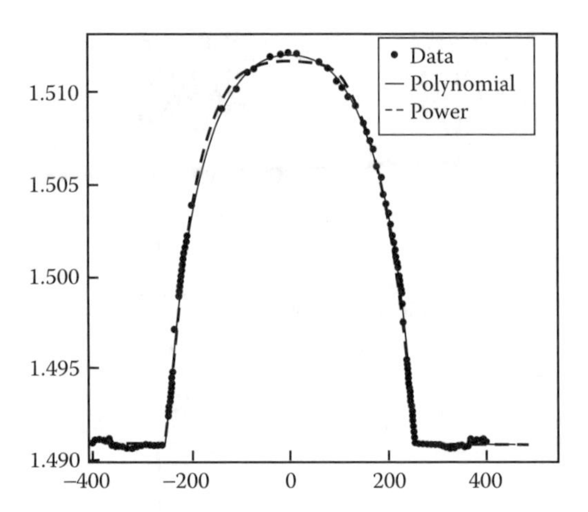

FIGURE 5.22 The refractive index profile measured with the near field refractive technique (points) and fits to a polynomial (solid curve) and single exponent (dashed curve). Adapted from Ref. [79] with permission of the Optical Society of America.

fit to a simple polynomial. Instead, the fit was best with a piecewise continuous polynomial of the form

$$n(r) = \begin{cases} n_{f_1} - \delta_{10}(r/a)^2 - \delta n_{11}(r/a)^4 \text{ for } r \le a_1 \\ n_{f_1} - \delta_{20}(r/a)^2 - \delta n_{21}(r/a)^4 \text{ for } a_1 \le r \le a \\ n_c \qquad\qquad\qquad\qquad\qquad\quad \text{ for } r > a \end{cases}, \tag{5.46}$$

where n_c is the refractive index of the cladding, a is the radius of the core, and the other parameters are varied to fit the data under the condition that the refractive index and its derivative are both continuous at $r = a_1$.

The data was also fit to a variable power law of the form

$$n(r) = \sqrt{n_f^2 - \left(n_f^2 - n_c^2\right)(r/a)^\alpha}, \tag{5.47}$$

where n_f is the refractive index at the center and α is the power parameter. The two fit functions in Figure 5.22 are given by Equations (5.46) and (5.47). Clearly, the power law gives a bad fit near the fiber center.

5.1.7 Summary of Refractive Index Measurements

Many methods are available for measuring the refractive index profile of a polymer fiber or preform. The interphako technique is well suited from measuring the profile from the side of the fiber, but assumptions about the refractive index profile must be made (such as a radial dependence). The DDM technique, on the other hand, can give highly accurate two-dimensional profiles, but a thin polished disk sliced from the end of the preform must be used. Small-diameter fibers, though, would not be amenable to this technique because of the fairly large beam diameters that are required to keep collimation. The refracted near field technique can be used on optical fibers, but only works when the fiber core is large enough to support many modes. The Fresnel reflection technique's strong suit is that it can measure the refractive index *tensor* in a sample even if it is highly absorbing at the measurement wavelength, but is highly sensitive to surface imperfections and surface damage.

5.2 OPTICAL LOSS

As we saw in Section 4.2 the optical loss is an important fiber property that is a measure of the amount of light that is absorbed by the fiber. There are two broad categories of loss — intrinsic and extrinsic loss. An intrinsic loss is due to the fundamental property of a material while the extrinsic loss depends on how the material is prepared. For example, the amount of light absorbed by a pure material is the intrinsic loss while optical scattering from sample imperfections is an extrinsic property. As such, the extrinsic loss can be changed through material processing. Since most applications require low loss fibers, materials that have low intrinsic loss should be used and the materials should be judiciously processed during the fiber drawing process to remove imperfections and impurities.

For both intrinsic and extrinsic loss mechanisms, the amount of light lost is proportional to the length of the material, L, the incident intensity, I_0, and the material loss. The optical loss per unit length of material, A, as given by Equation 4.56, is commonly expressed in a base 10 logarithm as

$$A = -\frac{10}{L} \log \left(\frac{I}{I_0} \right), \tag{5.48}$$

where I is the intensity exiting the material. The top portion of Figure 5.23 shows the parameters.

This definition of loss is a convenient one because the losses of multiple elements add. Consider the bottom portion of Figure 5.23. The sum of optical losses

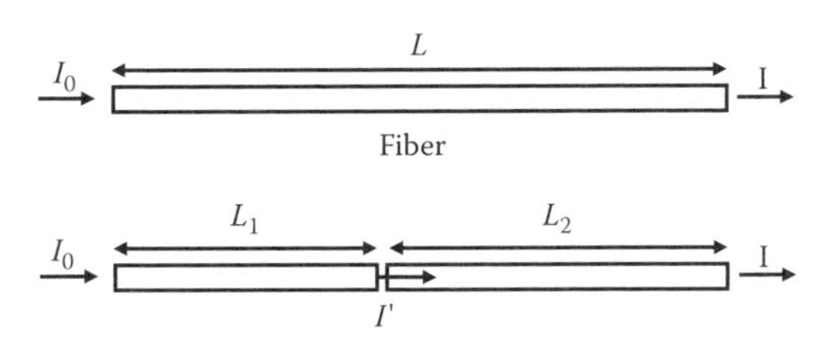

FIGURE 5.23 The optical loss in a fiber (top); and the loss in the same fiber when it is split into two sections.

of the two sections are

$$A_1L_1 + A_2L_2 = -L_1\frac{10}{L_1}\log\left(\frac{I'}{I_0}\right) - L_2\frac{10}{L_2}\log\left(\frac{I}{I'}\right) = -10\log\left(\frac{I}{I_0}\right) = AL, \tag{5.49}$$

where we ignore the surface reflections at the interfaces. Clearly, the loss for a stack of different materials of differing lengths and absorption per unit length would be given by

$$AL = \sum_{i=1}^{N} A_i L_i. \tag{5.50}$$

Given these definitions, the following subsections describe how the optical loss is measured using various techniques.

5.2.1 Cutback Technique

In the cutback technique, the transmitted intensity through a fiber is measured, a piece is removed, and the transmitted intensity is again measured. This technique can be used to measure the loss accurately, or to estimate the loss quickly. Keep in mind when reading the following sections that in both versions of the cutback method, the incident intensity does not need to be measured. This is an important issue, because it is almost impossible to measure with certainty how much light actually couples into the fiber.

5.2.1.1 Quick Cutback Method

To estimate the loss quickly, only two transmitted intensities are measured. First, the transmitted intensity, I, is measured for a fiber of length L. The output end of the fiber is carefully polished so that no light is scattered away from the detector. A section of fiber ΔL is cut from the output end while the input end remains fixed and the output end is re-polished. The intensity of the output, I', for this shorter length of fiber is measured. Figure 5.24 shows the two measurements. Before describing how these measurements yield the loss, we discuss some experimental issues that require some attention.

First, it is important that the input side of the fiber be left untouched during the whole measurement process so that the intensity coupled into the fiber does not change between the two measurements. The best method for protecting the input coupler side is to use a long length of fiber so that the cutting and polishing process is as far from the input end as possible.

Second, if the light source used is not stable, the drift of the intensity over time needs to be taken into account. This is done by placing a detector in front of the beam prior to each measurement. Alternatively, a small amount of the incident light can be deflected by a surface reflection from a piece of glass to monitor drift. The second method is shown in Figure 5.24.

Third, the intensity of light that is measured at the output of a fiber can be small and room light can add to the measured signal. This is especially true if the optical

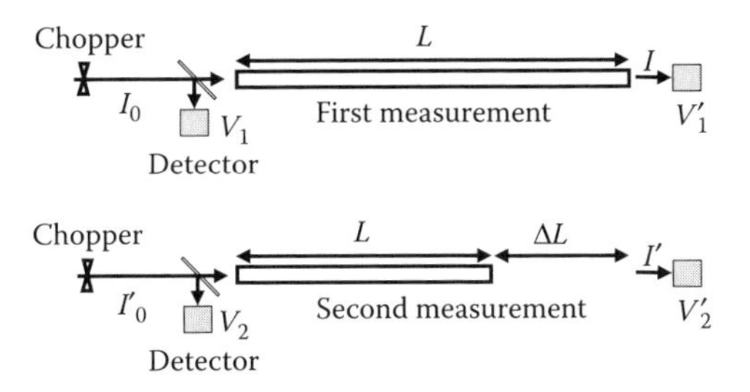

FIGURE 5.24 The quick optical loss measurement.

loss in the fiber is high. To eliminate background, the incident beam is chopped so that the light intensity is modulated at some frequency Ω and a lock-in amplifier is used to determine the signal at that frequency. Since the background light is not modulated, it is not measured with the lock-in amplifier. Note that the lock-in is not shown in the figure.

To measure the loss, we first recognize that Equation (5.48) is expressed in terms of a ratio of intensities. Since a detector converts intensity to a voltage (which can be measured by a computer interface or a simple voltmeter), Equation (5.48) can be expressed in terms of voltages. For the two experiments, we get

$$AL = -10\log\left(\frac{V_1'}{\beta V_1}\right), \tag{5.51}$$

and

$$A(L - \Delta L) = -10\log\left(\frac{V_2'}{\beta V_2}\right), \tag{5.52}$$

where β is related to the reflectivity of the glass used to divert part of the beam to the detector at the incident side of the fiber. This factor also includes the coupling efficiency into the fiber. So, βV_i is the actual intensity (measured in units of voltage) that enters the fiber. Note that β will not need to be determined.

If we subtract Equation (5.51) from Equation (5.52) and solve for loss A we get

$$A = -\frac{10}{\Delta L}\log\left(\frac{V_1'}{V_2'}\cdot\frac{V_2}{V_1}\right). \tag{5.53}$$

This result is independent of the factor β and is expressed in terms of those parameters that are simple to determine.

Note that the above approach assumes that the two detectors have the equivalent response. If they do not, they need to be calibrated. This is a simple process, in which a beam of light is launched on the first detector, then on the second one. Assuming that the light intensity is constant during this measurement, the calibration factor is just the ratio of the measured voltages. An example that illustrates the calibration process as applied to Fresnel reflection is described in Section 5.1.4. Thus, the calibration factor, when multiplied by the voltage read by one of the detectors converts it to the same scale that the second detector would measure if it were placed in the beam.

While this method is quick because it requires only two sets of simple measurements, it is inaccurate because of the variability in the amount of light that exits the fiber due to the polishing process. Since it is never possible to make a perfect polish, for a fixed input, the output intensity can easily vary by 20%.

5.2.1.2 Accurate Cutback Method

The large uncertainty of the measured loss that is introduced by the polishing process can be reduced if many measurements are averaged from multiple cutback experiments using a sequence of successively shorter fibers. The approach is more clear if we rewrite Equation (5.48) as

$$10 \log I = 10 \log I_0 - AL. \tag{5.54}$$

Equation (5.54), when plotted as $10 \log I$ as a function of L, yields a line of slope A and intercept $10 \log I_0$. Note that since the slope yields the loss, A, there is no need to know the intercept.

The cutback process is repeated several times by successively cutting small pieces of fiber from the detector end and measuring the transmitted intensity after each slice is removed to determine the transmittance as a function of the length of remaining fiber. A linear fit to the log plot of the data yields the loss. The uncertainty in the slope so determined decreases as the cutback process is repeated more times for a given fiber, leading to more data points and better statistics.

Figure 5.25 shows an example of data collected by Garvey and coworkers[19] for a multimode fiber. In these measurements, a laser diode provides light at $1.3\mu m$, which is modulated by a chopper and launched into the fiber with a lens. The transmitted intensity is measured with a Germanium detector and lock-in amplifier. A $1cm$ piece of the fiber end nearest the detector is cut and polished, and the intensity remeasured. (The cutting/polishing/measuring process is repeated about 50 times.) Multimode fiber is much more forgiving with regards to alignment on the input side so the data is less sensitive to the quality of the cut than for a single-mode fiber. All samples measured at $1.3\mu m$ fall within the 0.3dB/cm range, consistent with the neat PMMA value.[81,157]

From Figure 5.25, the loss of dye-doped PMMA at $1.3\mu m$ is found to be independent of which dopant is used and is identical to the neat PMMA loss. Furthermore, the fact that the dye does not contribute to the loss in a dye-doped polymer when the light wavelength is far from the dye's resonant wavelength is consistent with measurements of Skumanich and coworkers,[82] who use the photothermal deflection technique. This is significant, because it implies that the loss of a dye-doped fiber in the near *infrared* can be lowered simply by decreasing the loss of the host polymer, making possible low-loss nonlinear-optical switching devices.

Garvey also measured the loss in single-mode fibers and multimode fibers at $1.06\mu m$ using this technique. Even though the light intensity leaving a single-mode fiber is a sensitive function of the quality of the fiber end, the losses in the single-mode PMMA/ISQ fiber and the multimode PMMA/ISQ guide at $1.06\mu m$ are consistent with each other and with Kaino's result in neat PMMA of about 0.2*dB/cm*[81]. This is also the case for TSQ loss in both fibers at $1.06\mu m$. The

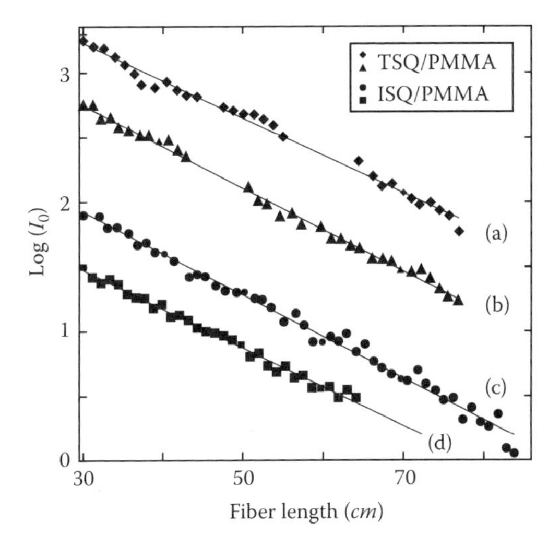

FIGURE 5.25 Absorbance in multimode fiber as a function of length for TSQ/PMMA ((a) and (b)) and ISQ/PMMA ((c) and (d)) multimode fibers. Doping levels are about 0.1% by weight. Adapted from Ref. [19] with permission of the Optical Society of America.

conclusion is that the dyes do not contribute to the loss in either of these materials. For BSQ, however, the loss of the single-mode fiber is much higher than the intrinsic loss as determined from the multimode measurement. This implies that the loss must be extrinsic (most likely due to inhomogeneity of the core), and can be made lower through material processing. It is encouraging that none of the dyes seem to contribute appreciably to the loss even at $1.06\mu m$. Table 5.1 summarizes Garvey's results.

TABLE 5.1
Optical loss in selected fibers[49]

λ	Dopant Dye	Diameter of Waveguide	Type	Loss (dB/cm)
$1.3\mu m$	ISQ	$510\mu m$	Multimode	0.31 (± 0.02)
$1.3\mu m$	TSQ	$440\mu m$	Multimode	0.30 (± 0.02)
$1.06\mu m$	ISQ	$790\mu m$	Multimode	0.19 (± 0.03)
$1.06\mu m$	BSQ	$790\mu m$	Multimode	0.15 (± 0.04)
$1.06\mu m$	TSQ	$790\mu m$	Multimode	0.28 (± 0.06)
$1.06\mu m$	HSQ	$790\mu m$	Multimode	0.17 (± 0.03)
$1.06\mu m$	PSQ	$790\mu m$	Multimode	0.22 (± 0.04)
$1.06\mu m$	ISQ	$16\mu m$	Single mode	0.23 (± 0.05)
$1.06\mu m$	BSQ	$15\mu m$	Single mode	0.42 (± 0.08)
$1.06\mu m$	TSQ	$8\mu m$	Single mode	0.18 (± 0.04)

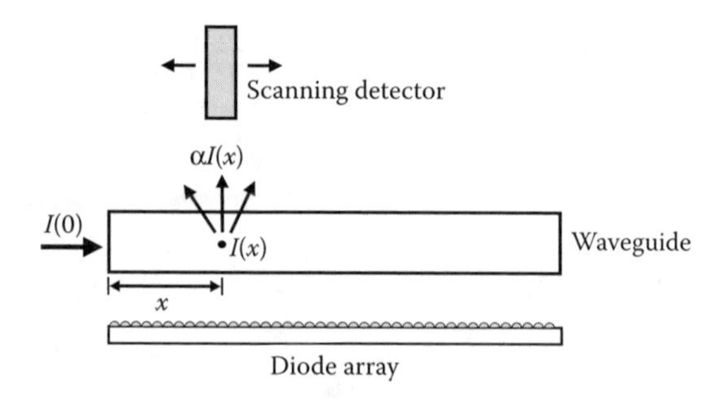

FIGURE 5.26 Light scattering from a waveguide.

5.2.2 Transverse Scattering Loss Measurement

Even in the best waveguides, some of the light leaves the waveguide through scattering. In a fiber waveguide, scattering from the core appears from the side as a bright line at the core. The light intensity that scatters from the waveguide at point x, $I_s(x)$ is proportional to the light inside the waveguide, $I(x)$. Figure 5.26 shows light that is coupled into the fiber from one end and scattering from a point x within the waveguide region. For a uniform waveguide, the constant of proportionality, α, is independent of x, so the scattered light is simply

$$I_s(x) = \alpha I(x). \tag{5.55}$$

Taking the logarithm of Equation (5.55),

$$\log\left(I_s(x)\right) = \log(\alpha) + \log\left(I(x)\right). \tag{5.56}$$

Recall from Figure 5.25 of Section 5.2.1 that the optical loss is given from a plot of the log of transmitted intensity versus position. According to Equation (5.56), a plot of $\log I_s(x)$ versus x gives the same slope as a plot of $\log I(x)$ versus x, except the intercept is different. As such, the value of the constant α is immaterial with regards to getting the loss from the scattered intensity.

Any method that records the intensity of light scattering from a waveguide as a function of position can be used to measure the loss using transverse scattering. As shown in Figure 5.26, either a scanning detector or a diode array are the most common. The advantage of the diode array is that the light streak can be measured quickly. The scanning detector, on the other hand, can cover more distance. In waveguide applications where the loss is on the order of 1dB/cm, a diode array works well. In the case of a fiber with low loss, a detector may need to be scanned over several meters of the fiber.

The advantages of the transverse scattering technique is that it is non-destructive and can be used to measure the loss at any point in the fiber provided that light can be launched into one end. For low loss fibers, where the scanning detector is used, it is difficult to keep the fiber straight and the detector at a fixed distance from the fiber. Sagging of the fiber is a common problem. In addition, scattering from the

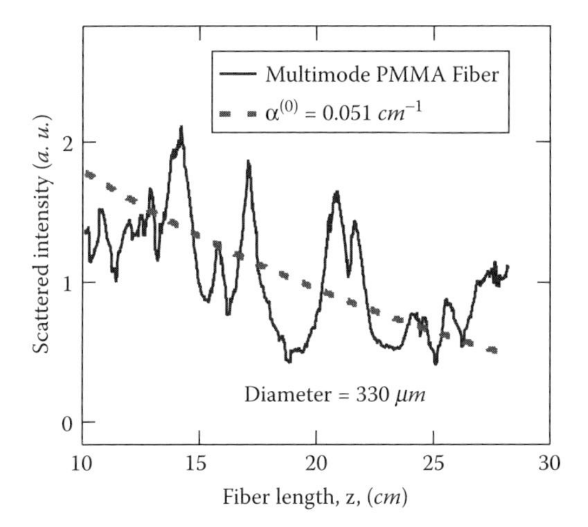

FIGURE 5.27 Data showing light scattering from a polymer fiber. Adapted from Ref. [84] with permission of the Optical Society of America.

fiber may not be uniform, so the transverse intensity profile may not accurately reflect the loss.

Skumanich and coworkers used the diode array detection scheme with a planar polymer waveguide to get reasonably accurate values of the loss.[82] In 1999, Kruhlak measured the loss in polymer fibers with the scanning technique.[83,84] Data with the largest variations in scattering intensity are shown in Figure 5.27 to illustrate the extreme case where the determination of the loss is difficult. In this fiber, several scattering centers along the length measured are responsible for the large intensity fluctuations. This data, however, can be used to study the scattering properties of the fiber. Indeed, by calculating correlation functions, Kruhlak was able to determine correlation lengths that are associated with defects that are originally fabricated into the preform and stretched along the fiber during the drawing process. Even so, the fit to the data gives a good approximation to the fiber loss if the data near the ends are ignored. This is justified because light scatters strongly from the collettes that hold the fiber in place at both ends.

5.2.3 The Eyeball Technique

While the performance of the human eye is somewhat subjective, the perception of intensity is known to be logarithmic. The ancient Greeks, who gazed upon the heavens, defined a brightness scale for stars which is approximately logarithmic with base 2.5. On this scale, the range of brightness that can be observed is about 5 to 6 orders of magnitude, or a dynamic range in intensity of over 100. To give the eyes a broad visual experience, makers of digital cameras and computer monitors make instruments whose dynamic range exceeds this value. Good cameras and displays provide 8 bits of shading, or an intensity range of 255; 24-bit color displays provide 8 bits for each of the three primary colors.

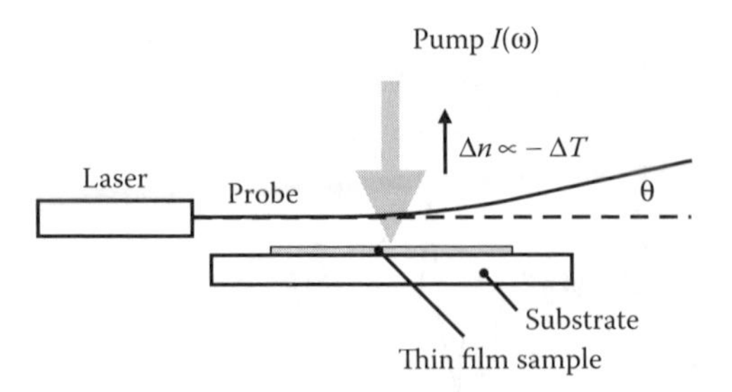

FIGURE 5.28 The photothermal deflection experiment.

Loss is usually given on a log scale in base 10. In the standard loss definition, the dynamic range of the human eye is about $21dB$. So, to estimate the loss of a fiber waveguide, all one has to do is to couple light into the core and observe the length of the fiber that appears to be illuminated, L. The loss is then simply given by $21dB/L$. Clearly, for this measurement to work, the length of the fiber must exceed the loss.

In reality, the eyeball technique is just the transverse scattering method where the eye is used as the detector. I am often surprised by the high degree of accuracy afforded by the eye and use this trick to awe the graduate students who might spend several hours to set up and measure a fiber loss, only to find that my estimate of the loss is within 25% of the more sophisticated method. As usual, a word of caution is necessary. **Never use the eyeball technique when the light source intensity is high enough to damage your eyes!**

5.2.4 Photothermal Deflection

While photothermal deflection measurements do not apply to optical fibers, it is worthwhile to mention this technique because it allows one to measure small optical loss in thin film samples. Because thin film measurements are usually dominated by intrinsic loss, this is a good complementary method to measure materials in a thin film geometry. The technique dates back to 1980, when Boccara and coworkers first measured thin film samples.[85] A detailed theory of both the transverse and collinear geometries, with experimental verification, was later presented by the same group.[86] Murphy and Aamodt also developed a similar theory and showed that photothermal deflection is related to photoacoustic spectroscopy,[87] where the absorption of a pulse of light yields a sound, whose strength is related to the amount of light absorbed.

Figure 5.28 shows a schematic diagram of the photothermal deflection experiment in the transverse geometry. A probe beam is launched near the surface of the sample and the pump beam is launched perpendicular to the sample. The pump beam is set to the wavelength at which the absorption loss is to be measured. Heat is generated in the sample when the pump light is absorbed, leading to a temperature gradient in the air around the sample. This temperature gradient results in

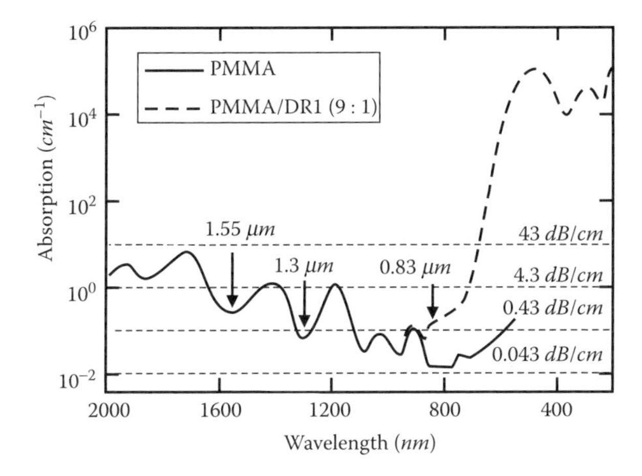

FIGURE 5.29 Loss in PMMA and DR1-Doped PMMA as measured with photothermal deflection. Adapted from Ref. [82] with permission of the American Institute of Physics.

a refractive index gradient that deflects the pump beam (analogous to the direct deflection method for measuring the refractive index gradient in a sample, as described in Section 5.1.2). The angle of deflection, θ, and the length of sample that is exposed to the pump is related to the absorption loss. The pump wavelength can be varied to get a spectrum. For this technique to work, air currents need to be kept to a minimum.

Skumanich and coworkers[82] used this technique to measure the loss of dye-doped polymer in the near *infrared*. Figure 5.29 shows a plot of the loss they measured in DR1-doped PMMA (10% by weight) and neat PMMA. In the near infrared, the neat PMMA loss and the DR1-doped PMMA loss are the same within experimental uncertainties until about $850nm$. The loss due to the dye dominates in the visible part of the spectrum.

For applications where prorogation distances are on the order of tens of centimeters, then active dye-doped optical fiber-based devices that operate in the near IR fall within the loss budget for most applications, such as fiber amplifiers and electrooptic switches. Note that the transparency windows in PMMA (and in any other polymer with C-H bonds) coincide with the important application wavelengths of $1.55\mu m$, $1.3\mu m$, and $830nm$. Note that the losses measured by Garvey and coworkers in dye-doped polymers (as presented in Section 5.2.1) are consistent with these results.[19]

More recently, Pitois and coworkers measured the optical loss of a series of crosslinked polymers[88] by using photothermal deflection (PD) and direct waveguide loss measurements. Since (PD) measures only the absorption loss, and the waveguiding studies measure the sum of scattering and absorption loss, they were able to separate the two mechanisms. Figure 5.30 shows the structure of the copolymer, and how the polymer becomes optically crosslinked.

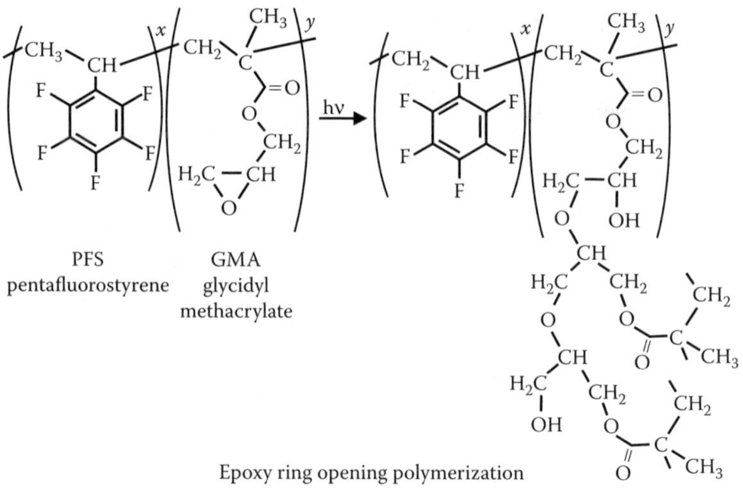

FIGURE 5.30 Photocrosslinking reaction in *p*(PFS-*co*-GMA) using epoxy groups.

The total loss in the waveguide was found to be greater than the value determined by PD — as expected, and the difference is attributed to scattering. To separate extrinsic scattering losses (for example, due to impurities) and intrinsic loss (due to polymer inhomogeneity due to shrinkage in the polymerization process), Pitois and coworkers assumed that the extrinsic losses for all the polymers are the same but the intrinsic loss depends on the ratio of monomers used to make the polymer. The reasoning behind this assumption is based on the observation that the degree of shrinkage during polymerization is correlated with crosslinking, so copolymers with a higher GMA content should suffer from higher scattering. In the limit of neat PFS, the authors assume that intrinsic scattering vanishes and that all the scattering loss is due to extrinsic sources. Since the PFS monomer contains fewer C-H bonds than GMA, the authors used the percentage of C-H bonds in the polymer as the variable accounting for scattering. This is a convenient variable to adjust for the study of intrinsic loss because it is well known that the oscillations in the absorption spectrum (see, for example, Figure 5.29) are due to overtones of the C-H natural oscillation frequency. Indeed, Pitois's data looks very similar to that of Skumanich.

When two monomers are copolymerized into a polymer, the relative proportion of each one depends on the initial ratio of monomers, or the ratio of x to y, as shown in Figure 5.30. Since the PFS monomer has fewer hydrogen atoms than GMA (see Figure 5.30), controlling the proportion of each component effectively controls the number of hydrogen atoms in the polymer. As such, the intrinsic loss increases with the number of C-H bonds, or equivalently, with the proportion of GMA.

Pitois found that both the intrinsic absorption loss and the scattering loss increase with GMA concentration. Table 5.2 summarizes their results. Note that

TABLE 5.2
Optical loss in four crosslinked copolymers of *p*(PFS-*co*-GMA)

C-H (bond %)	GMA (mol %)	PD loss Loss (dB/cm)	Waveguide (dB/cm)	Intrinsic Scattering Loss (dB/cm)
29.64	40	0.45	0.80	0.24
25.74	27	0.36	0.65	0.18
20.94	11	0.29	0.45	0.05
17.65	0	0.24	0.35	≈ 0

they found a constant extrinsic scattering loss of $0.11 dB/cm$, which is not listed in the table.

5.2.5 Side-Illuminated Fluorescence

The cutback technique is best suited for measuring losses that are very low because long lengths of fiber can be used. Photothermal deflection is sensitive to low losses in thin films. Thin film absorption measurements — where the light passes perpendicular to the plane of the sample, on the other hand, are best suited for strongly light-absorbing materials. The regions of the spectrum where the losses make a transition from large to small are often the most difficult to handle because separate measurements need to be combined where the measurements overlap. Side-illuminated Fluorescence Spectroscopy (SIF) fills this gap.[83,84]

To use SIF, the fiber material needs to be fluorescent. A light beam exposes the fiber from the side, and some of the fluorescence so generated guides down the fiber. The spectrum of light that exits the fiber is then measured with a spectrometer. The beauty of the technique is that the light source is effectively inside the fiber at whatever point the pump light enters. As such, the "light source" can be moved through the fiber so that the spectrum can be measured as a function of propagation distance, yielding an absolute value of the loss. In a sense, SIF is the inverse process of the transverse loss measurement described in Section 5.2.2; in SIF, light is scattered into the waveguide and the intensity of the guided mode is measured while in the transverse loss measurement, light is coupled into the fiber from one end and the light that is scattered out of the side of the waveguide is measured.

5.2.5.1 Theory

There are certain subtleties that need to be considered when calculating the propagation distance in the fiber. Figure 5.31 shows two possible geometries: a) when the light generated travels parallel to the fiber axis, and b) if the fluorescence acts as a point source. Clearly, the average distance traveled by the light in the point source case is longer than for the axial case. In reality, the excitation light will not be uniform, and the source is inhomogeneous.

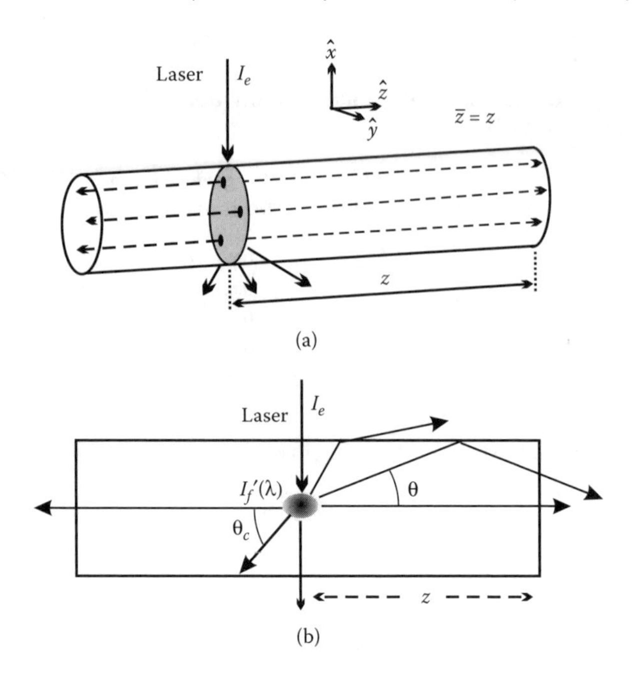

FIGURE 5.31 Side-illuminated fluorescence light that a) travels along the fiber axes and b) is a point source in which many rays below the critical angle are possible.

For illustration, Kruhlak considered the simple but important case of the point source. All other excitation profiles can be expressed as a sum over a distribution of point sources. If a monochromatic excitation source at wavelength λ_m produces a fluorescence spectrum through a one-photon absorption process, the intensity of the fluorescence spectrum at any wavelength λ is proportional to the excitation intensity, I_e. The fluorescence intensity, $I_f(\lambda)$, will then be of the form

$$I_f(\lambda) = C(\lambda_m, \lambda) I_e, \tag{5.57}$$

where $C(\lambda_m, \lambda)$ is the fluorescence yield at wavelength λ due to excitation by a source at wavelength λ_m. If the fiber diameter is small compared with the e^{-1} absorption length at the excitation source wavelength, λ_m, and the diameter of the source beam is larger than the fiber, the illuminated portion of the fiber can be approximated as a homogeneous light source.

Kruhlak also considered the simple case where all the rays of the generated fluorescence spectrum are parallel to the fiber axis, as is shown in Figure 5.31a. (While this is unphysical, it yields the shortest possible optical path length of light traveling down the fiber.) The light intensity a distance z away from the source, $I_f(\lambda, z)$, is then given by

$$I(\lambda, z) = I_0(\lambda) \exp[-\alpha(\lambda) z], \tag{5.58}$$

where $I_0(\lambda)$ is the intensity at the source which is given by $I_0(\lambda) = I_f(\lambda)$:

$$I_f(\lambda, z) = C(\lambda_m, \lambda) I_e \exp(-\alpha(\lambda)\, z) = I_f(\lambda) \exp(-\alpha(\lambda)\, z). \tag{5.59}$$

For the isotropic point source, Kruhlak considered the source at the center of the fiber. All the rays that are beyond the critical angle of the fiber remain in the waveguide and therefore must be traced. Since the refractive index of a neat polymer fiber or dye-doped fiber does not vary appreciably where the loss is low, we can assume that the index of refraction is roughly constant in the fluorescence wavelength range so the critical angle θ_c is independent of wavelength. Each ray within the critical angle cone travels different distances before reaching the fiber end, so the amount of absorption of each ray is different. Note that we can ignore all the rays that prorogate beyond the critical angle because the amount of light reflected at each bounce is typically low, so this light will be negligible after $1cm$ of propagation.) The fluorescence intensity of light at the plane z due to all the light rays inside $\theta = \theta_c$ is given by

$$I_f(\lambda, z) = C(\lambda_m, \lambda) I_e \frac{\int_0^{2\pi} \int_{-\theta_c}^{\theta_c} \exp(-\alpha(\lambda)\, z/\cos\theta) d\theta d\phi}{\int_0^{2\pi} \int_{-\frac{\pi}{2}}^{\frac{\pi}{2}} d\theta d\phi}, \tag{5.60}$$

where the denominator is the normalization factor, and $z/\cos\theta$ is the distance traveled by a ray at an angle θ if the projection of the ray along the axis is z (see Figure 5.31b). Integrating, we get

$$I_f(\lambda, z) = I_f^{'}(\lambda) \int_{-\theta_c}^{\theta_c} \exp(-\alpha(\lambda, \theta)\, z) d\theta, \tag{5.61}$$

where $\alpha(\lambda, \theta) \equiv \alpha(\lambda)/\cos\theta$ and $I_f^{'}(\lambda) = C(\lambda_m, \lambda) I_e/\pi$ is the initial fluorescence intensity at the point source.

To illustrate the effect of assuming a specific model for the ray properties of the source, Kruhlak compared the calculated intensity as a function of z for the point source and plane wave models for three values of the material loss.[83] Figure 5.32

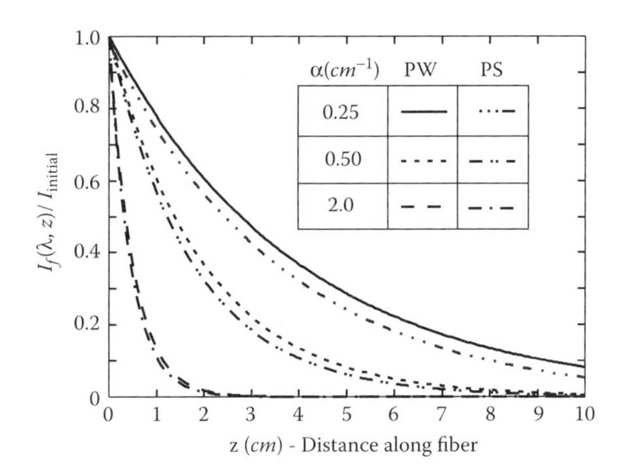

FIGURE 5.32 Intensity in a fiber as a function of z when the rays are parallel to the axis (PW) and originate from a point source (PS). Adapted from Ref. [84] with permission of the Optical Society of America.

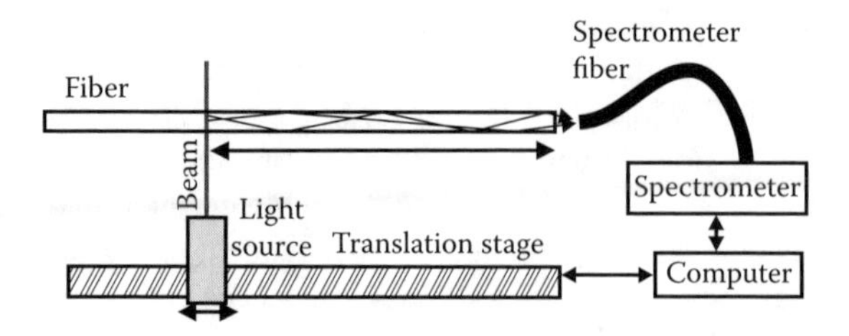

FIGURE 5.33 The SIF experiment. The light is collected with a spectrometer fiber after it has traveled a distance z in the sample fiber.

shows the results. The plane wave model predicts a lower loss in the fiber than the point source model because the rays, on average, travel further for the point source than for the plane wave.

5.2.5.2 Experimental Results

The experiment used by Kruhlak is diagramed in Figure 5.33. The light source excites fluorescence, the transmitted fluorescence spectrum guides down the fiber and exits from the fiber end. A spectrometer fiber collects the light and sends it to a diode array that measures the spectrum. A computer is used to control the translation stage and to record the spectrum. The translation stage is stepped over a range of z that is at least comparable to the e^{-1} loss length. The advantage of this setup is that the SIF spectrum is quickly obtained over a broad range of colors. Furthermore, an inexpensive laser diode light source such as a laser pointer can be used.

Figure 5.34 shows a series of SIF spectra as a function of z as measured in a PMMA fiber doped with PSQ dye. The amplitude of the spectrum decreases as light travels down the fiber. To get a loss at a given wavelength, the intensity at

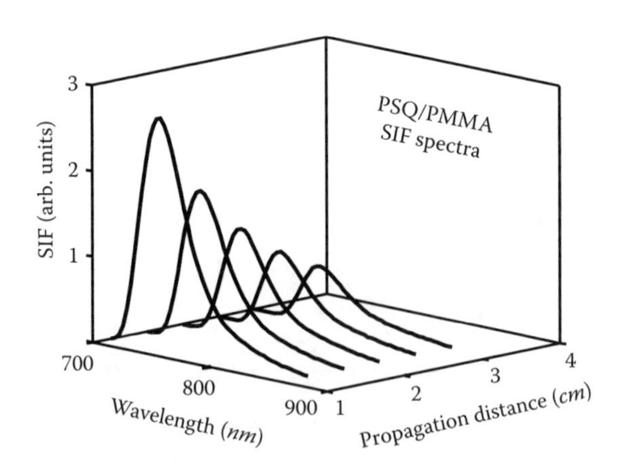

FIGURE 5.34 SIF spectrum as a function of propagation distance for PMMA/PSQ. Adapted from Ref. [84] with permission of the Optical Society of America.

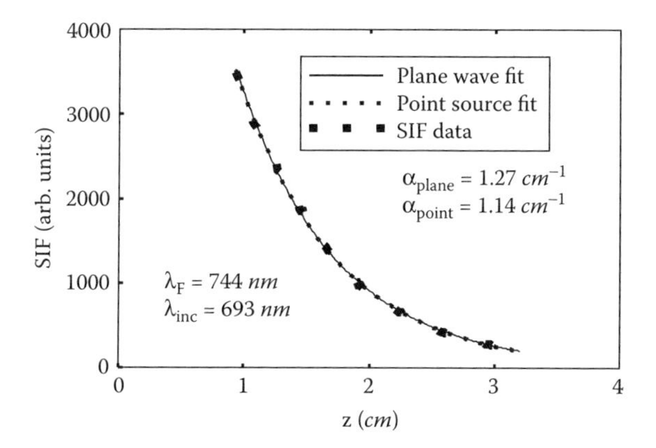

FIGURE 5.35 Measured intensity as a function of prorogation distance z for PSQ/PMMA-d8. The fit to the plane wave model and point source model look indistinguishable. Adapted from Ref. [84] with permission of the Optical Society of America.

that wavelength is determined as a function of z, which directly yields the loss. Figure 5.35 shows the SIF intensity as a function of z at 744*nm* as excited by a 693*nm* laser diode source. The fiber is deuterated PMMA (PMMA-d8) doped with PSQ dye. Note that while the fits to the plane wave and point source models look identical, the resulting loss that is calculated varies by 10%. Since the point source model is more realistic, using it is the most accurate. The uncertainty introduced with the point source model is probably less than 5%.

The rest of the SIF wavelengths are fit to a point and plane wave model to yield the loss spectrum. Figure 5.36 shows the results for PSQ/PMMA-d8.

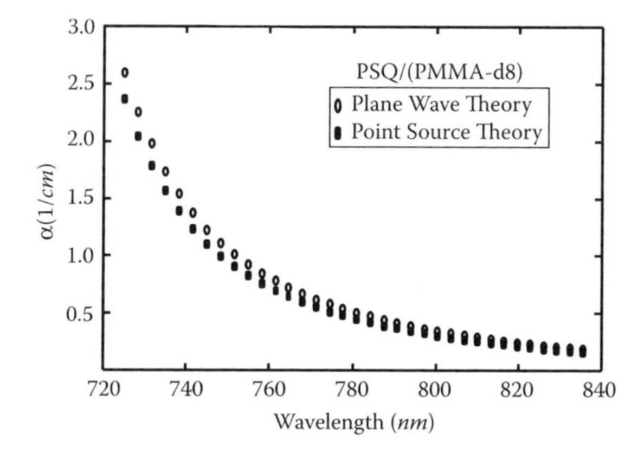

FIGURE 5.36 The loss spectrum of PSQ/PMMA-d8 as determined from a fit of the data to the plane wave and point source models. Adapted from Ref. [84] with permission of the Optical Society of America.

To conclude, SIF is a useful nondestructive technique that is used to measure the loss in a polymer fiber. In addition to characterizing polymer fibers for specific applications, the technique can also be used to study the properties of molecules that are embedded in the fiber. Indeed, Kruhlak conducted such experiments to study dye aggregation and excited state absorption from states with small absorption cross-sections.[84]

5.3 NUMERICAL APERTURE

The numerical aperture is a measure of the cone angle of the light that is incident on a fiber that will couple into the fiber. Similarly, the numerical aperture of a lens is defined as the cone angle of the focused light for a source at infinity. If the cone angle is θ_A, the numerical aperture is

$$N.A. = \sin\theta_A. \tag{5.62}$$

For a lens of diameter d and focal length f, the numerical aperture is

$$N.A. = \sin\left(\tan^{-1}\frac{d}{2f}\right) = \frac{1}{\sqrt{1+4\left(\frac{f}{d}\right)^2}}. \tag{5.63}$$

For an optical fiber, the numerical aperture is

$$N.A. = \left(n_1^2(r) - n_2^2\right)^{1/2}. \tag{5.64}$$

For a step-index fiber, $n_1(r)$ is a constant so the numerical aperture is independent of where the light enters the fiber. For a gradient-index fiber, the "local" numerical aperture depends on the refractive index at the entry point r of the light.

If the numerical aperture of an optical fiber is larger than the numerical aperture of a lens used to couple light into the fiber, then all of the light will enter the guide. If the numerical aperture of a lens is larger than the fiber's numerical aperture, some of the light will not couple into the fiber core. Ideally, an optical system made of multiple components is made with matched numerical apertures. Figure 5.37 shows an example of a lens and two fibers that each have the same numerical aperture. All the light from the lens is coupled into the first fiber then into the second fiber.

This picture is somewhat idealized. In a multimode optical fiber, the light in the core can be viewed as a collection of geometric rays. Each ray can be deflected by a refractive index inhomogeneity or scattering center. As such, energy will be transferred between the rays, which can result in a net redistribution of the energy between modes. In practice, this results in a loss of the rays near the critical angle. As such, the light leaving the end of the optical fiber will be confined to a tighter cone, resulting in a smaller numerical aperture than for the ideal fiber in which the modes are eigenstates. This situation turns out to be ideal for splicing together fibers. If a series of identical fibers are coupled together in series, the input numerical aperture will be larger than the output numerical aperture of the previous

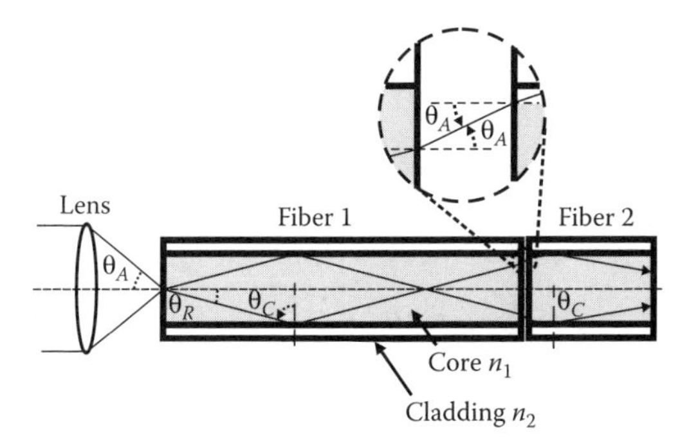

FIGURE 5.37 A lens of numerical aperture sin θ_A couples light into a fiber with numerical aperture sin θ_A whose output is coupled to another fiber of numerical aperture sin θ_A.

fiber — leading to efficient coupling of the light at each connection. Nature is not often so kind.

It is thus important to be able to measure both the input and output numerical aperture of a fiber. Each of these cases is described below.

5.3.1 Input Numerical Aperture Measurements

There are several methods for determining the numerical aperture. If the refractive index profile is known, then it can be calculated from Equation (5.64). The advantage of using the measured refractive index profile is that the "local" numerical aperture can be calculated based on the refractive index of the waveguide at the entry point of the beam of light. Other methods described below include objective lens coupling and ray angle determination.

5.3.1.1 Objective Lens Coupling

To get a quick (and rough) estimate of the numerical aperture of a multimode fiber, a known numerical aperture lens is used to couple light into the fiber core. The amount of light guiding down the fiber and exiting out the other end is related to the amount of light that was coupled at the incident end. For a Gaussian light source, the radial intensity distribution is given by

$$I(r) = I_0 \exp\left[-2r^2/w_0^2\right], \tag{5.65}$$

where w_0 is the beam waist at the focal point of the beam. The power within the beam waist can be determined by integrating Equation (5.65). It is straightforward to show that 86.5% $(1 - e^{-2})$ of the power is contained within the beam waist.

Beyond the focal point, the beam waist increases as a function of distance along the beam axis. For a Gaussian beam, far away from the focal point where geometrical optics is an appropriate picture, the waist as a function of distance,

$w(z)$, is given by

$$w(z) = \frac{\lambda z}{\pi w_0}, \tag{5.66}$$

where λ is the wavelength of the light. Note that the beam waist increases linearly as expected from geometric optics. As such, the angle of divergence, θ_L, of the beam is given by

$$\tan\theta_L = \frac{\lambda}{\pi w_0}, \tag{5.67}$$

so the numerical aperture of the lens is

$$\sin\theta_L = \frac{1}{\sqrt{\left(\frac{\pi w_0}{\lambda}\right)^2 + 1}}. \tag{5.68}$$

Provided that the beam spot size at the focus is smaller than the waveguide diameter, the light will couple into the fiber if the numerical aperture of the lens is smaller than the numerical aperture of the fiber. Figure 5.38a shows the light in a Gaussian beam coupling into a fiber. The solid line shows a beam waist with a small enough numerical aperture to couple light into the fiber while the dashed curve shows a smaller beam waist with high numerical aperture that scatters out of the fiber.

If the numerical aperture of a lens matches the numerical aperture of the fiber, the rays within the beam waist are below the critical angle so 86.5% of the power will couple into the fiber. If the fiber is lossless, detection of 86.5% of the light power coupled into the fiber at the output is a sign of matched numerical apertures, yielding the fiber numerical aperture. Unfortunately, many polymer fibers have

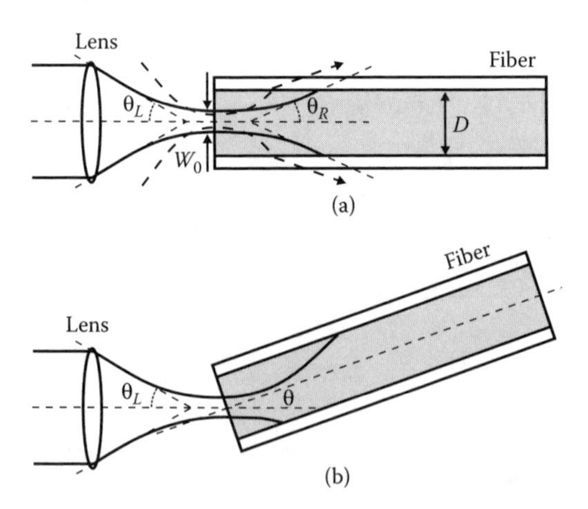

FIGURE 5.38 (a) A lens of numerical aperture $\sin\theta_L$ couples light into a fiber with numerical aperture $\sin\theta_A$ (solid line) while some of the light from a higher numerical aperture lens is radiated away (dashed line.) The lens is not shown. (b) The fiber is tilted.

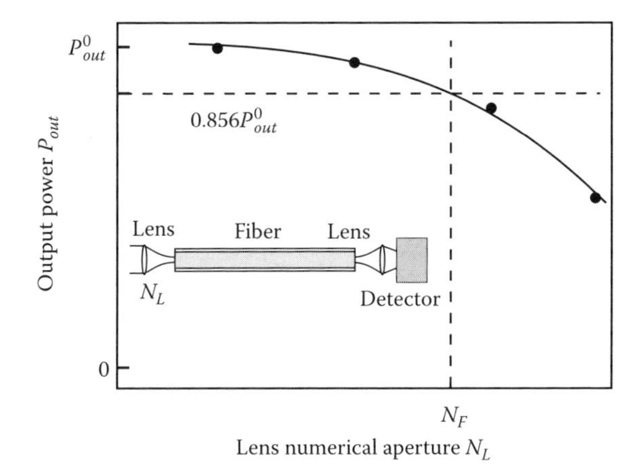

FIGURE 5.39 Simulated measured output as a function of numerical aperture of the input coupling lens. The inset shows the experiment.

large enough losses that loss must first be measured and taken into account. The procedure to take loss into account is straightforward, as follows.

A low numerical aperture lens is used to couple light into the fiber and the output power P^0_{out} is recorded. (The cladding modes need to be eliminated by dipping the clad fiber into index matching liquid or painting the cladding black.) The numerical aperture of the lens used is chosen to be much smaller than the expected numerical aperture of the fiber so that most of the light couples into the core. The numerical aperture of the coupling lens is changed and the intensity re-measured. The process is repeated for a series of lenses and the numerical aperture of the coupling lens for which the intensity drops by 15% is the numerical aperture of the fiber. Figure 5.39 shows the experiment (inset) and simulated data.

5.3.1.2 Ray Angle Method

The simplest approach to measuring the input numerical aperture is based on ray tracing. If a pencil of light is launched into a fiber at some angle θ and it guides, then we know that the numerical aperture is greater than $\sin\theta$. When the angle is increased until the light no longer couples into the core, then the numerical aperture is smaller than $\sin\theta$. As such, the numerical aperture can be determined by the value that straddles these two cases.

To get the most precise measurement of the input numerical aperture, it would be best to use an infinitesimal beam diameter light source with a small numerical aperture. Unfortunately, having simultaneously a small beam waist and a small numerical aperture are inconsistent with each other because diffraction rears its ugly presence. The more a beam width is confined at one point in space, the more it spreads from that point. The best experimental design balances the numerical aperture of the beam with its size. The beam waist should be smaller than the core size to be measured so that the light couples in easily but with as large a numerical aperture as possible. The optimum condition can be calculated as follows.

If the fiber has a numerical aperture defined by a cone angle θ_A, then the experiment will require that the output intensity be measured as the fiber axis is rotated from $\theta = 0$ to just beyond $\theta = \theta_A$. As we can see from Figure 5.38b, the projection of the beam waist on the face of the fiber must be smaller than the core diameter, or

$$w_0 < D \cos \theta_L. \tag{5.69}$$

For this size beam waist, the divergence angle of the incident beam will be given by Equation (5.67), so Equation (5.68) becomes

$$\sin \theta_L > \frac{\lambda}{\pi D}. \tag{5.70}$$

It is instructive to evaluate Equation (5.69) for a typical multimode polymer fiber. Consider a fiber with a diameter $D = 62.5\mu m$ and a light source of wavelength $\lambda = 640nm$. This yields $\sin \theta_L > 0.00325$. For a fiber with numerical aperture N.A.$= \sin 25^\circ = 0.42$, the acceptance angle is 25° and the divergence angle of the beam is $\sin^{-1}(0.00325) = 0.19^\circ$. When the light is launched on the axis of the fiber, all the rays are well within the acceptance cone angle. As the angle θ is increased, all the light couples into the waveguide. Once the angle reaches $\theta = 25^\circ - 0.19^\circ$, some of the light radiates away from the core. Once the angle exceeds $\theta = 25^\circ + 0.19^\circ$, all the rays in the guide will be beyond the critical angle and no light will be found to propagate in the waveguide. The numerical aperture will thus be given by the angle at which the intensity falls to half its value. Given that the light intensity falls over a 0.38° range, the input numerical aperture can be accurately determined. In this example, the numerical aperture can be easily determined to within less than 1% uncertainty.

As the fiber diameter gets smaller, the measurement gets less accurate. A single-mode fiber has the smallest core. In this case, since the light can not hop between modes (because there is only one mode), the exit and the entrance numerical apertures are the same, so the exit numerical aperture measurement can be used to determine both.

5.3.2 Output Numerical Aperture

A measure of the output numerical aperture is the most direct. Light is coupled into a fiber with a uniform illumination so that all the modes are excited. In the geometric optics limit, the output light will be contained within a cone angle defined by the total internal reflection condition. In reality, the cone is not sharply defined. As such, it is customary to define the cone as the region that contains most of the light. The point at which the light falls to $1/e^3$ of its peak intensity (i.e., 5%) is usually chosen to define the perimeter of the base of the cone.

A CCD camera or scanning pinhole/detector is used to image the light spot in the far field of the fiber output. The numerical aperture is then simply calculated from the $1/e^3$ cone angle according to Equation (5.62).

5.4 BANDWIDTH

Since the most important application for optical fibers is the transmission of data, the amount of data that can be packed into a fiber — or the bandwidth — is an important property that needs to be measured. While there are many methods for encoding information on a beam of light — such as Time Division Multiplexing (TDM) and Wavelength Division Multiplexing (WDM) — the simplest way to think about data density is to consider a binary system where a series of pulses carries the information.

As the individual pulses are made shorter, more such pulses can be crammed into a train of pulses yielding a higher bandwidth. As the pulses travel down a fiber, they spread. Once the pulses spread to the point where their widths exceed their spacing, it becomes impossible to resolve them — resulting in a loss of information. Figure 5.40 shows an example of a set of pulses launched into a fiber separated by Δt and the output pulses after traveling over a length L of fiber. The dashed lines show the original pulse train and the solid curve shows the sum of the pulses as would be seen by a detector. Clearly, the pulses are becoming inseparable at this point. The bandwidth thus depends on the length of the material and is expressed in units of frequency · length and, for the parameters in Figure 5.40, is given by

$$B = \frac{1}{2}\frac{1}{\Delta t} \cdot L. \tag{5.71}$$

There are two properties of a fiber that limit the bandwidth: material dispersion and modal dispersion. Modal dispersion depends on both material dispersion and the geometry of a fiber.

5.4.1 MODAL DISPERSION — RAY PICTURE

To illustrate modal dispersion, we start by ignoring material dispersion and consider a step-index fiber using simple geometric optics. Figure 5.41 shows a multimode step-index fiber with a ray that travels parallel to the axis of the core and a second ray at the critical angle. These two rays represents two modes — the one parallel to the axis being the lowest-order mode and the ray at the critical angle the highest-order mode. If a pulse of light is coupled into a fiber so that the energy

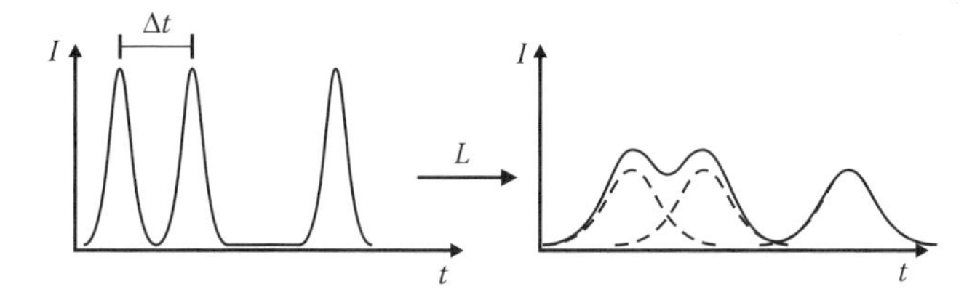

FIGURE 5.40 A train of pulses enter a fiber (left) and spread after traveling a distance L (right).

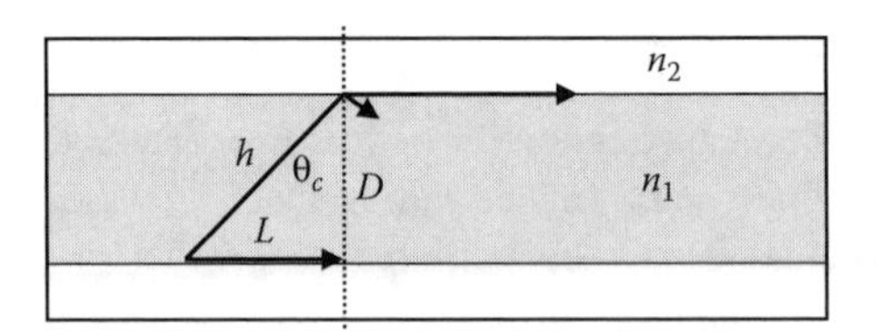

FIGURE 5.41 Geometrical optics method for calculating fiber bandwidth. The lowest-order mode travels parallel to the axis of a fiber while the highest-order mode hits the interface at the critical angle.

is equally shared between these two modes, the lowest mode travels a shorter distance and so exits the fiber first. The time delay between this pulse and the highest order one is a measure of the bandwidth.

The time delay between the two pulses along a length L of fiber is given by

$$\Delta t - \frac{n_1 h}{c} - \frac{n_1 L}{c}. \tag{5.72}$$

But using Snell's Law for the critical angle — that is $n_1 \sin\theta = n_2$ and $\sin\theta = L/h$ — eliminating h in Equation (5.72), we get

$$\Delta t = (n_2 - n_1)\frac{L}{c}, \tag{5.73}$$

yielding a bandwidth, according to Equation (5.71), of

$$B = \frac{c}{2(n_2 - n_1)}. \tag{5.74}$$

As an example, consider a multimode fiber with a numerical aperture of 0.3 and core refractive index of about 1.5. From the definition of the numerical aperture, this yields $n_2 - n_1 = 0.03$, which gives a bandwidth of $5 \times 10^9 m/s$ or $10MHz \cdot km$. For high-speed applications that require GHz bandwidths, only lengths of less than $10m$ would work for this type of fiber. Clearly, a fiber with these parameters is not suitable for long-haul applications.

There are methods for decreasing or eliminating modal dispersion by using a compensating fiber geometry. For example, in a graded-index fiber, the light ray that travels down the central axis is in a region of high refractive index while rays that oscillate about the fiber axis spend more time near the cladding. So, even though these oscillating rays travel further, their net speed is higher, resulting in no time delay between the two modes. A single-mode fiber, on the other hand, has no modal dispersion because such a fiber only supports one mode.

In a perfect, defect-free fiber, when light is launched into a given mode, it will stay in that mode. If that were so, if light were to be launched into one mode, there would be no modal dispersion. Since no fiber is perfect, small perturbations cause modes to couple. Thus, over a long enough fiber, all the modes will be excited at the output even if only one mode is launched into the fiber.

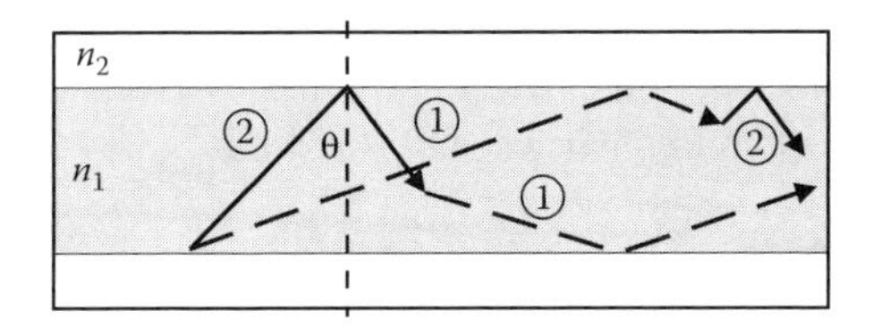

FIGURE 5.42 Geometrical picture of coupling between two modes in a fiber.

If mode coupling is strong, all the modes mix with each other after a short distance of propagation, which effectively increases the fiber bandwidth. Consider, for example, a fiber with two modes, which we will call 1 (dashed ray) and 2 (solid ray), as seen in Figure 5.42. Mode 1 has a larger component of the velocity along the fiber axis than mode 2, so the speed of the light beam down the fiber is faster for mode 1.

If mode 1 scatters into mode 2, the speed of the wave decreases and when 2 scatters into mode 1, the speed down the fiber increases. As a result, the bandwidth does not fall off as quickly with propagation distance as it would for an ideal fiber. In the absence of mode coupling, the pulse spreading increases linearly with propagation distance. With mode coupling, pulse spreading increases as the square root of the propagation length.

One might imagine making a fiber with strong mode coupling to make high-bandwidth fibers. Unfortunately, guiding modes can also couple into radiation modes, leading to higher optical loss. In applications where optical loss needs to be low, mode coupling should be avoided.

5.4.2 Material Dispersion

Even in single-mode and graded-index fiber, material dispersion can result in pulse broadening. A pulse of light can be made by summing over plane waves of differing frequencies. The broader the range of frequencies, the shorter the pulse. A wave with a single frequency is a plane wave that spans all of space. If the refractive index of the material depends on the frequency, ω, each plane wave that makes up a pulse travels at a different speed. These speeds are called the phase velocities. As a result, the pulse will spread. The velocity of the peak of the pulse is called its group velocity, v_g, and is defined by

$$v_g = \frac{d\omega}{dk}, \tag{5.75}$$

where k is the wavevector of the plane wave. A rigorous treatment of the group velocity can be found in many textbooks.[71] Here, we use the concept of the group velocity to present an intuitive argument of the role of material dispersion on bandwidth.

The dominant frequency component in a pulse is called the center frequency ω_c. If the spread in frequencies is $\Delta\omega$, the frequencies that make up a pulse will be found in the range between $\omega_c - \Delta\omega/2$ and $\omega_c + \Delta\omega/2$. The degree of spreading in a pulse after it travels through a fiber length L is given by the time delay in arrival between these two frequencies, and is given by

$$\Delta t = \frac{L}{v_g(\omega_c - \Delta\omega/2)} - \frac{L}{v_g(\omega_c + \Delta\omega/2)}. \tag{5.76}$$

If the spread in frequencies is small compared with the center frequency, we can approximate the group velocity as

$$\frac{1}{v_g(\omega_c \pm \Delta\omega/2)} \approx \frac{1}{v_g(\omega_c)}\left(1 \mp \left.\frac{dv_g(\omega)}{d\omega}\right|_{\omega_c} \frac{\Delta\omega}{2}\right). \tag{5.77}$$

Substituting Equation (5.77) into Equation (5.76) yields the time delay

$$\Delta t = -L\Delta\omega \frac{1}{v_g^2(\omega_c)} \left.\frac{dv_g(\omega)}{d\omega}\right|_{\omega=\omega_c} = L\Delta\omega \frac{d}{d\omega}\left(\frac{1}{v_g(\omega_c)}\right). \tag{5.78}$$

Substituting Equation (5.75) into (5.78), and substituting the result into Equation (5.71) yields the bandwidth

$$B = \frac{1}{2\Delta\omega \frac{d^2k}{d\omega^2}} = \frac{1}{2\frac{\Delta\omega}{c}\left(2\frac{dn}{d\omega} + \omega\frac{d^2n}{d\omega^2}\right)}, \tag{5.79}$$

where we have used the definition $k = n(\omega)\omega/c$, leading to

$$\frac{d}{d\omega}\left(\frac{dk}{d\omega}\right) = \frac{d}{d\omega}\left[\frac{1}{c}\left(n - \lambda\frac{dn}{d\lambda}\right)\right]. \tag{5.80}$$

It is useful to express Equation (5.79) as a function of wavelength. Using Equation (5.80), we get

$$\frac{d^2k}{d\omega^2} = \frac{1}{c}\left[\frac{dn}{d\omega} - \left(\frac{d\lambda}{d\omega}\right)\left(\frac{dn}{d\lambda}\right) - \lambda\left(\frac{d^2n}{d\lambda^2}\right)\left(\frac{d\lambda}{d\omega}\right)\right] = -\frac{\lambda}{c}\left(\frac{d^2n}{d\lambda^2}\right)\left(\frac{d\lambda}{d\omega}\right). \tag{5.81}$$

Recall that $\Delta\Omega$ is the spread in frequencies. Similarly, $\Delta\lambda$ will be the corresponding range in wavelengths, so if we approximate $(d\lambda/d\omega) \approx (\Delta\lambda/\Delta\omega)$ and substitute Equation (5.81) into Equation (5.79), we get

$$B = \frac{c}{2\lambda_c \cdot \Delta\lambda \cdot \left.\frac{d^2n}{d\lambda^2}\right|_{\lambda=\lambda_c}}, \tag{5.82}$$

where λ_c is the center frequency.

The Sellmeier Equation is used to approximate the refractive index dispersion of a material and is of the form

$$n^2(\lambda) \approx \sum_i \frac{A_i\lambda^2}{\lambda^2 - \lambda_i^2}, \tag{5.83}$$

where the set of parameters A_i and λ_i are the Sellmeier parameters for a given material. If the Sellmeier parameters are known for a given material, the bandwidth limit due to material dispersion can be calculated.

5.4.2.1 Numerical Example of Material Dispersion

When off resonance, the dispersion in a neat transparent polymer is low. Assuming that the refractive index is 1.51 at $500nm$ and 1.50 at $600nm$, and that only one term is dominant in the sum in Equation (5.83), we can solve for A_1 and λ_1. This yields $\lambda_1 = 102.4nm$ and $A_1 = 2.164$.

Clearly, in the range between $500nm$ and $600nm$, these wavelengths are long compared to λ_1. As such, we use Equation (5.83) to approximate the refractive index when $\lambda^2 \gg \lambda_1^2$:

$$n(\lambda) \approx \sqrt{A_1}\left(1 + \frac{1}{2}\frac{\lambda_1^2}{\lambda^2}\right). \tag{5.84}$$

Equation (5.84) leads to a dispersion in the refractive index given by

$$\frac{d^2 n(\lambda)}{d\lambda^2} = 3\sqrt{A_1}\frac{\lambda_1^2}{\lambda^4}. \tag{5.85}$$

To get the bandwidth, Equation (5.85) is substituted into Equation (5.79), yielding

$$B = \frac{c\left(\frac{\lambda_c}{\lambda_1}\right)^2}{6\frac{\Delta\lambda}{\lambda_c}\sqrt{A_1}}, \tag{5.86}$$

where the bandwidth is calculated at the center wavelength $\lambda = \lambda_c$. For a $0.1nm$ wavelength spread and a center wavelength of $\lambda_c = 600nm$, we get a bandwidth of

$$B = 7.0 \times 10^1 2m/s = 7GHz \cdot km. \tag{5.87}$$

This bandwidth is clearly higher than the limit given by modal dispersion. As such, the modal dispersion will limit the bandwidth.

For the case of a dye-doped polymer, the material dispersion is much higher so the bandwidth is much lower. For active device applications, though, only short lengths of fiber (typically less than one meter) are required so bandwidth is not an issue. Note that for dye-doped polymers, the material bandwidth can be calculated from the refractive index models in Chapter 4.

If material dispersion becomes an important issue, it is possible to find operational wavelengths where the second derivative of the refractive index is zero. In silica fibers, for example, the zero dispersion point is located at $1.3\mu m$, making this wavelength popular for telecommunications applications.

5.4.3 Modal Dispersion — Wave Picture

In the limit of geometric optics, we understood modal dispersion in terms of a set of geometrical rays that carry energy down the fiber at a set of distinct speeds. As the transverse dimension of a waveguide approaches the wavelength of the propagating light, interference effects result and only a discrete set of the rays will guide.

In the limit of a small waveguide, we saw in Chapter 2 that it is more convenient to express the electric and magnetic fields in terms of modes. Each mode of a fiber is an eigenstate, meaning that the transverse intensity profile does not change as it propagates. As we saw in Chapter 2, a ray is made up of a superposition of many modes. As such, in the short wavelength limit, the mode and ray pictures describe the propagation of the fields in terms of a different set of fundamental building blocks, i.e., rays versus modes. One is related to the other by taking a superposition over the corresponding basis.

In the mode picture, the electric field in a waveguide that propagates energy along the $\hat{z}$-direction can be expressed as

$$\vec{E} = \vec{E}_0(x, y) \exp\left[i(\omega t - \beta z)\right], \tag{5.88}$$

where β is the wave vector of the mode, ω the angular frequency, and $\vec{E}_0(x, y)$ the electric field profile in the plane perpendicular to the propagation direction. The mode profile can be obtained by substituting Equation (5.88) into Maxwell's wave equations, and solving the resulting differential equation with boundary conditions that are defined by the shape of the waveguide. Associated with each mode is a distinct value of β and ω. From these, one can obtain thc phase velocity ($v = \omega/\beta$) and group velocity ($v = \partial\omega/\partial\beta$) of each mode.

As we saw in Chapter 2, the process of calculating the modes even for simple geometries such as a cylindrical waveguide is complicated. So, it is useful to think about the correspondence between the ray and mode pictures. For a ray traveling at an angle θ to the z-axis, the mode wavevector is

$$\beta = n_{core} k_0 \cos\theta, \tag{5.89}$$

where $k_0 = \omega/c$ is the free-space wavevector and n_{core} the refractive index of the waveguide. So, the phase velocity is given by

$$v = \omega/\beta = \frac{c}{n_{core}} \cos\theta. \tag{5.90}$$

As such, we find that the effective refractive index of the waveguide mode is $N = n_{core}/\cos\theta$. We can imagine that each mode is like a ray traveling at its own unique angle with a corresponding unique phase velocity. (However, the analogy cannot be taken literally because the intensity profile remains unchanged as the wave propagates.) Modal dispersion and its effect on bandwidth can therefore be calculated by replacing the refractive index by the mode index and treating the problem in the same way as if the bandwidth depended on material dispersion, as described in Section 5.4.2.

5.4.4 Measurement Techniques

There are two broad measurement techniques that are used to determine the bandwidth of a fiber, which can be separated into time domain and frequency domain experiments. In a time domain experiment, a short pulse is launched into one end of a fiber and the broadened pulse is measured at the other end. In the frequency domain measurement, a high-bandwidth signal is launched into one end of the fiber and the power spectrum is measured at the other end.

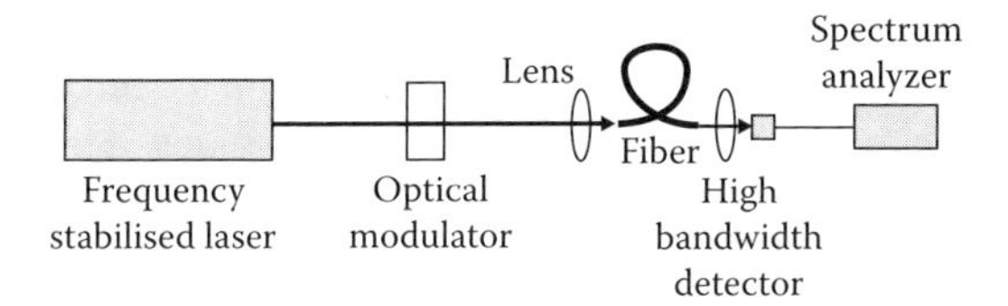

FIGURE 5.43 A typical frequency domain bandwidth experiment.

5.4.4.1 Frequency Domain Measurements

Figure 5.43 shows a schematic diagram of an experiment used to measure the bandwidth using the frequency domain technique. A frequency stabilized source that produces a highly monochromatic beam at the carrier wavelength of interest passes through an optical modulator. The modulator provides a high-frequency intensity modulation to the beam. The output of the modulator is first measured without the fiber to determine the source's intensity amplitude as a function of the modulation frequency. The light is then launched into fiber and the frequency-dependence remeasured. (Note that the optical loss due to the fiber must be carefully taken into account.) The ratio of these two signals gives the frequency response function, and the $1/e$ roll-off frequency observed is a direct measure of the bandwidth.

5.4.4.2 Time Domain Experiments

Figure 5.44 shows a schematic representation of a time domain bandwidth experiment. A series of properly collimated short optical pulses are launched into a fiber. After traveling the length of the spool, the pulse's intensity profile is recorded by an oscilloscope. Because the pulse widths used are typically less than $1ns$, a fast sampling oscilloscope is used to record the data.

To obtain an accurate determination of the bandwidth, several experimental issues need to be addressed. First, it is important that the peak power of the optical pulse train be small enough so that nonlinear optical processes — such as optical

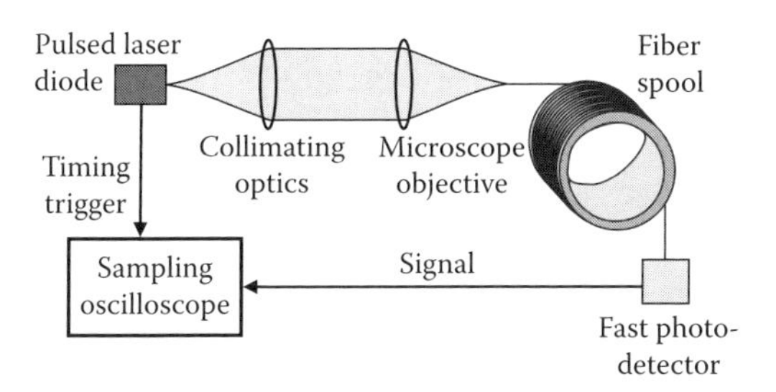

FIGURE 5.44 A typical time domain bandwidth experiment.

phase modulation, do not significantly add to pulse broadening. As such, the energy per pulse should be high enough to be measurable with a good level of signal to noise, yet lower than the threshold for nonlinear processes. Second, to keep the peak power low, the pulse width should be as broad as possible but shorter than the expected amount of broadening.

The light leaving the fiber is converted to an electrical signal with a fast photodetector, such as a semiconductor photodiode. All of the detector's time constants (such as its rise time) must be smaller than the pulse width of the laser source. A sampling oscilloscope takes a short time slice of each of a series of signal pulses and from these reconstructs the pulse. As such, fluctuations in the pulse shape and energy should be as small as possible. Furthermore, the repetition rate of the pulse train should be high so that the data can be accumulated over a reasonable time. Note that an accurate trigger pulse with minimal time jitter must be provided to the sampling oscilloscope.

The sampling scope is used to determine the FWHM pulse width of the laser source, τ_{in}, and the width of the pulse exiting the fiber, τ_f. The impulse response of the signal, τ, is then given by

$$\tau = \sqrt{\tau_f^2 - \tau_{in}^2}. \tag{5.91}$$

For pulses with a Gaussian profile in time, the bandwidth of a fiber of length L is then given by[79]

$$B = \frac{0.44}{\tau} L. \tag{5.92}$$

Note that for multimode and graded-index fibers, the measured bandwidth can depend on the launch conditions if the fiber length is short compared to the mode coupling length.

A sampling scope works well for nanosecond pulses that result from long lengths of fiber with moderately high bandwidth. For ultra-high bandwidth fibers, the ultra-fast pulses that are used may be too short in duration for the sampling scope to resolve. In these cases, an optical Kerr gate can be used, which is made of a nonlinear optical material that is placed between crossed polarizers. When the material is isotropic, no light is transmitted through the gate. Through the optical Kerr effect, if a "pump" pulse illuminates the sample from the side, it makes the sample birefringent. Since many Kerr materials have fast response times (less than $1 ps$), the birefringence of the sample follows the time profile of the pulse. Figure 5.45 shows an experiment used to measure the temporal width of a laser pulse or pulse broadening effects using a Kerr gate.

Consider the laser pulse in Figure 5.45, which is split into two parts by the beam splitter. One part takes a detour as it is reflected from four mirrors, which provides a variable time delay when two of the mirrors are translated (as shown by the double-sided arrow). Upstream from these mirrors, a sample (labeled "x") may be placed in the beam, which leads to a change in the temporal profile of the pulse. We label the pulse intensity profile leaving the laser as $I_p(t)$ and the time-delayed and broadened pulse as $I_s(t - \tau)$. Through the nonlinear-optical interaction of the

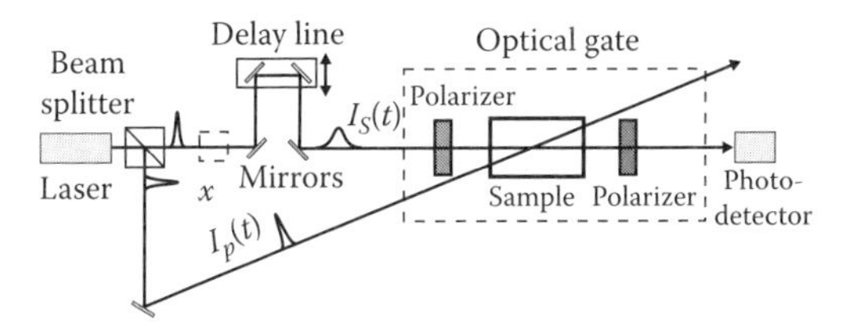

FIGURE 5.45 An optical Kerr cell acts as an autocorrelator to determine the temporal pulse width.

two pulses, the light transmitted by the cell is proportional to the product of the two pulse intensities, so the total energy measured by the detector is

$$U(\tau) = \int_{-\infty}^{+\infty} I_p(t) \cdot I_s(t - \tau) dt, \tag{5.93}$$

where τ is the time delay imparted to the signal by the delay line. $U(\tau)$ is somctimes called the correlation function.

The task at hand is to determine $I_s(t)$ from a measurement of $U(\tau)$. This requires two steps. First, the laser beam profile, $I_p(t)$ is determined by removing the sample and measuring $U(\tau)$. Subsequently, the sample is placed in the experiment and $U(\tau)$ is remeasured. With $I_p(t)$ known, $I_s(t)$ can be determined using the mathematical technique of de-convolution. Mathematically, these two steps are equivalent; so in what follows, we assume that $I_p(t)$ is known and show how $I_s(t)$ is determined.

As a prerequisite, we define the Fourier transform of the function $f(t)$ to be $\bar{f}(\omega)$,

$$\bar{f}(\omega) = \frac{1}{\sqrt{2\pi}} \int_{-\infty}^{+\infty} f(t) e^{-i\omega t} dt, \tag{5.94}$$

and the inverse Fourier transform of $\bar{f}(\omega)$ to be given by

$$\bar{f}(t) = \frac{1}{\sqrt{2\pi}} \int_{-\infty}^{+\infty} f(\omega) e^{+i\omega t} dw. \tag{5.95}$$

The Fourier transform of $U(\tau)$ in Equation (5.93) is therefore given by

$$\bar{U}(\omega) = \int_{-\infty}^{+\infty} I_p(t') \bar{I}_s(-\omega) e^{-i\omega t'} dt', \tag{5.96}$$

where have made the substitution $t' = t - \tau$ in order to get this integral into the above form.

Writing Equation (5.96) as

$$\bar{U}(\omega) = \bar{I}_s(-\omega) \sqrt{2\pi} \frac{1}{\sqrt{2\pi}} \int_{-\infty}^{+\infty} I_p(t) e^{-i\omega t} dt, \tag{5.97}$$

we can immediately evaluate the Fourier transform in Equation (5.97) to get

$$\bar{U}(\omega) = \sqrt{2\pi} \cdot \bar{I}_s(-\omega)\bar{I}_p(\omega). \tag{5.98}$$

Making the substitution $\omega \rightarrow -\omega$ and solving for $\bar{I}_s$, we get

$$\bar{I}_s(\omega) = \frac{1}{\sqrt{2\pi}} \frac{\bar{U}(-\omega)}{\bar{I}_p(-\omega)}. \tag{5.99}$$

It is important to point out that all of the intensities $I_p(t)$, $I_s(t)$, and $U(\tau)$ are real so that $\bar{I}_p(-\omega) = \bar{I}_p(\omega)$, $\bar{I}_s(-\omega) = \bar{I}_s(\omega)$, and $\bar{U}(-\omega) = \bar{U}(\omega)$. We can solve for the signal intensity $I_s(t)$ under the condition that the intensities are real by taking the inverse Fourier transform of Equation (5.99)

$$\boxed{\bar{I}_s(t) = \frac{1}{2\pi} \int_{-\infty}^{+\infty} \frac{\bar{U}(\omega)}{\bar{I}_p(\omega)} e^{i\omega t} d\omega.} \tag{5.100}$$

To summarize the process used to measure the temporal profile of an arbitrary pulse shape using an all-optical shutter, we enumerate the sequence of events required to evaluate Equation (5.100), as follows:

- Determine $\bar{I}_p(\omega)$
  - **Experiment:** The sample is removed and the pump mixes with itself ($I_s = I_p$) in the Kerr gate as a function of delay provide by the translation stage, τ, yielding $U_p(\tau)$, the correlation function of the pump with itself (called the autocorrelation function).
  - **Analysis:** With $I_p = I_s$, solving Equation (5.98) for the pump intensity and using the reality condition, we get

$$\bar{I}_p(\omega) = \sqrt{\frac{\bar{U}_p(\omega)}{\sqrt{2\pi}}}, \tag{5.101}$$

    where $\bar{U}_p(\omega)$ is the Fourier transform of the measured autocorrelation function as determined by Equation (5.94).
- Determine $U(\omega)$
  - **Experiment:** The sample is placed in the beam and the intensity passing through the optical gate is again measured as a function of delay, τ, from which the correlation function $U(\tau)$ is determined.
  - **Analysis:** The Fourier transform of the cross-correlation function, $\bar{U}(\omega)$ is determined using Equation (5.94).
- $I_s(t)$ is determined from $\bar{I}_p(\omega)$ and $U(\omega)$ using Equation (5.100).

Using a Kerr gate, it is thus possible to measure temporal profiles of an optical pulse with a resolution that is limited by the material response time. The electronic response time of organic materials such as dye-doped polymers is in the femtosecond regime; however in liquids, other mechanisms such as molecular

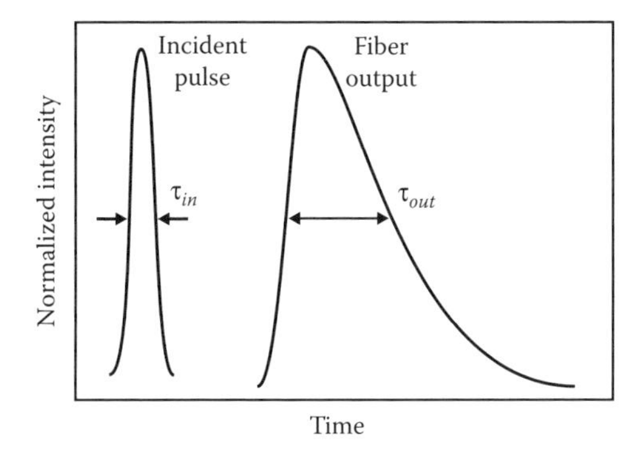

FIGURE 5.46 A qualitative rendering of the expected temporal profiles of an input and output pulse in a time domain bandwidth experiment.

reorientation have a response time of about a picosecond. It is therefore reasonable to get sub-picosecond resolution in a Kerr gate.

An alternative to a Kerr gate is the parametric mixing process, where the two beams interact through a second-order nonlinearity to provide a beam whose frequency is given by the sum of the frequencies of the two input beams. (In a parametric optical gate, the polarizers are removed and an interference filter is placed in front of the detector to pass only the sum frequency.) Such second-order parametric processes have femtosecond response times and can be used for resolving some of the shortest optical pulses known to humanity.

5.4.4.3 Time Domain Measurements Applied to Graded-Index Polymer Fibers

Garito and coworkers have applied the time domain measurements to graded-index polymer optical fibers.[79,80,89,90] Figure 5.46 shows a sketch of the data expected from such time domain experiments. The input pulse is sharp and symmetric while the output pulse is broadened and deformed. Indeed, this is what is observed. For a gradient-index fiber made with a 20% benzyl benzoate and 80% PMMA core (with $<175\ dB/km$ loss at 650nm), they launched $45ps$ pulses into a $98.8m$ length of fiber and observed an output pulse width of $145ps$. This yields a bandwidth of about $300MHz \cdot km$. This bandwidth is suitable for high-speed local area network applications where network components are separated by tens of meters.

5.4.4.4 Mode Mixing in Polymer Fibers

As discussed in Section 5.4.1, the effect of modal dispersion on bandwidth can be decreased if modes couple to each other. In polymer optical fibers, mode mixing has been shown to lead to higher bandwidths.[89] Figure 5.47 shows the measured degree of pulse broadening, τ, as a function of length of a step-index multimode fiber with a numerical aperture of 0.51.

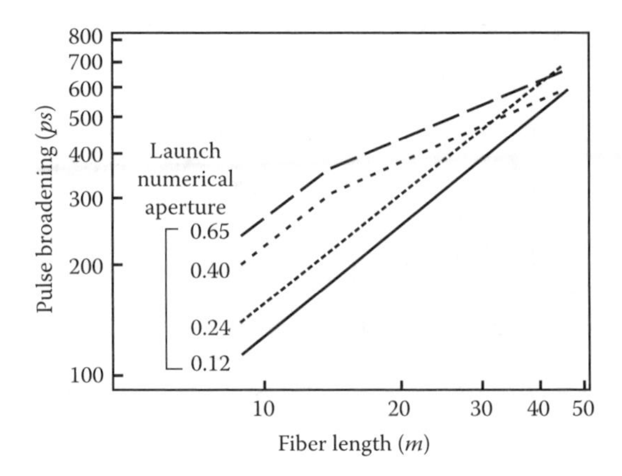

FIGURE 5.47 Pulse broadening as a function of fiber length for a 0.51 numerical aperture fiber. The launch numerical aperture is varied from 0.12 to 0.65. Adapted from Ref. [89] with permission of IEEE (©(1997) IEEE).

In these studies, the launch numerical aperture is varied between 0.12 and 0.65. On the log plot, the data appears linear — implying a power law. A fit of the data to the function $\tau \propto L^{\alpha}$ for the 0.12 and the 0.25 launch numerical aperture yields $\alpha = 0.99 \pm 0.01$ and $\alpha = 0.97 \pm 0.02$, respectively. The other two plots show two distinct slopes — a short length slope that is higher than the long length limit. For the short length section of the 0.40 and 0.65 launch numerical aperture, the data fit yields $\alpha = 0.97 \pm 0.03$ and $\alpha = 0.96 \pm 0.10$, respectively. The slope of one is consistent with modal dispersion.

Beyond about $15m$ of propagation, the large numerical aperture data yields $\alpha = 0.54 \pm 0.01$ for a launch numerical aperture of 0.40 and $\alpha = 0.57 \pm 0.01$ for a launch numerical aperture of 0.65. In this region, mode mixing is established and follows the square root of the length law as predicted by modal diffusion models. The explanation of these two distinct slopes is as follows: For high-input numerical aperture, all the modes are excited so mixing occurs over shorter lengths than for a small numerical aperture source, where only a few of the modes are excited. With fewer propagating modes, it takes a longer distance for them to scatter and couple into all the other modes. This hypothesis is supported by the observed intensity profile at the output. In the low launch numerical aperture case, the profile shows spatial oscillations. Once mode mixing is established, though, a smooth uniform profile is observed.

Using the refractive index profile (measured as explained in Section 5.1.6), the calculated bandwidth of the fiber due to modal dispersion is about one order of magnitude smaller than the observed bandwidth for $100m$ lengths of fiber. Clearly, the effects of mode mixing dominates, leading to higher bandwidth propagation than would be expected based on material dispersion alone.

6 Transmission, Light Sources, and Amplifiers

This chapter focuses on the use of polymer fibers for passive applications such as in transmission systems and displays, light sources, and amplification. Transmission applications include optical interconnects between chips and circuit boards, which are tens of centimeters in length, to long-distance applications, which require kilometers of fiber. Amplification is important in telecommunications applications such as in repeaters, which amplify the signal that is lost due to transmission losses. Fiber amplifiers and light sources, while important for telecommunication, are also important all-purpose stand-alone light sources. We also discuss more whimsical applications such as polymer optical fiber displays in textiles.

6.1 TRANSMISSION

6.1.1 LOSS CONVENTIONS

As we saw in previous chapters, optical loss is defined on a logarithmic scale and is expressed in units of decibels per kilometer (*dB/km*). For high-loss systems, the loss may be expressed in dB/cm. The usefulness of defining a logarithmic loss scale is the fact that it is additive. Hence, if a device has an input coupling loss of 1 dB, a transmission loss of 0.4 dB, and an output coupling loss of 0.8 dB, the total device loss is 2.2 dB.

Similarly, the power of a light beam is often defined in terms of a logarithmic scale. Since typical powers in a fiber are on the order of milliwatts, 0 dBm is defined as 1 mW of power. So, the logarithmic power, P_{dBm}, is related to the absolute power, P, according to

$$P_{dBm} = 10 \log \left[\frac{P}{1mW} \right]. \tag{6.1}$$

As such, a 1 W beam corresponds to 30 dBm, 1 μW corresponds to $-30\,dBm$, etc. Thus, to calculate the loss of a component, one need only subtract the output power from the input power in units of *dBm*, or

$$Loss = P^{in}_{dBm} - P^{out}_{dBm}. \tag{6.2}$$

6.1.2 FIBER MATERIALS

There are a variety of fiber types that are each designed for specific applications. The earliest fibers were large-core multimode fibers while more recently, graded-index fibers made with proprietary materials are common. It is useful to review the

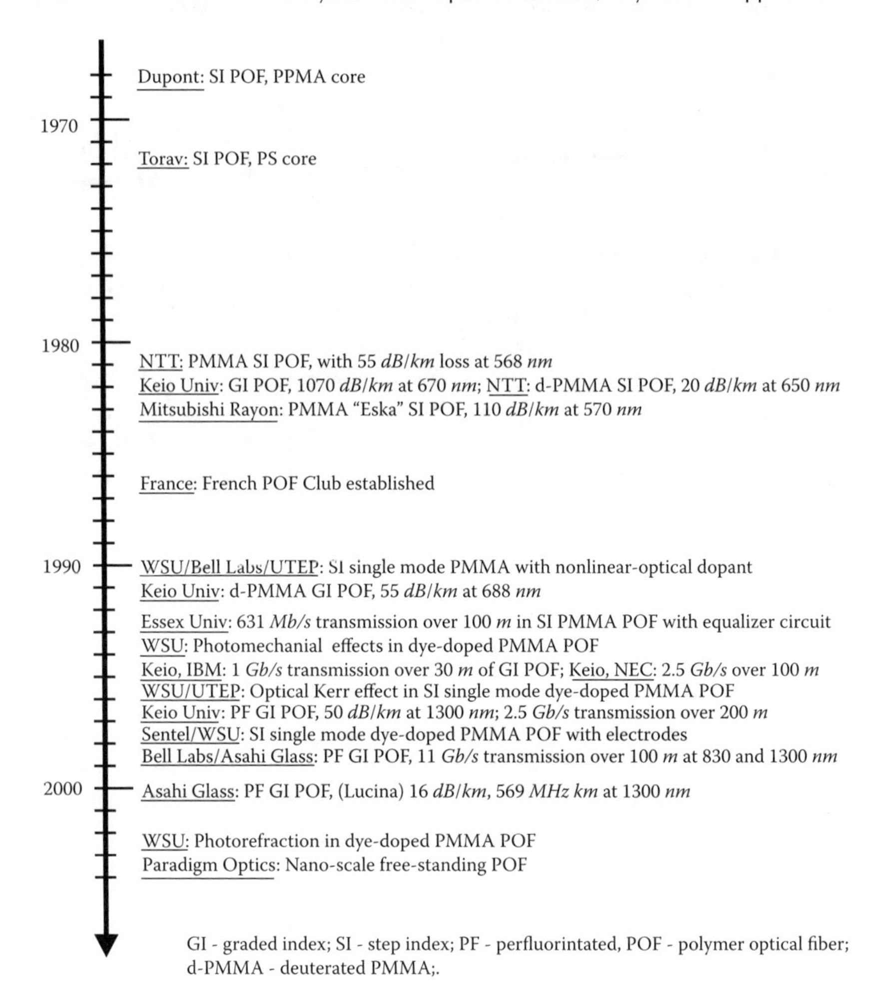

FIGURE 6.1 A nonexhaustive survey of the development of polymer optical fibers.

major developments leading to today's cutting-edge technology. Figure 6.1 shows a survey of the developments in the study of polymer fibers in the form of a time line. Note that there are far too many important contributions to list in this limited space, so the time line is not meant to be exhaustive. Rather, it should be used as an approximate gauge of when certain new ideas or demonstrations were first reported. Clearly, developments in recent years have grown exponentially, and few of these are listed.

From the late 1960s to 1990, research focused on passive fibers that were made with neat polymers or dopants that controlled the refractive index. The 1990s saw a series of reports of active polymer fibers that were doped with various molecules to provide functionality. For example, dye-doped fibers were reported with an intensity-dependent refractive index that can be used for all-optical switching.

A fluorescent dye-doped fiber was reported for optical amplification and rare-earth-ion-doped fibers were demonstrated to have properties similar to silica glass with higher pumping efficiency and potentially with broader gain curves. Electrooptic modulation was demonstrated in dye-doped fiber with electrodes, and photomechanical effects in dye-doped fibers were shown to act as active elements in all-optical position-stabilization circuits. (These applications are discussed in later chapters.)

The 1990s also showed continued progress in the development of passive materials with greatly lowered loss by fluorination and increased bandwidth in gradient-index POFs due to better control of modal dispersion. Many of these high-tech materials have found their way into commercially available fiber.

The late 1990s and early 2000s were times of continued reports of novel effects in fibers such as the demonstration of photorefractive effects, the writing of Bragg gratings, the demonstration of holey fiber, and papers on using POFs as light sources with reversible photodegradation. Additionally, fibers reached the nano-realm with the demonstration of free-standing fibers with a diameter of less than $1\mu m$.

There are three broad classes of fibers for transmission applications, which include large-core multimode fiber that is of low bandwidth but is easy to splice; graded-index fiber with high bandwidth and ease of splicing; and small-core single mode fiber with high bandwidth but also higher intolerance to core misalignments at a splice. Table 6.1 summarizes properties of the large-core fibers that are relevant for transmission applications. Small-core polymer optical fibers, with cores typically smaller than $10\,\mu m$, are not used in long-distance telecommunications applications, which is presently dominated by silica glass. However, small-core POF could be used for active inline devices, such as amplifiers, in long-distance transmission systems.

TABLE 6.1
Properties of some polymer optical fibers

Material	Loss (dB/km)	Bandwidth ($GHz \cdot km$)	Applications	N.A.	Core D (μm)
PMMA	55 (538 nm) 230 (660 nm) 2000 (850 nm)	0.003	LANs, industrial communications, and sensing	0.47	250–1000
PS	330 (570 nm)	0.0015	high T industrial short-haul communications and sensing	0.73	500–1000
PC	600 (670 nm)	0.0015	high T communications and sensing	0.78	500–1000
CYTOP	16 (1310 nm)	0.59	LANs	0.40	125–500 GRIN POF

6.2 DISPLAYS

Polymer optical fiber can be woven like any other textile to form a light-guiding cloth. Figure 6.2a shows a photograph of a polymer fiber fabric display on a jacket made by Koncar and coworkers. [91] Polymer fibers are lightweight, inexpensive, and flexible, making them ideal for such a product. In this particular demonstration, the fabric was made with optical fibers for wefts and silk in chain.

A closeup of the fabric is shown in 6.2b. The red-, green-, and blue-emitting PMMA fibers run parallel to each other and are illuminated with light-emitting diodes. The pattern to be displayed is etched into the cladding by making a series of pits as shown in 6.2c. Each pit is a pixel, and scatters light out the fiber. Given the losses in the system, many LED light sources are usually used to power the garment.

The pits can be made using various methods, such as micro-perforation by chemical treatment with a solvent. In this method, the image is permanent. However,

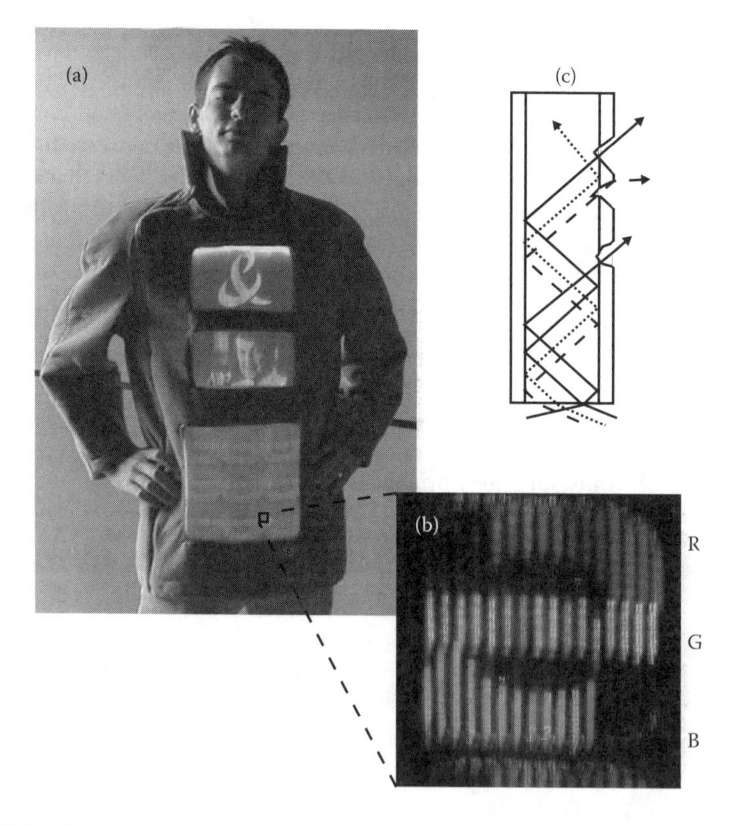

FIGURE 6.2 (a) A garment with a built-in POF display, (b) a closeup of the fiber "threads" (notice how the R,G, and B channels are interlaced), and (c) a schematic diagram of an individual fiber. All photographs are provided by and courtesy of Vladan Koncar, and used with his permission.

multiple images can be made and interlaced. In this way, one of several images may be selected by illuminating the appropriate set of fibers. The initial uses for such garments, as proposed by Koncar, include applications in which the wearer needs to be seen in the dark, a means for making a fashion statement, or for marketing purposes.

6.3 OPTICAL AMPLIFICATION AND LASING

There is a long history of light-emitting polymers. The next step is to make a polymer optical fiber light source that can be connected directly into a fiber-optics system. In this chapter, we discuss progress in both fiber-optic light sources and amplifiers. Given the great flexibility in controlling the optical properties of a dye-doped polymer, it is possible to make polymer fiber sources and amplifiers that act over a broad range of colors. In this section, we discuss the principles of light emission, materials for lasing and amplification, and polymer fiber-based sources.

6.3.1 PRINCIPLES OF STIMULATED EMISSION

In order to produce light, the material needs to be prepared in a way that stores energy that can be released as electromagnetic radiation. Since the lasing process can only be understood with a quantum mechanical picture, we start with a brief tutorial.

One of the important results of quantum mechanics is that the energy levels of a system are not continuous. In atoms and molecules, the energies of the excited states are discrete. In crystalline materials where the electrons are delocalized over many atoms (like metals or semiconductors) or molecules (such as organic polymer crystals), the energy levels are replaced by bands that define a range of energies that the system can occupy. Forbidden energy ranges are called gaps. In these systems, we usually consider only the electron states since their energy level differences correspond to photon energies in the near infrared to the ultraviolet part of the spectrum — where most lasers operate.

To store energy in a system, the electrons must be in an excited state. More importantly, more electrons need to be in the excited state than in the ground state. Let's first see if this can happen in a two-level system (i.e., the atoms or molecules in the material each have only one excited state and a ground state). Consider what happens when heat is fed into the system. The ratio of the number of atoms in the excited state, N_e, divided by the number of atoms in the ground state, N_g, is given by statistical mechanics to be

$$\frac{N_e}{N_g} = \frac{\exp\left[-\frac{E_e}{kT}\right]}{\exp\left[-\frac{E_g}{kT}\right]} = \exp\left[-\frac{(E_e - E_g)}{kT}\right], \tag{6.3}$$

where E_e is the energy of the excited state, E_g the energy of the ground state, T the temperature, and k is Boltzmann's constant. Since $E_e \geq E_g$, the population ratio given by Equation (6.3) is always less than or equal to unity. So, the population of the excited state can never be greater than the ground state population; and the two populations are only equal at infinite temperature.

We know from experiments that in lasing materials, the excited state population is greater than the ground state population. In the thermodynamic model, this corresponds to negative temperature, which is clearly nonsensical. Whenever a system appears to have a negative temperature, it is because the states considered are a subsystem of a larger number of excited states. Once these other states are included, the temperature is found to be positive and finite. So, heating a two-level system will not lead to an inverted population.

Another method for exciting a two-level system is by using light of an energy (per photon) that matches the energy difference between the two levels. This approach also fails because of stimulated emission: when a molecule is in its excited state, its rate of de-excitation to the ground state is proportional to the number of photons impinging on the molecule with energies that match the transition energy (defined as $(E_e - E_g)$). So, as we pump more and more light into the system to drive all of the molecules to their excited state, the same light also stimulates the molecules that are already in their excited state to decay back to the ground state. Calculations show that at best, the number of molecules that can be excited is equal to the number that remain in their ground state. So, it appears that nature does not like to provide conditions that are conducive to lasing. Luckily, this is not true for a three-level system.

Figure 6.3 shows an energy level diagram of a three-level system. For now, let's assume that the material is excited with light to achieve an inverted population. If the energy of the photon is equal to $(E_1 - E_g)$, then the system will be excited to State 1. Once in the excited state, the system can either de-excite back to the ground state; or, through a non-radiative process, relax into State 2. If the lifetime of State 2 is longer than State 1 ($\tau_2 > \tau_1$), then there is a high probability that the electron will be found in State 2. Furthermore, because the exciting photon energy does not match the energy difference $(E_2 - E_g)$, there will be no stimulated emission to the ground state. For these reasons, the electron can stay in this excited state for a long time. In a collection of such molecules, an inverted population can be reached where most of the molecules are found in State 2.

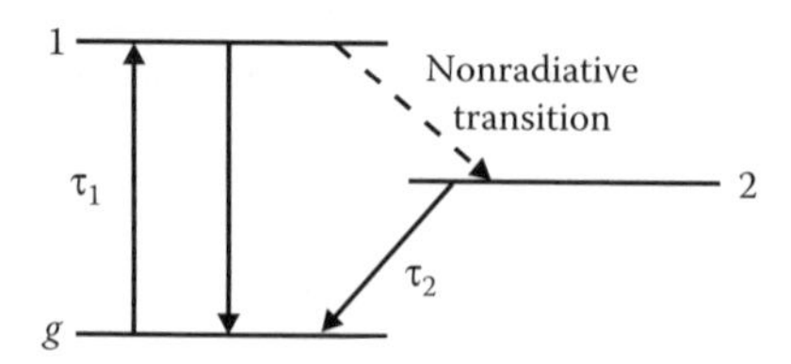

FIGURE 6.3 A three-level system is required to make an inverted population.

An inverted population can be reached by several means. Ion and gas lasers are usually pumped with an electrical current while pulsed lasers such as Neodymium Yttrium Aluminum Garnet lasers are optically pumped with flash lamps. Alternatively, the output of one laser can be used to pump another one.

Let's now assume that we have an inverted population. Consider one of the excited molecules. If an isolated molecule decays over a characteristic time τ_2, then its decay rate is $1/\tau_2$. For a collection of molecules, then, the decay rate is given by

$$\frac{dN_2}{dt} \propto -\frac{N_2}{\tau_2}, \tag{6.4}$$

where N_2 is the number of molecules in State 2. The decay rate can be calculated from quantum mechanics and is given by

$$\frac{1}{\tau_2} \propto \left| \int \psi_2^*(\vec{r}) \cdot e\vec{r} \cdot \psi_g(\vec{r}) dV \right|^2 = \left| \langle 2 \left| e\vec{r} \right| g \rangle \right|^2, \tag{6.5}$$

where ψ_i is the wavefunction of State i, $e\vec{r}$ is the dipole operator, and the expression to the righthand side of the equality is a shorthand notation for the integral on the left. $\langle 2 \left| e\vec{r} \right| g \rangle$ is called the transition dipole moment. So, if the wavefunctions for a given system are known, the transition rate can be calculated. For the calculations that follow, knowledge of quantum mechanics is not required. The important result is that the excited state has a characteristic decay rate.

To understand stimulated emission, consider a population of excited molecules as depicted in Figure 6.4. (Even though the material has three levels, we concentrate only on the two states that are responsible for emitting light.) At time t_1, a photon of light is launched into the material. As the photon passes a molecule in its excited state, it stimulates the molecule to de-excite and emit a photon in phase with the original one. At time t_2, two of the atoms have emitted photons, which are now all

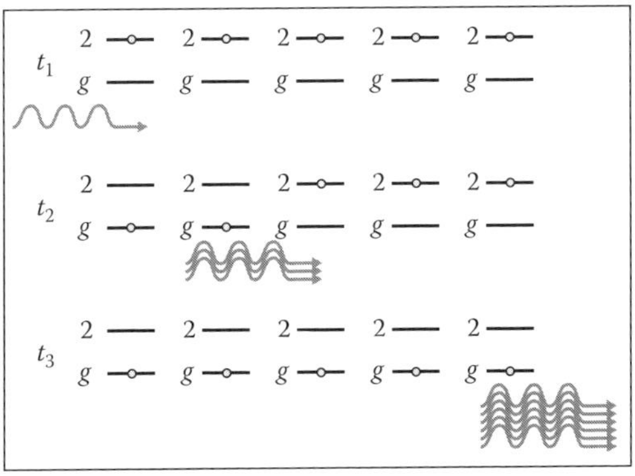

FIGURE 6.4 The presence of a photon near an atom stimulates it to de-excite and emit a photon of the same color and in phase with the first one. Shown are three time slices in the process of stimulated emission.

traveling in phase with each other. At time t_3, all of the molecules have de-excited due to stimulated emission, leading to a bright coherent beam.

In the presence of light, the transition rate in a molecule between State 2 and the ground state is given by

$$\frac{1}{\tau_2} \propto (n+1) \left| \langle 2 \left| e\vec{r} \right| g \rangle \right|^2 , \tag{6.6}$$

where n is the number of photons that are interacting with the molecule. So, the more photons that are present, the greater the probability of an emission. When $n = 0$, no light is present, so the probability of an emission is given by Equation 6.5. This type of decay is called spontaneous emission. When $n > 0$, the process is called stimulated emission. Therefore, as light travels through a material with an inverted population, it gets amplified as intensity is added from emitting molecules. This is the principle behind an optical amplifier. A flash of light pumps a material to cause a population inversion and the beam carrying a signal (such as a fiber-optic telecommunications line) is amplified in a way that preserves the phase and intensity profile of the signal.

Lasing is also a stimulated emission process. The material, though, is placed between two mirrors that are highly reflecting. After the material is pumped to form an inverted population, some of the atoms or molecules decay through spontaneous emission. A photon that is formed by stimulated emission gets amplified if it happens to pass by an excited molecule, though even then, the probability is small. If a photon is emitted in a direction that causes it to strike one of the mirrors at normal incidence, it bounces back and forth between the mirrors, increasing the probability that it will cause stimulated emission in one of the molecules. Once a stimulated photon joins the original one, the probability of that pair inducing a stimulated emission is higher. As more and more photons join the process, the probability of a stimulated emission grows until eventually, the material gets depleted of its inverted population.

In a continuous wave (CW) laser, the material is constantly excited to keep the population inverted and the mirror reflectivity is typically greater than 99%, so that the average photon bounces back and forth about 50 times before the probability of it exiting through the mirror is 40%. Since this is a statistical process, at every bounce, there is a 1% chance that a photon will pass. So, about 1% of the energy in the laser cavity exits at each bounce. To compensate, energy must be pumped into the cavity at least at that rate to keep the rate of light emission fixed. Figure 6.5 shows a spontaneous photon being amplified in a laser cavity. (The rays are not drawn normal to the mirror to make it easier to follow the time sequence of bounces.) The light leaving the cavity is not shown.

There are many ways to make a pulsed laser. In a diode laser system, a pulse of light can be made by applying a time-dependent voltage to the laser, causing it to turn on, then off, over a short time interval. To make a very short duration high-power pulsed laser, an externally controlled loss element is placed in the cavity, which allows energy to build, then to be released quickly. Otherwise, the design is the same as the continuous wave laser.

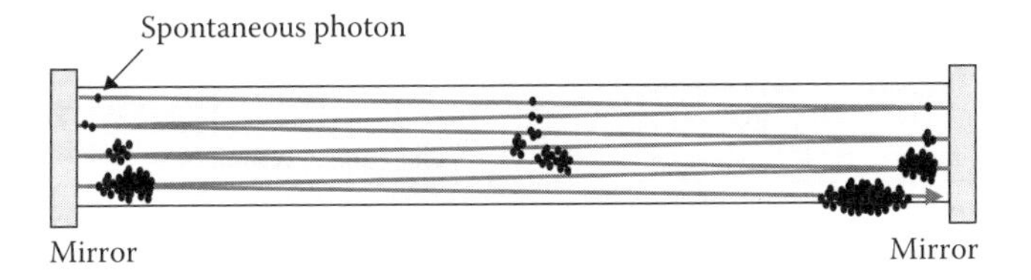

FIGURE 6.5 Lasing in a cavity.

For illustration, we focus on high-power pulsed lasers, which operate as follows. The lasing material is pumped while the loss element is turned on to prevent lasing. When the loss element is turned off for a short period of time, lasing occurs and all of the energy leaves the cavity in a short time. In a Q-switched laser, the loss element is a Pockels cell, in which the birefringence in the refractive index of the material in the cell is controlled with a high voltage. So, the Pockels cell rotates the polarization of the laser beam; and, in combination with a polarizer, the light beam can be blocked when the polarization is perpendicular to the axis of the polarizer. The combination of a Pockels cell and polarizer is often called a Q-switch because it can be used to drastically decrease the quality of the cavity, Q.

In one design of a Q-switched laser, an electrical current pulse energizes a flashlamp which in turn excites the lasing medium. By storing lots of charge in a large capacitor, a bright short flash of light can be generated when the capacitor is discharged through the flashlamp. Just after the lasing material is fully excited by the flashlamp, the Q-switch is pulsed to start the lasing process. A large fraction of the energy in the cavity is then dumped out as laser radiation while the Q of the cavity is high.

A mode-locked laser, on the other hand, contains a loss element that is turned off and on over a time period that matches the round-trip travel time of light inside the laser cavity. In this way, a pulse is amplified as it bounces back and forth in the cavity. Since the mode locker allows lasing only for that one pulse, it gets all of the energy at the expense of all the other modes. An acousto-optic mode-locker is made from a quartz crystal that is excited by an acoustical wave that periodically diverts the light. This is an example of active mode locking, where an external source controls the process. The frequency of the crystal needs to be accurately tuned to the cavity length for a mode locker to work properly.

Passive elements, on the other hand, are controlled by the light beam itself. Examples of passive mode lockers include materials that photobleach (i.e., become transparent) at high intensity or self-focusing materials, which change the wave propagation parameters that determine whether or not the light remains in the cavity. Such passive mode lockers only transmit light when a threshold intensity is exceeded. As such, they require enough light to make the mode locker pass light to initiate the lasing process. During each pass through the laser cavity, the pulse picks up more energy and is therefore more easily transmitted through the mode locker. Ironically, some lasers of this type require a kick start, where they literally need to be jarred to start made locking. In the early days of laser development,

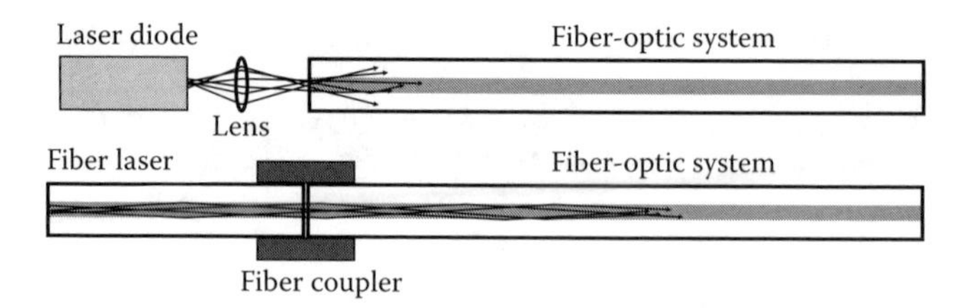

FIGURE 6.6 Coupling a laser diode to a fiber (top) and fiber-laser-to-fiber coupling (bottom).

this jarring was provided by a tap of the hand. More high-tech models include a piezoelectric crystal that provides the jolt. Q-switched lasers usually provide nanosecond pulses while mode-locking yields pulses in the femto- to picosecond range.

Commercial laser diodes emit a very messy mode pattern. When the emitted light from a laser diode is focused into the end of a single-mode fiber, only a small fraction of the light makes it into the core, as shown in the top part of Figure 6.6. Even with the introduction of specialized cylindrical lenses that clean up the spatial profile dramatically, much of the light is still lost in the coupling process. In addition, permanently connecting a laser source to a fiber is labor intensive and adds greatly to the cost of the final package.

A fiber laser source solves many of these problems. If the numerical aperture of the fiber laser matches the numerical aperture of the fiber-optic system, coupling between the two is straightforward and the amount of light lost, minimal. The bottom part of Figure 6.6 shows that a fiber laser with a fiber coupler can be easily attached to a fiber in a manner similar to how passive fibers are connected to each other.

Two devices that are of importance to telecommunications applications are amplifiers and lasers. An amplifier is like a laser without mirrors. When an inverted population is excited in the material, any pulse of light that enters the material at the correct range of frequencies will induce stimulated emission and the pulse will be amplified, preserving both the color and phase of the signal. To make a polymer fiber laser, on the other hand, mirrors need to be defined around the active region. One way to do so is to define Bragg gratings as shown in Figure 8.1, Section 8.1.

Both devices can be excited either from the side or with a pump beam that travels down the waveguide. The top part of Figure 6.7 shows how an amplifying fiber can be pumped either from the side (a) or down the waveguide where the pump is launched into the fiber using a dichroic mirror (which reflects the pump and passes the signal). The fiber laser, on the other hand, has Bragg grating reflectors that form a cavity and is pumped the same way, as shown in the middle portion of Figure 6.7.

In real fiber devices, a directional coupler is used. The signal light is launched into the bottom port and exits the bottom port of the first coupler. The pump beam, being a different color, enters the top port and exits the bottom port. In the amplifier, the pump beam and signal travel together and through stimulated emission, the

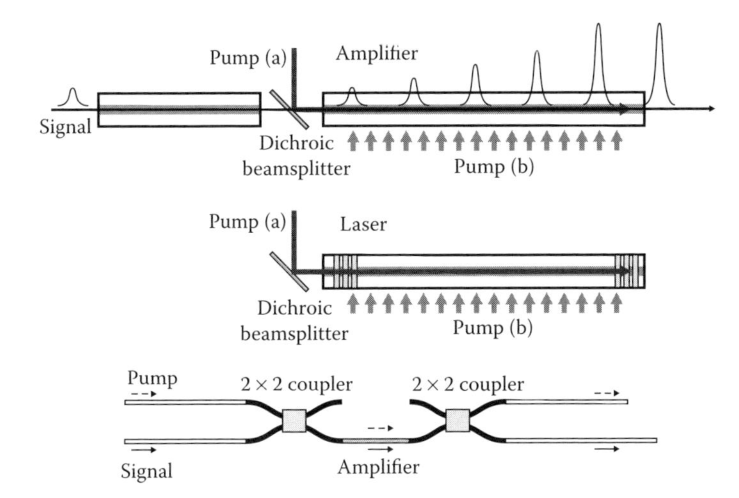

FIGURE 6.7 A fiber amplifier (top) and a fiber laser (middle). The all-fiber version of an amplifier (bottom).

energy is transferred to the signal beam. In a well-designed fiber device, most of the pump light is depleted in the process. The signal then enters the bottom port of the second 2×2 coupler and exits out the bottom port. Whatever pump light remains is removed by the second coupler to the upper port so that the signal light is not contaminated with the pump. The same idea can be applied to make an all-fiber laser.

The sections that follow describe examples of demonstrations of the fluorescence and stimulated emission processes that have been observed in polymer optical fiber.

6.3.2 Fluorescence

We begin by presenting a brief description of the fluorescence process. To understand fluorescence, we must first understand the energy level diagram of a typical molecule. A molecule is made of two or more nuclei that are arranged in some fixed relative positions with respect to each other when the molecule is in its lowest energy state. The electron cloud is smeared around these nuclei.

There are two types of excitations that occur in a molecule that is embedded in a polymer matrix. In a purely electronic excitation, the electron cloud is deformed while the nuclei stay fixed. A typical transition energy for such an excitation is on the order of a couple electron volts. In a purely vibrational excitation, the nuclei oscillate about their equilibrium positions in a way that keeps the center of mass of the molecule constant. The energies of these nuclear vibrations are much smaller than typical electronic energies. Since the electrons are much lighter than the nuclei, the electron cloud readjusts itself in response to the nuclear motion on a time scale that is much smaller than the time it takes the nuclei to undergo an oscillation.

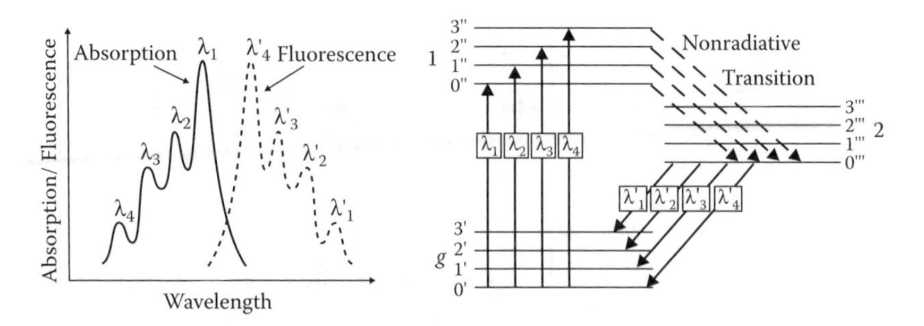

FIGURE 6.8 An energy level diagram showing the important features of absorption and fluorescence.

In reality, when a molecule is excited, both electronic and vibrational motions result. The electronic excitation occurs instantaneously, followed by oscillations of the nuclei. Figure 6.8 shows a simplified diagram of the process. Both the electronic ground state, g, and electronic excited state, 1, correspond to the same positions of the nuclei. The finer structure shows vibrations of the nuclei. But, the nuclear coordinates in the first excited electronic state do not yield the lowest possible energy state, so the shape of the molecule will relax into the state labeled 2. In State 2, the electronic energy is lower.

Let's follow the excitation and de-excitation process. While the molecule is usually found in its ground electronic state, at non-zero temperature, the molecule can be found in one of the higher vibrational states (labeled $1'$, $2'$, and $3'$). For the sake of argument, we consider that the molecule starts out in the ground state, which is a reasonable approximation for typical molecules. When a light beam's energy per photon matches the transition energy to an excited state, those photons will be absorbed. In the energy level diagram shown in the righthand part of Figure 6.8, the absorbed colors will be given by λ_1, λ_2, λ_3, and λ_4. The strength of the absorption will depend on the amount of overlap between the vibronic state's wavefunctions.

To calculate the transition rate between these two states, we use the fact that the wavefunction can be written as the product of the electronic wavefunction and the nuclear wavefunction. This is true if the energy of the molecule can be expressed as the sum of an electronic and nuclear contribution. If there is some coupling between the two, as long as the coupling is small, the wavefunction can be approximated by the product of the two wavefunctions. So, the transition dipole moment between States $0'$ and $1''$ is given by $\langle 1''|0'\rangle\langle 1|e\vec{r}|g\rangle$, where r is the electronic coordinate. Since $\langle 1|e\vec{r}|g\rangle$ is a common factor to all transitions between States g and 1, the transition between States i' and j'' is proportional to $\langle j''|i'\rangle$. In the absorption spectrum shown in the left part of Figure 6.8, the transition at λ_1 is shown to be the strongest and at λ_4, the weakest. The strengths of these transitions is thus proportional to $|\langle j''|i'\rangle|^2$.

Once the molecule is excited to State 1, it will nonradiatively decay to the lowest vibrational energy of State 2, which is labeled $0'''$. From this state, it can de-excite to any one of the vibrational levels in the electronic ground state. In the decay process, a fluorescence photon is emitted with a wavelength between λ'_1 to λ'_4. The resulting fluorescence spectrum is shown by the dashed curve.

Since the transition strength between States $0'$ and i'' in the absorption process is similar to the transition strength between States $0'''$ and i', the fluorescence spectrum will look like a mirror image of the absorption spectrum. The shift between the two spectra, called the Stokes shift, is due to the energy lost in the nonradiative decay from State 1 to State 2. Note that there are many potential mechanisms for the Stokes shift, which include relaxation of the nuclear positions and electrostatic interactions between the the molecule and surrounding medium.

To some degree, most dye-doped polymers fluoresce; so, it is not surprising that dye-doped polymer optical fibers also fluoresce. The important issue is that when light is generated inside a fiber, all of it that is contained within the cone defined by the critical angle will guide down the fiber, yielding an efficient method for collecting the light. Indeed, this idea was used by Kruhlak (as described in Section 5.2.5) to measure the optical loss.

Once fluorescence is generated in a fiber, its small diameter and long length enables the high-intensity light to interact with lots of material. Observation of fluorescence in a material is the first step in observing optical amplification, making light sources, and observing lasing. Fluorescence *in a fiber* can enhance the effect due to transverse confinement, making it possible to build efficient optical amplifiers, light sources, and lasers.

6.3.2.1 Decay and Recovery of Fluorescence

Peng and coworkers reported on fluorescence decay *and recovery* in three dye-doped modified poly (methyl methacrylate) fiber materials shown in Table 6.2. [92] Figure 6.9 shows the experiment. An argon laser beam at 514*nm* wavelength passes through a filter (F1) to remove background fluorescence in the laser medium. A lens focuses the light through a chopper into the end of the fiber that is part of a 2×2 coupler. Part of the light is diverted to a detector (D2) which monitors the laser power while the remaining light is launched into the polymer fiber sample that is butt-coupled and epoxied to the coupler. The fluorescence is sent to a scanning monochromator whose output is converted to an electrical signal by a detector.

TABLE 6.2
Dye-doped fibers measured by Peng and coworkers

Dopant	Concentration (*ppm*)	Cladding diameter (μm)	Core diameter (μm)	Core to cladding Δn
Rodamine B	2500	300	30	0.0035
Pyrrometheme	350	150	15	0.01
Fluorescein	175	200	20	0.01

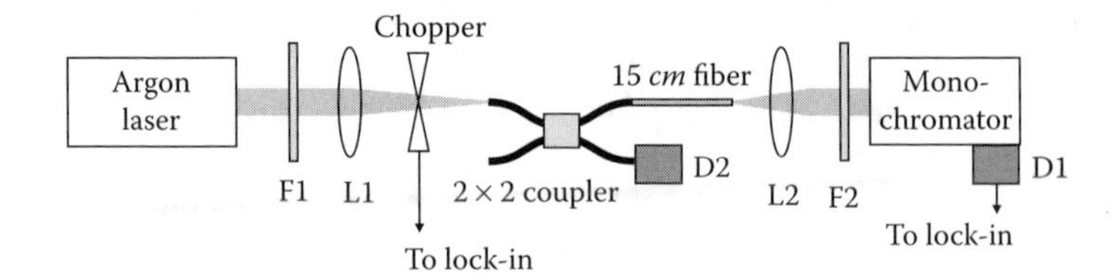

FIGURE 6.9 Peng's fluorescence decay experiment. Adapted from Ref. [92] with permission of IEEE. (©(1998) IEEE.)

A lock-in amplifier is used to determine the magnitude of the fluorescence signal at each wavelength.

Figure 6.10 shows a plot of the fluorescence spectrum of the three dye-doped fibers as a function of time under 514*nm* wavelength exposure. Note how the fluorescence spectrum from each different molecule decays as a function of time very differently from the others both with regard to the intensity of exposure and the shape of the spectrum. The qualitative form of degradation observed is common, though the physical mechanisms responsible are not well understood (and may even be different).

For example, the energy deposited into the material when light is absorbed is turned into heat that raises the temperature of the material. The rise in temperature can cause the molecules to change shape or break up into smaller ones. In both processes, this leads to a decrease in the fluorescence intensity. The fact that the temperature rise is responsible can be investigated by measuring the amount of fluorescence as a function of the temperature in an oven chamber.

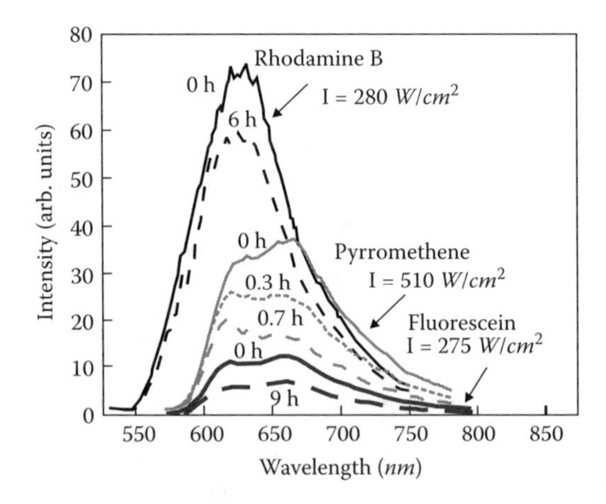

FIGURE 6.10 Fluorescence decay data for rhodamine B- (top two curves) pyrromethene- (middle three curves) and fluorescein-doped polymer fiber (bottom two curves). The exposure time, in hours, is labeled near each curve. Note that while the intensity is labeled with arbitrary units, the relative intensities are accurate. Adapted from Ref. [92] with permission of IEEE. (©(1998) IEEE.)

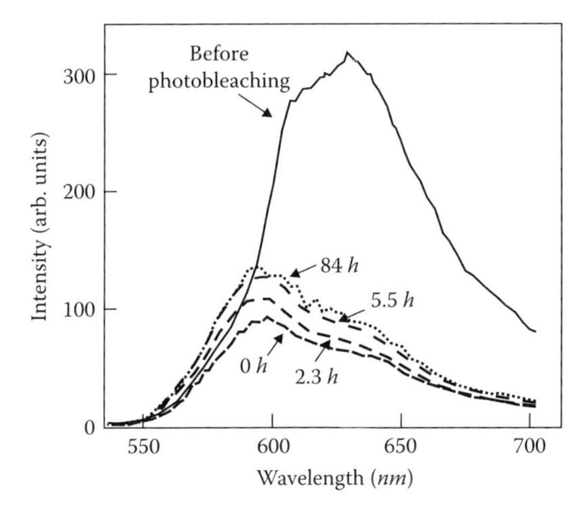

FIGURE 6.11 Partial fluorescence recovery for rhodamine B-doped polymer optical fiber. The time that the sample is kept in the dark after photobleaching (in hours) is shown on each fluorescence spectrum. Adapted from Ref. [92] with permission of IEEE. ((©)(1998) IEEE.)

Chemical reactions with a molecule that is in an excited state is a second process that results in molecule degradation. The long-lived state shown in Figure 6.8 is often a triplet state, which is known to be highly reactive. [92] Triplet oxygen is also known to be a catalyst that promotes photo-induced chemical reactions that lead to the fragmentation of larger molecules into smaller ones. Indeed, when oxygen is kept away from a molecule, it degrades much more slowly. For example, dye-doped cores that are surrounded by cladding material survive temperatures that would normally be catastrophic in open air.

Bleaching is a familiar household term, and refers to the decrease of the intensity of color in a molecule when it is exposed to harsh chemical agents such as chlorine. Usually, the process is irreversible. Photobleaching is a general term that refers to the change in color of a molecule due to light. There are many ways that molecules can change color. For example, Section 9.2.2 discusses the reversible process of photoisomerization. In a nonreversible process, bonds may be broken. In all the degradation measurements reported by Peng, they also observed a decrease in the amount of light absorbed through the dye-doped polymer. Such a decrease of absorption will cause a decrease in the depth of color of the material.

Peng found that under high enough intensities and exposure time, the degree of photobleaching was irreversible so that once the fluorescence spectrum decreased, it never returned even when the sample was kept in the dark for a long period of time. However, at low intensities, the fluorescence spectrum was found to return to its original value. At medium exposures, the fluorescence was found to partially recover.

Figure 6.11 shows fluorescence recovery in the rhodamine B-doped polymer optical fiber for various times after the photobleaching beam is turned off. The initial spectrum is also shown. Peng found that irreversible photobleaching occurs once the fluorescence spectrum falls 30% below its original value. In Figure 6.11, the sample was irradiated for 2 hours at 2550 W/cm^2. The $t = 0h$ curve shows the fluorescence spectrum just before the photo bleaching laser is turned off to allow recovery.

Since most optical light sources and amplifiers are based on fluorescence, the fact that fluorescence can recover as long as the material is not pumped too hard and allowed to "rest" has promising reliability implications. Such recovery is not normally observed. Later in Section 6.3.3, we will see that amplified spontaneous emission can also recover.

6.3.3 Amplified Spontaneous Emission

Amplified Spontaneous Emission (ASE) can be viewed as a laser without the cavity. When many neighboring molecules are excited into an inverted population, a photon spontaneously emitted by one molecule will be multiplied by stimulated emission from other molecules that lie in the path of the originally emitted photon.

With low-intensity illumination, not many of the molecules in a material are excited so that stimulated emission is not likely. Under these conditions, only a fluorescence spectrum is observed. As the intensity is turned up, stimulated emission begins. Since many more photons are created at a wavelength corresponding to the fluorescence peak than at any other wavelength, and more of the inverted population corresponds to this energy, this wavelength will be favored for stimulated emission and amplification. If the photons that are produced are strongly absorbed, then stimulated emission will not be favored. The wavelength at which a sample will emit ASE is therefore determined from the competition between the material absorption spectrum and fluorescence spectrum.

The lefthand part of Figure 6.12 shows how the emission spectrum changes as the pump beam power is increased. At high intensities, the ASE peak grows and narrows until most of the energy from the sample is emitted within the ASE peak. If the sample is illuminated with a line of light from the side (righthand side of Figure 6.12 for low intensity), the light will be emitted in all directions. As the intensity increases, stimulated emission will favor a direction that contains the largest population of excited molecules. As such, the ASE output will be mostly collinear with the line and exits the sample out the ends, as shown in the right-hand side of Figure 6.12.

6.3.3.1 Gain

When light travels in a material that absorbs and scatters light, its intensity decreases. If the material is made of molecules with an inverted population, the intensity grows. When the gain due to stimulated emission exceeds the loss, the light beam will be amplified. To understand the process, we consider the one-dimensional case, where all the light travels in the same direction — as is the case in a fiber.

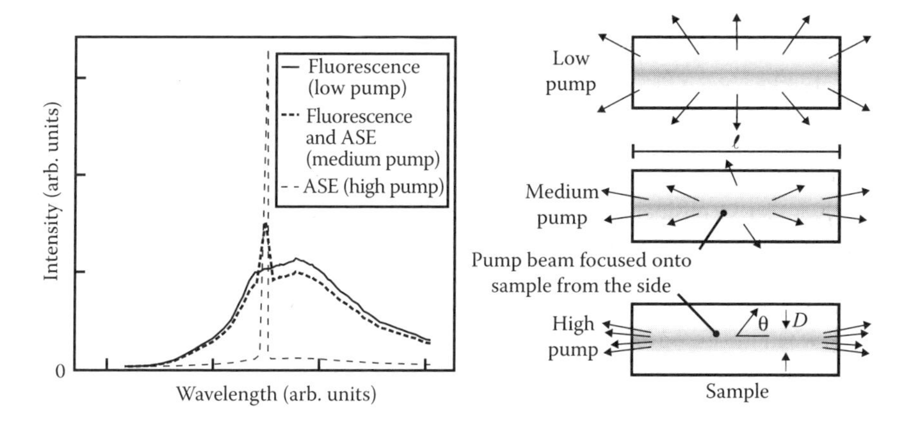

FIGURE 6.12 A schematic representation of the emission spectrum as a function of pump intensity (left) and the directionality of the fluorescence for a pump beam that is focused to a line in a sample (right).

The left-hand part of Figure 6.13 shows a thin slice in a material of width dx. The photon flux (number of photons per unit of area per unit of time) entering and leaving this element is ϕ and $\phi + \Delta\phi$, respectively. If the material gain is $\gamma(\lambda)$ at wavelength λ, and the loss is $\alpha(\lambda)$, then the change in the flux through the slice is

$$\frac{\phi_\lambda(x)}{dx} = \gamma(\lambda)\phi_\lambda(x) - \alpha(\lambda)\phi_\lambda(x), \tag{6.7}$$

where the subscript λ refers to the wavelength of the flux.

The solution to Equation (6.7) for the flux at any point x along the fiber, $\phi_\lambda(x)$, is

$$\phi_\lambda(x) = \phi_\lambda(0) \exp\left[(\gamma(\lambda) - \alpha(\lambda))x\right], \tag{6.8}$$

where $\phi_\lambda(0)$ is the flux entering the material. The gain depends on the population of excited molecules, which in turn depends on the intensity of the pump. As the pump intensity is increased, so does γ, until the gain overcomes the loss, or $\gamma(\lambda) - \alpha(\lambda) > 0$.

If a continuum of colors is launched into the material, the point at which gain exceeds loss occurs first at one particular wavelength. Since the process is highly

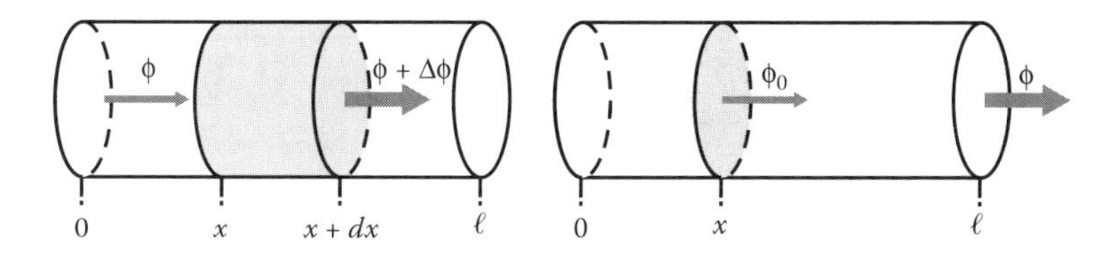

FIGURE 6.13 Gain in a small slice of material (left) and the total gain for a flux originating at point x and exiting the material at $x = \ell$ (right).

nonlinear (depending exponentially on the gain), the one dominant wavelength will win over the other ones, taking all of the energy. This is the source of the sharp spike shown in Figure 6.12. It is worthwhile to note that the above analysis only holds when the pump is not depleted. Clearly, energy must be conserved; so at best, the amplified light beam cannot gain more energy than provided by the absorbed pump light.

Now let's consider the case where only the pump beam is present. So, the light is generated by a spontaneous emission from the inverted population inside the material. Furthermore, we consider the case where the inverted population is confined to a line of width $D(\ll \ell)$, as shown in Figure 6.12 — which is the case when the pump is focused by a cylindrical lens. Furthermore, since only the dominant wavelength is present at high gain, we can neglect the loss, and do not explicitly display the wavelength-dependence of the gain.

First, we calculate the angular distribution of the emitted light. Let's consider one point that is isotropically fluorescing at the the end of the fiber as shown in the right-hand portion of Figure 6.14. A ray that leaves this point at an angle θ travels a distance $D/(2\sin\theta)$ through the material (Note that this holds only when $D/2\sin\theta < l_0$). If the intensity directed away from the fiber at angle θ is I_0 at the source, it will be amplified to an intensity $I(\theta)$ to yield an amplification factor A:

$$A(\theta) = \frac{I(\theta)}{I_0} = \exp\left[\frac{\gamma D}{2\sin\theta}\right] \tag{6.9}$$

where we have used Equation (6.8) and neglected the loss.

The left-hand part of Figure 6.14 shows a plot of the intensity ratio as a function of the material gain-diameter product. Clearly, with just small increases in the gain, the output becomes more highly directional. As long as the line width D is large compared with the wavelength of the light, most of the ASE light will exit the material in a cone angle given by $\tan\theta = D/2\ell$. The effect of gain narrowing is also due to the high nonlinearity of the amplification process. For a large gain,

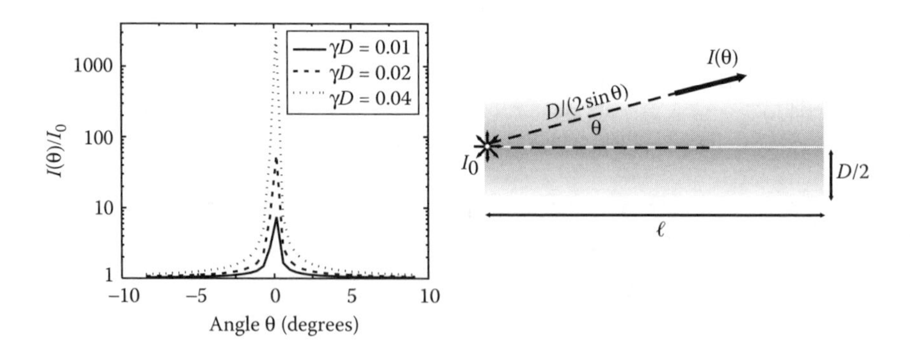

FIGURE 6.14 The inverted population distribution in a sample when excited by a line source of width D (right) and a plot of the angular dependence of the gain for various values of γD (left).

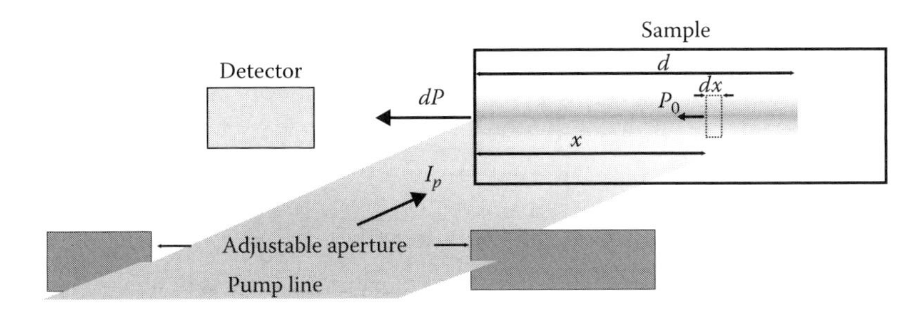

FIGURE 6.15 A common gain measurement that uses a line source of variable length d.

most of the light emitted will do so over a small range of wavelengths. Thus, the ASE light will be emitted over a narrow range of angles and wavelengths.

The material gain is commonly measured using the technique diagramed in Figure 6.15. A line pump source of intensity I_p passes through an adjustable aperture of width d to produce a line of fluorescence in the sample as shown. As described above, the ASE intensity for a strong pump is highly directional and exits mostly out the two ends of the line, so a detector placed at the edge of the sample catches about half the total ASE light that is generated.

To determine the gain from this measurement, we first need to calculate the total ASE intensity leaving the line out of the detector end. Consider a section of the line of thickness dx and a distance x away from the output end. The power of the light that is spontaneously emitted from this slice along the line axis is $P_0(I_P)dx$, where $P_0(I_P)$ is the power per unit thickness. (The emission power is proportional to the pump intensity.) Using Equation 6.8, the power from the slice, dP, that exits the fiber end is

$$dP = P_0(I_P)dx \exp\left[\gamma(I_P)x\right], \tag{6.10}$$

where we assume that the loss is negligible. This expression also assumes that the gain is uniform along the whole length of the line. If a large fraction of the pump beam is depleted, this assumption might not hold.

To get the total output power, we integrate over the length of the line:

$$P = \int_0^d P_0(I_P)\exp\left[\gamma(I_P)x\right]dx = \frac{P_0}{\gamma}\left(\exp\left[\gamma d\right] - 1\right), \tag{6.11}$$

where we have suppressed the pump-power-dependence of the gain and of the emitted power. If the aperture is reduced to $d/2$, the power density P_0 remains the same, so the ratio of powers is given by

$$\frac{P(d)}{P(\frac{d}{2})} = \frac{\left(\exp\left[\gamma d\right] - 1\right)}{\left(\exp\left[\gamma \frac{d}{2}\right] - 1\right)} = \frac{\left(\exp\left[\gamma \frac{d}{2}\right] - 1\right)\left(\exp\left[\gamma \frac{d}{2}\right] + 1\right)}{\left(\exp\left[\gamma \frac{d}{2}\right] - 1\right)}, \tag{6.12}$$

which can be inverted to solve for the gain,

$$\boxed{\gamma = \frac{2}{d}\cdot \ln\left(\frac{P(d)}{P(\frac{d}{2})} - 1\right)}\ . \tag{6.13}$$

Note that the gain is defined as the fractional increase of the light per unit length of propagation (see Equation 6.7); so, it has units of inverse length. The reciprocal of the gain gives the length over which the intensity increases by a factor of $e(\approx 2.7)$. Since the gain depends on the pump power, it is more useful to define the gain coefficient, G:

$$\boxed{G(P_p) = \frac{d\gamma}{dP_p}}, \tag{6.14}$$

where P_p is the pump power. The gain coefficient thus has units of inverse length and inverse power. The gain coefficient can be experimentally determined by measuring the gain as a function of power. For a given pump power, the gain is given by the slope of the line tangent to the gain curve. As the pump power is increased, the gain coefficient drops once all of the molecules are excited (called saturation). When doing a gain measurement, the largest slope is usually the best measure of the intrinsic gain.

In a dye-doped polymer fiber, light is absorbed and emitted by the dye molecules. As the concentration of molecules increases, so does the gain coefficient. When the concentration becomes so high that the molecules interact with each other or form crystals, there is no longer an advantage to making samples of higher concentration. To determine a molecule's intrinsic ability to amplify light, we divide the amplification coefficient by the number density N to get

$$\boxed{g(P_p) = \frac{G(P_p)}{N} = \frac{1}{N} \cdot \frac{d\gamma}{dP_p}}. \tag{6.15}$$

The concentration and power that yields the largest value is the intrinsic molecular gain coefficient. Note that it has units of area per unit of power.

In any application, the important quantity is the maximum gain, γ, that is achievable in a bulk device. Thus, we can define the gain per molecule, Γ, which we get from Equation (6.13):

$$\boxed{\Gamma = \frac{\gamma}{N}}. \tag{6.16}$$

Γ has units of area and represents an effective area, or cross-section, of a molecule for amplification. Note that this is not a real area, but a quantity that is proportional to the probability that a molecule will de-excite and provide amplification.

There is one other important factor that needs to be mentioned, and that is the effect of the electric-field polarization of the pump on the pattern of emission. If the molecule or piece of material is initially isotropic, the induced dipole moment, $\vec{p}$, will be along the polarization axis of the pump light. Figure 6.16 shows a pump beam that is incident on a molecule in the sample along the y-axis and polarized along $\hat{z}$. The figure eight patterns represent the amplitude of the instantaneous electric field lines of the induced dipole moment. Figure 6.16 shows the pattern only in the xz plane. The full pattern looks like a donut whose axis is coincident with $\hat{z}$.

The Poynting Vector, $\vec{S}$, which is perpendicular to the density of electric field lines, shows the direction of the radiation at that point and the magnitude of the intensity. (Only a few Poynting vectors are drawn in the figure.) So, in this picture,

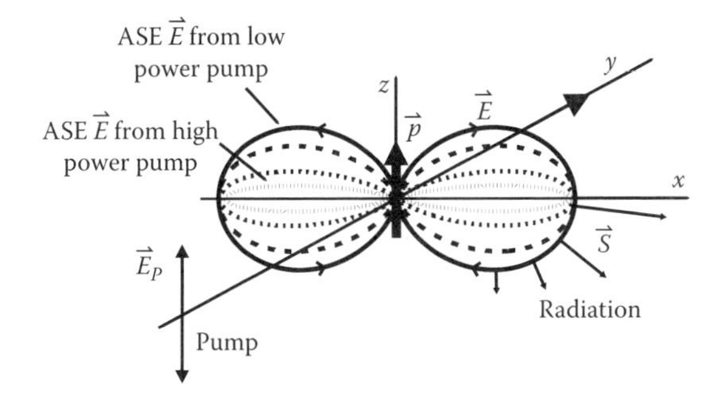

FIGURE 6.16 The dipole radiation pattern in the xz plane for an incident pump beam polarized along the z-axis. The figure eight patterns represent the electric field lines and the arrows on the bottom-right lobe show the Poynting vector at that point in space.

the light will radiate mostly in the zy-plane. In the case of a line excitation pump, in order for the ASE to be emitted along the line, the polarization of the pump must be perpendicular to it. If the sample is anisotropic, the radiation pattern can in general be emitted at any specific angle — depending on the susceptibility tensor. In such cases, the optimal polarization will not be perpendicular to the line source.

Due to stimulated emission, the pattern becomes more flattened as the pump power is increased (as shown in Figure 6.16), so at high enough intensity, the light will be emitted in the xy-plane. For the isotropic material, this effect reinforces the tendency of the light to be emitted along the line source.

6.3.3.2 Reversible Degradation in ASE

In Section 6.3.2, we discussed how Peng and coworkers demonstrated recovery of fluorescence provided that the degree of degradation was less than 30%. Howell demonstrated that ASE can also recover, even after a more substantial amount of degradation in the signal. [93]

The dopant chromophore used by Howell is sold under the trade name of Disperse Orange 11 (DO11). Its molecular structure is shown in the inset of Figure 6.17. The figure also shows the absorption spectrum and fluorescence spectrum of DO11-doped PMMA. There are two important features of this data to note. First, the two spectra are nearly mirror images of each other, as we would expect (see Section 6.3.2). Second, the separation between the peaks is large (about 125*nm*).

When the separation between peaks is large, this implies that the relaxation process (as shown by the transition between States 1 and 2 in Figure 6.8) results in an energy change that is larger than can be accounted for by the small readjustment of nuclear positions. The structure of the DO11 molecule is similar to the class of Coumarin dyes, which are known to undergo phototautomerization; a process where a proton hops from an adjacent ring to the double-bonded oxygen. In the case of DO11, the hydrogen nucleus on the NH_2 group hops to the upper oxygen

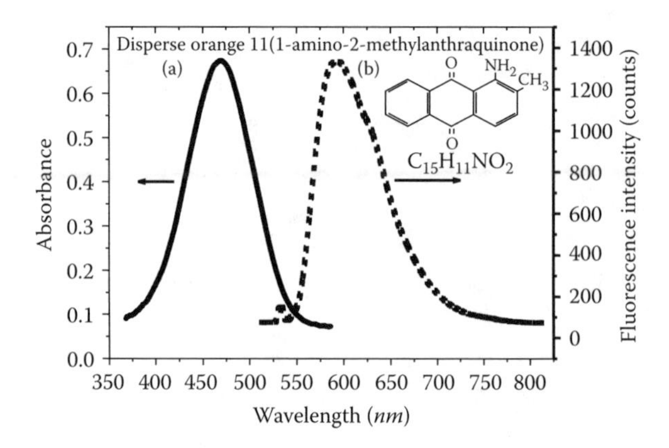

FIGURE 6.17 The absorption and fluorescence spectrum of DO11 dye in PMMA polymer. Adapted from Ref. [93] with permission of the Optical Society of America.

atom to form an OH^+ group. Since the remaining group is NH^-, the two interact strongly, resulting in lowered energy. In the Coumarin dyes, this process is known to protect the molecule from degradation.

Figure 6.18 shows the measured spectra of light emitted by a polished cube of DO11-doped PMMA polymer that is excited by a line source at three different intensities (0.5″ diameter cylinders were also used in such measurements). The pump line is focused on the side of the $1cm$ cube, and the concentration of the dye is so high that no significant penetration of the the pump into the sample is

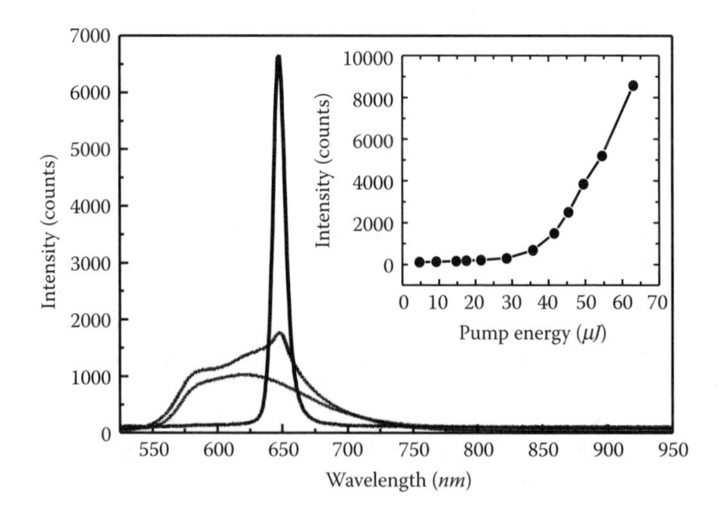

FIGURE 6.18 The spectrum of light emitted from a cube of DO11 dye in PMMA polymer when excited by a line source. The inset shows the power in the ASE peak. Adapted from Ref. [93] with permission of the Optical Society of America.

observed (the ASE line appears only on the surface). The pump light is polarized perpendicular to the line (when it is polarized parallel to the line, the output drops dramatically) and the spectra are all measured at the end of the excitation line, which coincides with the edge of the cube.

As the intensity of the pump is turned up, the spectrum makes a transition between a broad fluorescence peak and a sharp ASE peak. The inset shows the intensity at the peak wavelength of $650nm$ as a function of the pump energy, per pulse. In these experiments, the pump was a $25ps$-duration pulse at $532nm$. After a pump threshold of about 30 μJ, the ASE intensity is observed to increase quickly.

It is interesting to note that the ASE peak at $650nm$ is not coincident with the fluorescence peak at $620nm$. This is most likely due to optical absorption by the polymer. PMMA is known to have a window of transparency at about $650nm$ and a local peak absorption due to a C-H stretch overtone at about $625nm$. This clearly illustrates how competition between optical loss and gain determines the wavelength at which spontaneous amplified emission results. Interestingly, for these very reasons, $650nm$ is a standard telecom wavelength for polymer fibers. As such, DO11-doped PMMA makes a good gain medium for local area network applications that use polymer fiber.

Figure 6.19 shows the output power exiting the line out one side of the sample as a function of pump energy. The concentrations in the legend are given by the mass of dye divided by the volume of liquid MMA that were mixed together and polymerized to make the bulk dye-doped polymer. Above a pump energy of $500\mu J$, the ASE intensity starts to saturate. The highest observed conversion efficiency is about 12% and the gain saturates to around $0.6\,cm^{-1}$ at a pump power of $4.1\,MW$.

Figure 6.20 shows a plot of the ASE intensity as a function of time for DO11-doped PMMA. For the first 5 hours, the pump runs continuously at about $10Hz$, during which time the ASE signal degrades by about 60%. At 5 hours, the pump

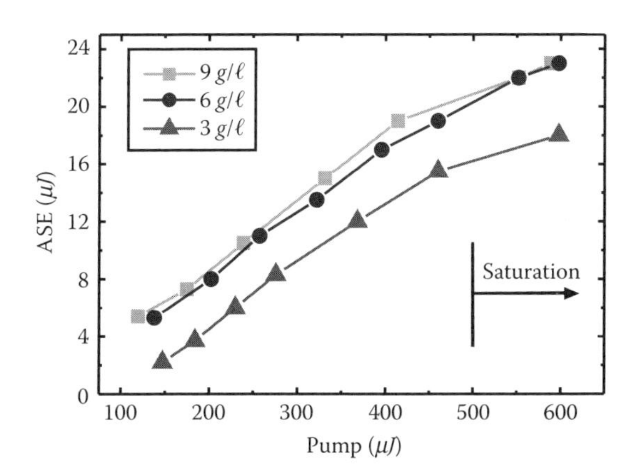

FIGURE 6.19 The ASE output as a function of pump for DO11-doped PMMA. Adapted from Ref. [93] with permission of the Optical Society of America.

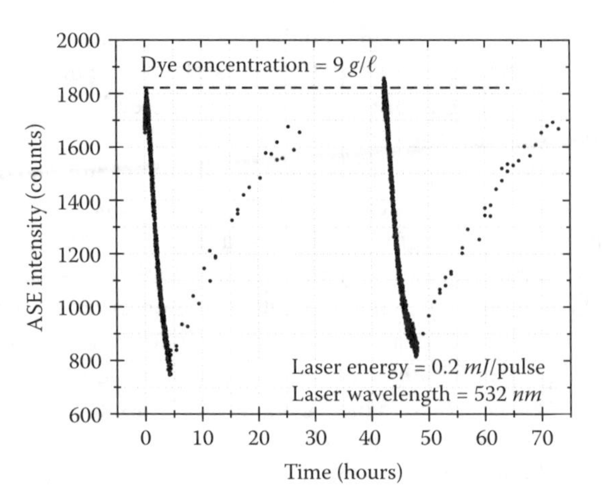

FIGURE 6.20 ASE output for DO11-doped PMMA as a function of time when the pump remains on (decay curve) and when the pump is only turned on to measure the degree of recovery of the ASE efficiency. Adapted from Ref. [93] with permission of the Optical Society of America.

laser is turned off and is only turned on for about one minute every hour to measure the ASE signal. During a period of about 30 hours, the signal recovers to its original value. More interestingly, Howell reports that the signal usually recovers to a value that is greater than 100% (as seen in Figure 6.20 using the dashed line as a reference) of the original value. In addition, there is evidence that the degree of recovery increases and the decay rate decreases with each cycle of degredation/recovery. [93] The implication is that materials can be hardened to have longer lifetimes if they are cycled at high pump power.

All of the decay curves observed by Howell exhibit the property that the ASE intensity, upon degradation, levels off to a nonzero value. This can be understood as being due to the competition between the decay and recovery mechanisms. Once the decay rate matches the recovery rate, an equilibrium ASE intensity is reached. Since the decay rate depends on the pump intensity, the state of equilibrium ASE output also depends on the intensity. Note also that the decay and recovery time constants are different. The pump intensity in the experiment that generated the data for Figure 6.20 was high, so the equilibrium intensity was found to be small. As such, the time constant of the decay curve is dominated by the decay rate, which is clearly much faster than the recovery rate.

While the mechanism of the degradation and recovery process is not yet well understood, Howell and Kuzyk proposed a model that is illustrated in Figure 6.21. [94] The proposed energy-level diagram is shown on the left-hand side of Figure 6.21. The absorption spectrum, shown on the right-hand side of the figure, yields an energy of the first excited state of 2.64 eV. The large separation between the singlet state and observed fluorescence led them to propose that the fluorescence spectrum is from the tautomer state. As such, the singlet state relaxes

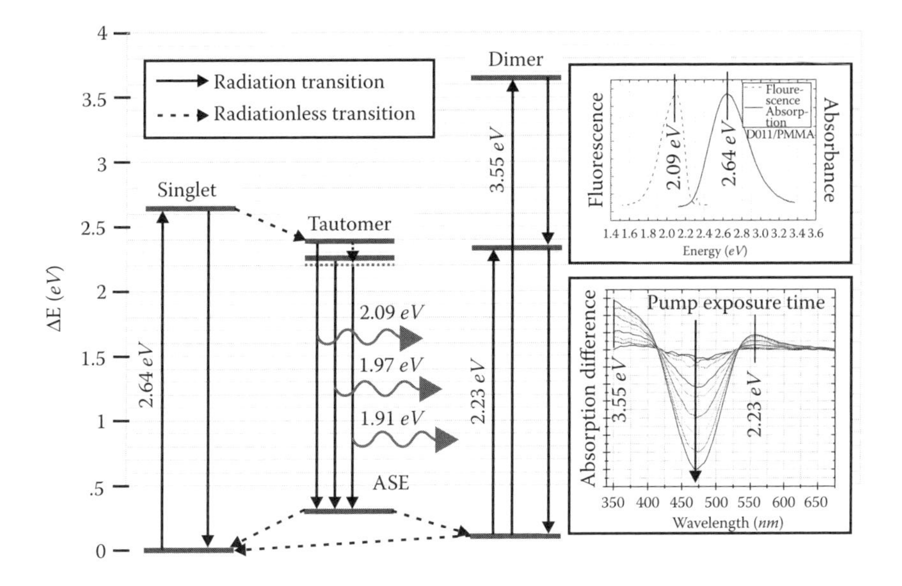

FIGURE 6.21 The left portion of the figure shows the energy level diagram of DO11 dye proposed by Howell and the right side of the figure shows some of the data used in getting to this result.

to the excited tautomer state, which itself can relax to the state from which fluorescence is observed. At low pump intensity, in the absence of ASE, the fluorescence spectrum peaks at $2.09\,eV$. It is well known that a broad fluorescence peak is composed of several narrow peaks that are unresolved. At higher intensities, an ASE peak is observed at $1.91\,eV$.

Recovery of the ASE intensity is only observed when the dye concentration is high. Furthermore in liquid solution, decay is observed, but not recovery. [95] As such, Howell and Kuzyk argued that the process is mediated by interactions between molecules. Since the tautomer has a large dipole moment, they argued that dimers are formed through dipole interactions. They did experiments that showed that during pumping the main peak in the absorption spectrum gets depleted while two other peaks form. The bottom inset on the right-hand side of Figure 6.21 shows the change of the absorption spectrum as a function of time. The observation is consistent with the formation of a dimer with two energy levels at $3.55\,eV$ and $2.23\,eV$.

To summarize, the decay/recovery mechanism can be reconstructed from the data as follows. When a DO11 molecule is excited, one possible outcome is non-radiative decay to a tautomer state, which upon de-excitation, yields the observed fluorescence. The tautomer, in its ground state, can form a dimer, which has an energy that is between the ground state of the single molecule and the tautomer molecule. The dimer does not show any fluorescence but can be observed with linear absorption spectroscopy. The dimer can decay entropically to the ground state of the single DO11 molecule, as was confirmed by Howell by showing that recovery of ASE and fluorescence accelerates at elevated temperatures. The DO11 molecule thus recovers its ability to fluoresce.

The observation of recoverable photodegradation is of interest scientifically because of the rich set of mechanisms that are responsible. While the model proposed by Howell and Kuzyk is still tentative, whatever process is found to be responsible is surely an interesting one. The idea that a material can recover its ability to generate light, and more importantly that it can be hardened through cycling, makes this material class promising for applications that require operation at high intensities. Material reliability is always an important issue in high-intensity applications, for which DO11-like materials may play a crucial part.

6.3.3.3 ASE in Fibers

Figure 6.22a shows a DO11-doped PMMA fiber that is excited with green light from the side. The ASE light guides down the fiber and exits out both ends. A mirror can be placed at one end so that all of the ASE light exits out the other end. For a sense of scale, note that 1/4–20 screws hold the fiber in place. Figure 6.22b shows the red spot on a screen generated by the ASE fiber. Figure 6.22c shows that the ASE light can be coupled to a passive fiber that is placed at the end of the fiber laser.

A large variety of molecules can be dispersed into a polymer, so devices can be designed from the ground up. A chemist can make a series of molecules for testing structure-property relationships. The best molecules can then be dispersed into a polymer and tested. The resulting polymeric materials with the desired

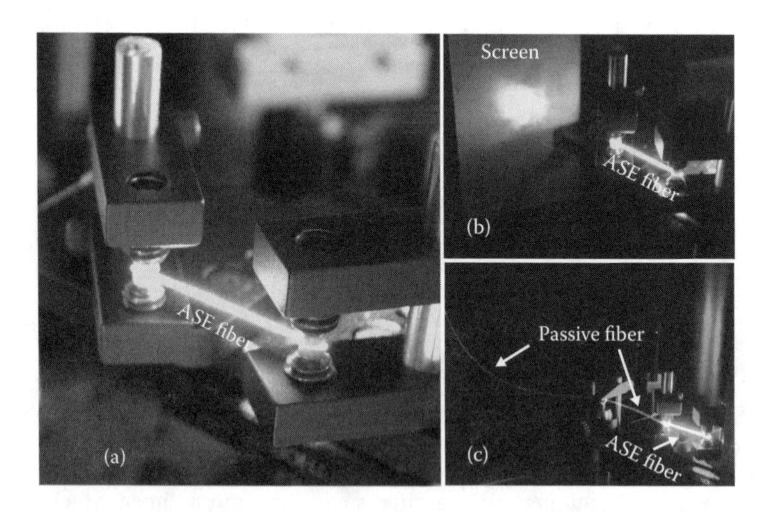

FIGURE 6.22 ASE in a fiber. (a) A closeup of the active fiber pumped with green (533nm) light. (b) The ASE fiber forms a red spot (650nm) on a screen. (c) The red ASE light is coupled into a passive fiber. Photograph courtesy of Brent Howell and Mark G. Kuzyk.

properties can then be made into fibers. In this way, it is possible to design a wide range of fiber light sources and amplifiers. Far too many material systems have been demonstrated to review here. Most importantly, however, all other reported materials suffer from permanent photodegradation — a problem that has made polymer fibers not generally accepted as sources or amplifiers for commercial applications. As such, molecules such as DO11 may provide the breakthrough required for general acceptance of polymers in commercial systems.[1]

The red sources described above are well suited for local area networks that operate in the $650nm$ transparency window of polymers. Other applications, such as optical memory, require shorter wavelengths. Since the area required to store one bit of information scales as the square of the wavelength, the shorter the wavelength, the better. Díaz-García and coworkers reported that the common hole-transporting agent, N,N'-Bis(3-methylphenyl)-N,N'-diphenylbenzidine (TPD), that is routinely used to control the conductivity of polymers in electro-luminescent applications is itself a material that produces blue spontaneous emission. [96]

In these studies, Díaz-García used films in the range of 300–$1000nm$ thickness of polystyrene or poly(N-vinylcarbazole) (PVK) doped with TPD in the range of 5% to 20% by weight. The films were pumped with a $355nm$ wavelength nanosecond pulse that was focused onto the sample with a cylindrical lens to form a line to produce ASE that is collected at the end of the line, similar to the diagram shown in Figure 6.15. The observed ASE spectrum peaks at about $420nm$, and, the polystyrene polymer is found to be much more stable to photodegradation.

Fifteen percent doping by weight of TPD in polystyrene was found to yield the most efficient ASE material. A gain between 3.5 and $10\,cm^{-1}$ was observed for pump energies of 66.5 to $140\,\mu J/pulse$ for pulses of $10nm$ duration. While this is not as efficient as is found for certain semiconducting polymers, the dopant concentrations are also lower. Since electrodes can be embedded in optical fibers, as illustrated in Figure 1.4.5 of Section 1.11, this material might be suitable for making an electrically powered light source rather than one that needs to be optically pumped.

Kobayashi and coworkers observed blue ASE (at $494nm$) in a step index fiber made of PMMA doped with 1,4-bis(4-diphenylamino-styryl)-benzene (SP35) (shown in Figure 6.23). [97] The fiber is fabricated by filling a glass capillary tube with a solution of polystyrene, SP35 dye, and orthoxylene solvent. The fiber is excited by a line source from the side using a $10Hz$ repetition-rate $10ns$ pulse train of wavelength $355nm$. A fiber fluence of $12\,mJ/cm^2$ was used throughout all experiments, and the observed gain was found to be $36\,cm^{-1}$. The loss of the fiber was determined using a SIF experiment (see Section 5.2.5), where the signal at the ASE wavelength from the end of the fiber is measured as a function of the position of the excitation line. Kobayashi and coworkers found a loss of $\alpha = 0.7\,cm^{-1}$, a value much smaller than the net gain.

---

[1] Polymers are being used in commercial thin-film electro-luminescent displays.

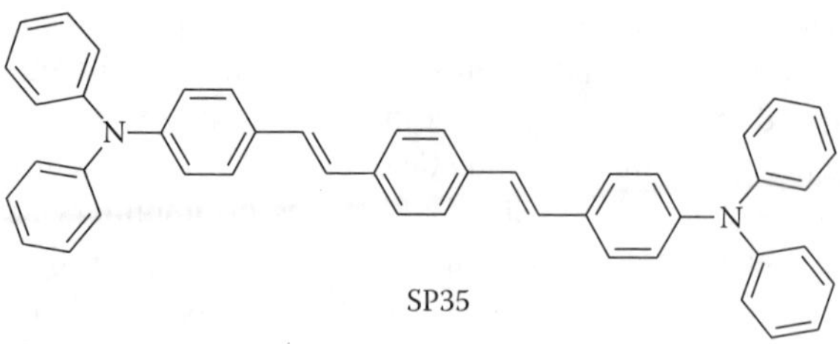

SP35

FIGURE 6.23 1,4-bis(4-diphenylamino-styryl)-benzene.

6.3.4 Polymer Optical Fiber Amplifiers

6.3.4.1 Laser Dyes

The first polymer optical fiber amplifier in a graded index core was reported by Tagaya and coworkers. [98] The fiber was made with PMMA host material doped with the organic laser dye rhodamine B and benzyl *n*-butyl phthalate (BBP), a molecule used to elevate the refractive index. The fiber was made using the interfacial gel techniques, which yielded a parabolic refractive index profile with the laser dye and BBP concentrated in the core region.

The inset of Figure 6.24 shows the absorption and fluorescence spectrum of the dye-doped polymer. A signal at 591*nm* (at the peak of the fluorescence spectrum)

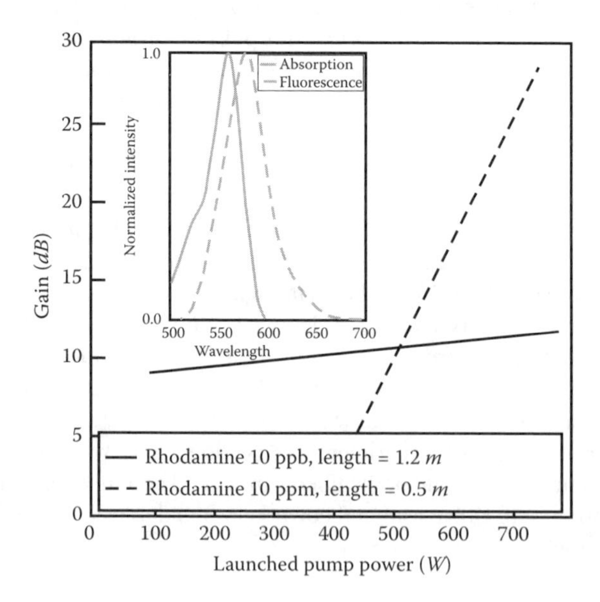

FIGURE 6.24 Gain as a function of pump power in two different fibers. The inset shows the absorption and fluorescence spectrum of the dye-doped polymer.

and a pump beam at $532nm$ are both launched into the core. Over a length of $0.5m$, the signal is amplified by $27\,dB$ for a pump power of 690 W in a fiber doped with $10ppm$ of laser dye. The energy conversion efficiency is about 10–15%. The gain in the lower concentration fiber is much less.

In the local area network (LAN), large fiber diameters are used to make installation easier. As we saw in Section 2.4.2, large-core step-index fibers with high refractive index differences, Δ, suffer from lowered bandwidth as the refractive index difference increases. So, graded-index fiber is the best solution. The graded-index fiber amplifier is thus best suited for applications that use such fiber; namely, in LANs, interconnects and fiber-to-the-home. Long-distance telecommunication, however, uses mostly small-core step-index single-mode fiber. Peng and coworkers first demonstrated optical amplification in a step-index fiber, [53] albeit in a larger core ($30\mu m$ diameter) than the $8\mu m$ diameters required by typical long-haul telecom applications.

Peng's fibers were made with PMMA cladding and rhodamine B(RhB)-doped cores (the same laser dye as Tagaya's demonstration). The preforms were fabricated by polymerizing the cladding around a Teflon wire, removing the wire, and pouring an unpolymerized mixture of MMA/dye into the hole. (See Section 3.2.4.) The dye concentration in the core is about $2,500\ ppm$, substantially higher than the $10\ ppm$ concentration used by Tagaya.

As was discussed in Section 5.2.5, self-absorption of fluorescence, as it propagates down a fiber, causes the spectrum to change. Peng observed that a fluorescence spectrum of RhB dye that is peaked at $572nm$ is shifted to about $633nm$ after propagating $73cm$ down the fiber. These spectra were determined by exciting fluorescence at one end of a fiber and observing it at the other end. The non-self-absorbed spectrum was determined by placing a silica fiber just outside the polymer fiber to collect the fluorescence near the excitation point. As such, for the high concentrations of dye and long lengths of fiber used, the peak fluorescence wavelength could be shifted to the telecom wavelength of $650nm$, right in the middle of the transparency window of hydrogenated polymers.

Because of the higher concentrations of RhB dye, a maximum of just under $25dB$ of gain was observed for a pump peak power of under $1kW$, and an energy conversion gain of almost 60%. This is in contrast to Tagaya's observation of comparable gain, which required an order-of-magnitude greater pump intensity. Table 6.3 summarizes the results of Peng and coworkers. Note that the gain peaks at $23.5\,dB$ for a fiber of $0.9\,m$ length.

6.3.4.2 Lanthanide Complexes

Rare-earth-doped inorganic glass fibers are commonly used for making commercial optical amplifiers. The concentrations that are attainable with such lanthanide ions in glass is relatively low compared with the concentrations that are possible in lanthanide-containing dyes that are dissolved into a polymer. As such, the higher concentrations can lead to higher gain. Unfortunately, lanthanide atoms can interact with the surrounding polymer, quenching the fluoresce.

TABLE 6.3
Gain and peak emission wavelength as a function of fiber length for PMMA fiber doped with RhB laser dye[53]

Fiber length (*m*)	Wavelength of maximum gain (*nm*)	Gain at (*dB*)
0.56	612	20
0.69	616	22.5
0.9	620	23.5
1.08	624	17
1.29	628	15

One approach for protecting the lanthanide molecule from the polymer is to incorporate it into a structure of several ligands. Such ligand structures are called chelates, where the rare-earth atom is held in place with ligands in the shape of pincers. As an example, the left part of Figure 6.25 shows a diagram of the molecular structure of a Europium chelate. [99] The middle part of the figure shows a view with only those atoms of the ligand that interact with the Europium ion while the right part illustrates the geometry of the interactions and how the ion is surrounded by the ligand to protect it from the environment.

Even in a chelate, rare-earth complexes will aggregate if the concentration gets high enough. For example, Xiaohong and coworkers found that the Europium complex $Eu(C_7H_15COO)_3$ doped in PMMA polymer can aggregate as determined from both fluorescence measurements and the more direct method of near field optical microscopy. [100] In these studies, the aggregates were observed to form fractal clusters.

Another issue is the efficiency of excitation. In a rare-earth-ion-doped glass, the efficiency is limited by the low cross-section of the ion. However, ligands typically have a large cross for excitation; and the absorbed energy can be transferred

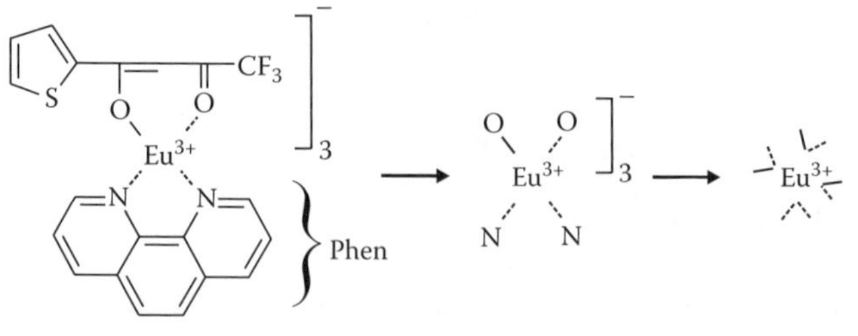

FIGURE 6.25 A typical rare-earth chelate structure of Europium (left), chelate that shows only the atoms interacting with the rare-earth ion (middle), and the geometry of the ligands in a chelate (right).

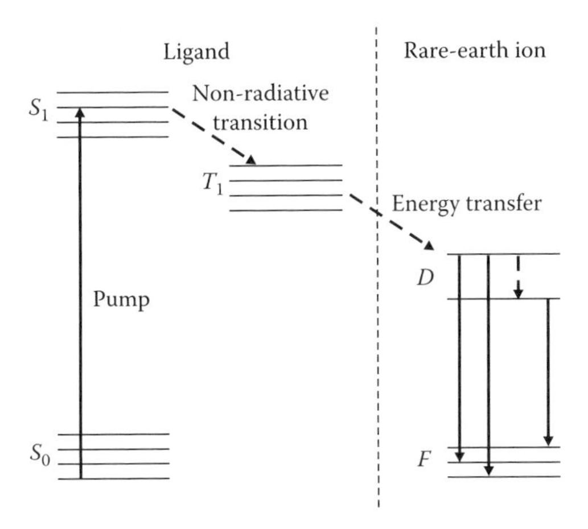

FIGURE 6.26 An energy level diagram that illustrates excitation of the ligand, transfer of energy to the rare-earth ion, and subsequent emission.

to the nearby rare-earth ion. Figure 6.26 shows an energy level diagram of the process.

A pump laser induces a transition from the singlet ground state to a singlet excited state. Through a nonradiative transition, the system relaxes into a long-lived triplet state, which transfers energy to an excited state of the lanthanide ion, provided that the ion's excited state energy is lower than the triplet state of the ligand. Subsequently, the ion emits light, and with feedback, stimulated emission results. For the Europium complex, emission from the manifold of ^5D excited states to the ^7F states is observed. The dominant wavelengths are around $538nm$, $592nm$, and $613nm$. [99]

Figure 6.27 shows a set of representative ligands that have been used to make lanthanide complexes. In 1964, Huffman was the first to report stimulated emission in a PMMA polymer fiber doped with $Tb(TTFA)_3$(work he started in 1963 [101]). [102] One of the polished ends of the $6cm$-long $750\mu m$-diameter fiber was silvered to provide feedback. When pumped with $335nm$ light, the fiber produced light at $545nm$ with a 225 J threshold pump energy. The fiber needed to be cooled to 77 K in order for stimulated emission to be observed.

There are other means for energy transfer such as Förster transfer. In this process, the rare-earth complex is doped into a semiconducting polymer, such as PVK. A current applied to the polymer can induce excitations whose energy is transferred to the rare-earth complex. In this way, electrically pumped sources or amplifiers can be made. Similarly, the polymer can be excited by light, followed by the same process of energy transfer. In both cases, the emission band of the polymer must overlap the absorption band of the ligand for energy to be efficiently transferred. If a polymer is not semiconducting, such as PMMA or polystyrene, it can be doped with hole-transporting molecules, such as 2-(4-biphenylyl)-5-(4-*tert*-butylphenyl)-1,3,4-oxadiazole (PBD), to make it semiconducting.

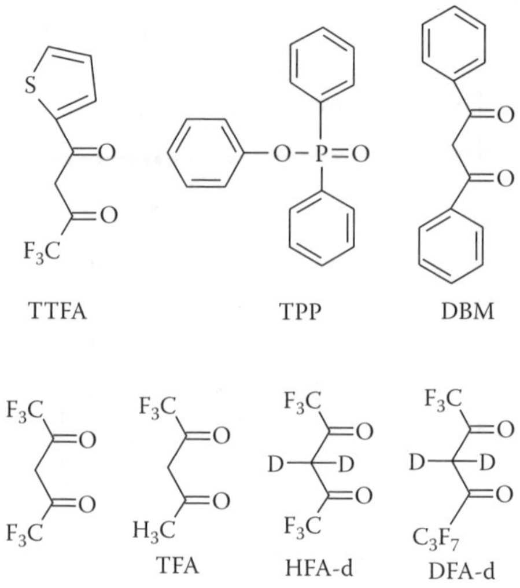

FIGURE 6.27 Several representative ligands that have been used to make rare-earth complexes that are doped into polymers.

Díaz-García and coworkers studied rare earth complexes of Europium and Samarium with 2,2,6,6-tetramethyl-3,5-heptanedionate (tmhd) and 3-(trifluoromethylhydroxymethylene)-(+)-camphorate (tfc) ligands. [103] The six materials studied were $Eu(tmhd)_3$, $Eu(tfc)_3$, and $Sm(tmhd)_3$ doped into either PVK polymer or polystyrene with 10% PBD to make it semiconducting. The strength of energy transfer from the semiconducting polymer to the rare earth complex was determined by comparing the integrated intensity under the emission peak of the complex with the same complex doped in the same concentration into the inert neat polymer polystyrene. It was found that the rare earth complexes in neat polystyrene emitted an order of magnitude less light than in PVK polymer. For example, $Eu(tmhd)_3$ and $Eu(tfc)_3$ at 15% doping levels in PVK had an integrated intensity that was a factor of 15 and 5 larger than in neat polystyrene. Polystyrene doped with PBD was found to have even a stronger Förster energy transfer. This is impressive given the fact that the concentration of the PBD molecule is an order of magnitude smaller than the concentration of PVK.

The efficiency of energy transfer can be quantified by the transfer rate parameter, ϵ_{TRP}, defined by [103]

$$\epsilon_{TRP} = \frac{I(A) - I(0)}{I(0)}, \tag{6.17}$$

where $I(A)$ is the ratio of the area of the rare earth's emission peak to the area of the broad fluorescence band of the PVK polymer for a rare earth concentration of A. Table 6.4 summarizes the results. [103]

More recently, researchers have been working on improving the efficiencies of emitting materials for applications in optical amplifiers and light sources. The

TABLE 6.4
Energy transfer efficiencies of several rare earth complexes doped into PVK or Polystyrene polymer

	PVK polymer			Polystyrene/PBD	
A%	$Eu(tmhd)_3$	$Eu(tfc)_3$	$Sm(tmhd)_3$	$Eu(tmhd)_3$	$Eu(tfc)_3$
5	0.52	0.15	0.034	0.61	0.68
10	0.63	0.42	0.43	0.80	0.80
15	0.72	0.68	0.59	0.91	0.83
20	0.75	0.70	0.62	0.91	0.78

wealth of materials made gives the device designer many choices with regard to the wavelengths that can be amplified (due to the chelate complex) and the mechanical properties of the fiber (due to the polymer). Figure 6.28 shows some of the polymers that have been used in making fibers.

The fluorescence efficiency is adversely affected by vibrations in the surrounding chelates and polymers, as are found in C-H bonds. By substituting a heavier-mass atom for the hydrogen atom, the vibration frequency can be decreased, leading to an increase in the florescence efficiency. HFA and HFA-d, as shown in Figure 6.27, are examples of chelates made with ligands in which hydrogen atoms are replaced by fluorine and deuteron atoms. Similarly, PPMA-d8 — as shown in Figure 6.28 — is a fully deuterated polymer analogous to PMMA and CYTOP is a fully fluorinated polymer. Combinations of such chelates and polymer hosts make improved materials.

For example, Gao and coworkers [104] studied Sm and EU chelate complexes. They included the ligand HFA in the form $Sm(HFA)_4N(C_2H_5)_4$ and

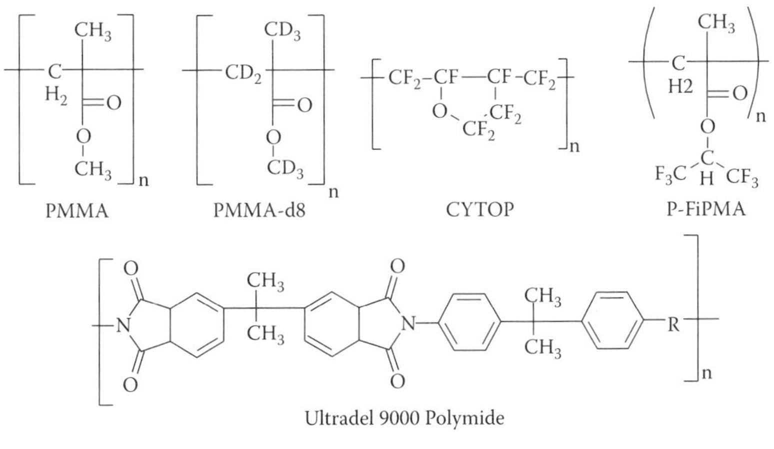

FIGURE 6.28 Several representative polymers that have been used to make rare-earth-doped fibers.

TABLE 6.5
Fluorescence properties of Europium Chelates doped in PPMA at a concentration of 800 *ppm* by weight

Material	Concentration of TPP (*ppm*)-wt	Fluorescence Wavelength (*nm*)	Fluorescence Intensity (relative)	Fluorescence Lifetime *ms*
$Eu\,(TFA)_3$	0	613	0.036	0.31
	20%	614	0.070	0.39
$Eu\,(HFA)_3$	0	614	0.208	0.39
	20%	615	1.000	0.61
$Eu\,(TTFA)_3$	0	614	0.083	0.31
	20%	614	0.110	0.40
$Eu\,(DBM)_3$	0	614	0.145	0.37
	20%	613	0.317	0.55

$Eu(HFA)_4N(C_2H_5)_4$. In PMMA-d8, $Sm(HFA)_4N(C_2H_5)_4$ has a fluorescence peak at $645nm$ due to the transition in Sm^{3+} between the states $^4G_{5/2} \rightarrow ^6 G_{9/2}$. The measured values of the lifetime, $\tau = 194\mu s$, and emission cross-section of $4.5 \times 10^{-21} cm^2$ yielded a theoretical waveguide gain of $20\,dB$ at a wavelength of $650nm$ with a pump power of $50\,mW$.

Kuriki and coworkers review the properties of several materials. [99] Table 6.5 summarizes some of the important parameters. Note that the best gain materials have the longest lifetimes. Table 6.5 shows a result reported by Kobayashi and coworkers that adding triphenylphosphate (TPP) to Europium chelates increases both the fluorescence efficiency and the lifetime. They found that a $70\,cm$-long GI-POF made with 20 wt % of TTP was found to have an attenuation loss of $0.4\,dB/m$ at $650nm$ with 1% chelate doping, by weight. [105]

Liang and coworkers reported on optical amplification of a Europium ion surrounded by DBM ligands (see Figure 6.27) and Phen (See Figure 6.25) — designated $Eu(DBM)_3$Phen, doped in a step-index PMMA polymer optical fiber at 4,000 ppm by weight. [106] The core diameter of the fiber is $400\mu m$ and was pulled from a $10mm$ rod and coated in a fluorine-containing resin as cladding.

The $613nm$ fluorescence peak, due to the $^5D_0 \rightarrow {}^7F_2$ transition, was observed to dominate the spectrum. From the theory of Judd and Ofelt, they find the emission cross-section is $4.72 \times 10^{-21}\,cm^2$ which is comparable to the value of Erbium-doped silica glass fibers of $5 \times 10^{21}\,cm^2$. The radiative lifetime is calculated to be $1.68\,ms$ and the measured lifetime is $0.49\,ms$, yielding a luminescence quantum yield of 28.9%. In the $30\,cm$-long fiber, a gain of $5.7dB$ is determined after accounting for the $3.2dB/m$ loss of the fiber. Over the duration of about 10^4 pump laser shots at $355nm$ (tripled YAG) with an energy flux of $3.02\,mJ/mm^2$, no measurable degradation of the fiber was reported. As such, the performance of this particular fiber is comparable to silica glass technologies.

6.3.5 Applications

As illustrated by the large number of demonstrations described above, the ASE process in a polymer optical fiber can be used to make light sources and amplifiers. Graded-index amplifiers that operate in the visible are well suited for fiber-to-the-home applications. Telecom applications, on the other hand, which use small-core single mode fibers need fiber amplifiers whose cores match the telecom fiber's diameter and numerical aperture.

Wavelength division multiplexing (WDM) is used in high bandwidth telecom systems to increase bandwidth. The underlying principle is simple. If one particular wavelength can carry a certain bandwidth, B, then N different wavelengths, when propagating together in a fiber core, will have a bandwidth NB. Wavelength division multiplexing is simple to implement because sources at different wavelengths provide the light while demultiplexing requires only passive components such as Bragg fiber gratings for spectral separation. So in a WDM system, an amplifier system must meet the additional criteria that the gain be broad enough to amplify all of the frequency components with approximately the same gain. (Dense WDM — or DWDM — refers to systems that pack the most number of wavelengths into a fiber.)

Single-mode dye-doped polymer optical fibers can meet the requirements as light sources, broad-gain amplifiers, and demultiplexers. Clearly, the large number of dyes available can meet the requirements of making sets of sources at specific ranges of wavelengths. Additionally, as we saw above, the wavelength of emission of a particular dye can be tuned. As we will see in Section 8.1, Bragg gratings can be made in fibers; and, they can be tuned mechanically.

Perhaps the most unique advantage of a dye-doped polymer is the fact that it can be impregnated with several different dyes simultaneously to make a broad-band amplifier. Figure 6.29 shows an example where the gain peaks of each dye overlap. The net gain is then given by the broader curve. The vertical dashed lines show the position of the wavelength division multiplexed channels. As such, a much broader

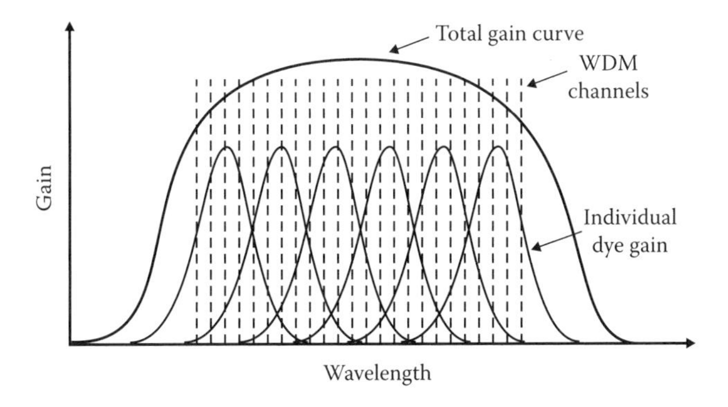

FIGURE 6.29 Several individual gain curves of different molecules add to yield a broader gain curve. Vertical dashed lines show WDM channels.

range of wavelengths can be amplified. In making a multiple-dye amplifier, the choice of dyes used and their concentrations need to each be carefully adjusted to make the gain curve flat over the wavelength range of operation. Given the broad range of dyes available, controlling the gain profile should be, in principle, straightforward.

Given the flexibility afforded by dye-doped polymers, they can be used to make a large number of fiber components. It is not difficult to imagine polymer optical systems that use polymers in many or all of their components. The biggest hurdle to using polymers in active device applications is photostability. Given the fast pace of improvements in polymeric materials, and reports of the resistance of dyes to photodegradation, polymer systems may one day be the material of choice for photonics applications.

7 Optical Switching

Optical fibers are universally used to transmit information in systems that operate at high bandwidths. It would therefore be an attractive option to have a pallet of active inline fiber devices available as components to build larger systems. In fact, all-optical switches and logic elements had been a popular and intense area of research in the 1980s. The lull in activity in the 1990s and through the middle of the 2000s is based on a combination of technical factors and the business cycle. Since all-optical fiber devices still offer the highest level of performance, they are bound to show up in fiber-optic systems in the future. The large nonlinear-optical susceptibilities of dye-doped polymers thus make single-mode polymer optical fibers a natural choice for inline devices.

In this chapter, we review the underlying principles of optical switching, and how polymers can play a role in making devices. Also discussed are measurement techniques for characterizing fibers and materials, device designs, and applications — such as optical multistability.

7.1 ELECTROOPTIC SWITCHING

The electrooptic effect is a second-order nonlinear-optical process where a static electric field induces a change in phase of an optical field. For historic reasons, the change in the refractive index is related to the field in a somewhat cumbersome way:

$$\Delta n = \frac{n_0^2}{2} r E, \tag{7.1}$$

where E is the applied static electric field, r is the electrooptic coefficient, and n_0 the zero-field refractive index. Figure 7.1 shows that the angle of deflection of a light beam in a prism made from an electrooptic material depends on the voltage applied.

The degree of bending depends on the wavelength of the light as well as its polarization. As such, the electrooptic coefficient is a wavelength-dependent tensor and the electrooptic effect should be expressed as

$$\Delta n_{ij}(\omega) = \frac{\left[n_{ij}^{(0)}\right]^2}{2} r_{ijk}(\omega) E_k(0), \tag{7.2}$$

where i, j, k are indices representing the Cartesian components and the k indices are summed (according to Einstein summation convention). The i and j indices are not summed. r is a third-rank tensor while the refractive index is a second-rank tensor. $E_k(0)$ is the static electric field that is applied to the sample.

The electrooptic prism is essentially an optical switch, where an applied voltage can redirect a beam of light to multiple optical fibers. The deflection angle,

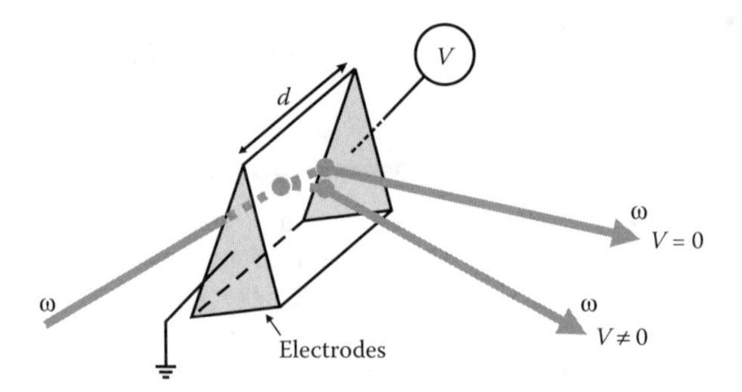

FIGURE 7.1 The refraction angle of a beam of light can be changed by applying a voltage across a prism made of an electrooptic material.

however, is so small that such a device would not be practical. Most designs for electrooptic switching devices rely on interference effects that leverage the small electrooptic-induced refractive index change.

It is instructive to estimate the change in angular deflection that one would expect for an electrooptic process. For example, consider a typical dye-doped polymer or inorganic crystal with an electrooptic coefficient of $10 pm/V$. (Note that $1\ pm/V = 10^{-12} m/V$). The angle of refraction, θ_r, of a beam exiting a material of refractive index n is given by Snell's Law,

$$n \sin\theta_i = \sin\theta_r, \tag{7.3}$$

where θ_i is the angle of the beam with respect to the surface normal inside the material. For this illustration, we will assume that the material is isotropic so that the tensor nature of r can be neglected. If the refractive index changes (but the incident angle is fixed), the angle of refraction changes, which we can calculate by differentiating Equation (7.3),

$$\Delta\theta_r = \Delta n \left(\frac{\sin\theta_i}{\cos\theta_r} \right) = \Delta n \left(\frac{\sin\theta_i}{\sqrt{1 - n^2 \sin^2\theta_i}} \right), \tag{7.4}$$

where the term to the right of the last equality is determined using Snell's Law, Equation (7.3).

Equation (7.4) shows that the change in the angle of the refracted ray depends on the angle of incident. The angle at which the angle of refraction is the greatest is simply the critical angle, where the denominator in Equation (7.4) vanishes. Unfortunately, at the critical angle, most of the light is reflected so that the intensity of transmitted light is small. So, for our estimate, let's pick an angle that is intermediate between the critical angle and normal incidence, or $n^2 \sin^2\theta_i = 0.5$. For a refractive index of $n = 1.5$, this yields a change in the angle of refraction on the order of $\Delta\theta_r = 0.67\Delta n$. For a 50V potential difference across a $2\mu m$ thick sample (an electric field of $2.5 \times 10^7 V/m$), which for $r = 10^{-11} m/V$ with Equation (7.1) yields $\Delta\theta_r = 1.87 \times 10^{-4} rad = 0.011°$. This is too small for making a practical device.

7.1.1 Electrooptic Modulation

An electrooptic modulator is one in which an applied voltage induces a phase shift in a beam of light. The best way to test such a modulator is to place it inside one arm of an interferometer and to observe the change in the intensity exiting the interferometer as a function of the applied voltage. Indeed, the electrooptic coefficient of polymeric materials is often characterized using an interferometer. In this section, we begin by presenting the theory behind electrooptic modulation in a Mach-Zehnder interferometer. Subsequently, an experimental technique to measure the electrooptic coefficients is introduced, culminating in an application of this technique to a polymer optical fiber. We note that the theory, as derived in Section 7.1.1, is rigorous and messy. If the reader has no interest in derivations but needs to learn only about the details of doing the experiments and measuring the electrooptic coefficients, (s)he should proceed to Section 7.1.1.2.

7.1.1.1 Theory of Electrooptic Modulation in a Mach-Zehnder Interferometer

Figure 7.2 shows a Mach-Zehnder interferometer experiment that is used to measure phase modulation in a thin film.[107–109] While there are other experimental techniques for doing so, [110] the Mach-Zehnder technique has been the only one used to measure electrooptic polymer optical fiber.

The light intensity exiting the laser is controlled with a half-wave plate and polarizer. The light passes through a chopper, then it is separated into two equal intensities with a beam splitter. One of these beams passes through the sample while the second beam passes through a wedge. The two beams are then recombined

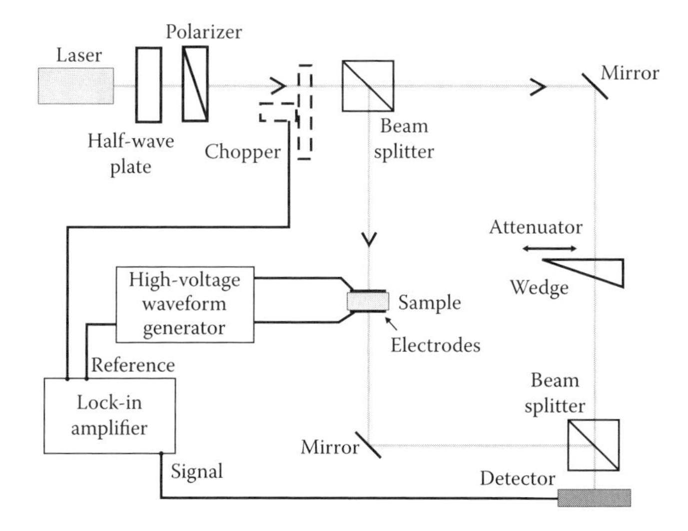

FIGURE 7.2 The Mach-Zehnder experiment for measuring phase modulation.

so that they are on top of each other. This yields a bull's-eye pattern of light and dark rings. The detector is placed at the central region of the pattern, where the intensity is brightest for the constructive interference condition. Any phase difference between the two arms results in a change in the intensity measured by the detector.

The experiment is performed in two steps. First, with the voltage off and the chopper on, the intensity is measured at the interferometer as a function of the postilion of the wedge. The lock-in amplifier is synched to the chopper to remove background noise from ambient light. This yields a sinusoidal intensity dependence. Next, the chopper is turned off and a sinusoidal voltage is applied to the sample. Through the electrooptic effect, this yields a sinusoidal phase modulation. When the phase-modulated beam interacts with the other beam, a sinusoidal modulation in the intensity illuminates the detector. The lock-in amplifier, which is synched with the high-voltage waveform, measures the amplitude of this oscillation. Since the phase shift is usually small, the amplitude of intensity modulation is small compared with the laser power. It is not uncommon for the signal to be 5 orders smaller than the background. As such, the lock-in amplifier is essential for picking out the signal. Note that because the signal is so small, care must be taken to prevent electromagnetic pickup in the cables leading to the lock-in amplifier. Even small amounts of pickup can overwhelm the signal.

While the intensities in both arms of the interferometer are ideally the same, experimentally this is not always true. The light path inside the interferometer that includes the sample we call the sample arm and the other path, the reference arm. Let's define the electric fields at the combining beam splitter in the reference arm as $\vec{E}_1$ and the field in the sample arm as $\vec{E}_2$ with incoming field amplitudes E_{01} and E_{02}, respectively. After passing through the sample, the field E_2 is given by

$$E_2 = E_{02} \exp\left[i\left(\phi_2 + \frac{2\pi d(n-1)}{\lambda}\right)\right], \tag{7.5}$$

where λ is the wavelength, d the sample thickness, ϕ_2 the phase of the light with no sample present, and n the refractive index of the sample. (ϕ_2 is real and n is complex.) The light in the reference arm is given by

$$E_1 = E_{01} \exp[i\phi_1], \tag{7.6}$$

where ϕ_1 is the phase in the reference arm, which can be adjusted by translating the wedge perpendicular to the beam. If the wedge is transparent, ϕ_1 is real. We assume this to be so.

The refractive index of the sample, n, is a function of the applied static field given by Equation (7.2). To use the lock-in amplifier, the applied voltage is modulated at a frequency Ω, which is much smaller than the optical frequency ω. As such, the applied electric field can be considered static. The magnitude of the quasi-static electric field, $E(0)$, is given by

$$E(0) = 2^{1/2}\frac{V_{rms}}{d}\cos(\Omega t), \tag{7.7}$$

where d is the sample thickness and V_{rms} the root-mean-square voltage. Since we will be dealing with poled polymers — where the dopant molecules are polar aligned inside the polymer — the samples are approximately isotropic provided that the degree of orientation is not too large. The zero-field refractive index tensor will also be isotropic.

If the two light beams remain similarly polarized after being recombined at the second beam splitter, they will add and result in interference. Using Equations (7.6) and (7.5), the square of the magnitude of the electric field gives the intensity,

$$|E_1 + E_2|^2 = E_{01}^2 + E_{02}^2 \exp\left[-4\pi n_I \frac{d}{\lambda}\right] + 2E_{01}E_{02} \exp\left[-2\pi n_I \frac{d}{\lambda}\right] \times \cos\left([\phi_2 - \phi_1] + \frac{2\pi d\,(n_R - 1)}{\lambda}\right), \tag{7.8}$$

where we have expressed the refractive index in terms of its real and imaginary parts, $n = n_R + i n_I$. The refractive index is of the form given by Equation (7.2), so Equation (7.8) includes the intensity dependent terms. All of the information that is required to determine the complex tensor component of r is given by Equation (7.8). The trick is to measure all of the other quantities in this equation so that we can solve for r.

The tensor nature of the electrooptic coefficient complicates the calculations unless the optical and static field are each along one of the principle axes of r. In such cases, only one tensor component will contribute, so the scalar form for Δn — as defined by Equation (7.1) — can be used. For example, if both fields are along the 3-axis, then r_{333} will be the only component that is probed. If the static and electric fields are perpendicular to each other, then only β_{iij} will need to be considered. (If the material has ∞_{mm} symmetry — as is the case for isotropic dye-doped poled polymers, all the tensor components of the form β_{iij} will be equal for $i \neq j$ when j is along the symmetry axis.) With this in mind, we will express all quantities as scalars with the understanding that the tensor component to be used depends on the geometry. Note that in Figure 7.2, E_0 (which, let's say, is along z) is perpendicular to the polarization of the electric field so r_{xxz} is being measured.

The electrooptic measurement is performed in two consecutive steps. First, the absolute intensity and contrast of the interferometer is determined from a zero-field phase sweep. Next, the chopper is turned off and a sinusoidal voltage is applied to the sample. The amplitude of the signal is measured with a lock-in amplifier as a function of phase difference between the two arms. The curves obtained in these two measurements are then used to determine all of the parameters required to solve Equation (7.8) for r. The theory behind the measurements follows.

STEP ONE — Chopper ON, Voltage OFF

The intensity, which is what we measure, is proportional to $|E|^2$ — where the constant of proportionality depends on the units. Multiplying Equation (7.8) by this constant and setting the applied voltage to zero, we get

$$I_0 = I_{01} + I_{02} \exp\left[-4\pi n_{0I} \frac{d}{\lambda}\right] + 2\sqrt{I_{01} I_{02}} \exp\left[-2\pi n_{0I} \frac{d}{\lambda}\right] \times \cos\phi, \quad (7.9)$$

where we have defined the phase,

$$\phi = [\phi_2 - \phi_1] + \frac{2\pi d\,(n_{0R} - 1)}{\lambda}, \quad (7.10)$$

where n_{0R} and n_{0I} are the real and imaginary parts of the zero-field refractive index.

With the chopper on and the voltage off, the signal measured by the lock-in amplifier as a function of ϕ is a sinusoidal function. We call this function an interferogram. From Equation (7.9), the maximum and minimum values of this interferogram are given by

$$I_{max,min} = I_{01} + I_{02} \exp\left[-4\pi n_{0I} \frac{d}{\lambda}\right] \pm 2\sqrt{I_{01} I_{02}} \exp\left[-2\pi n_{0I} \frac{d}{\lambda}\right]. \quad (7.11)$$

The amplitude of the oscillation is given by half the difference between the maximum and minimum intensity values:

$$\frac{I_{max} - I_{min}}{2} = 2\sqrt{I_{01} I_{02}} \exp\left[-2\pi n_{0I} \frac{d}{\lambda}\right]. \quad (7.12)$$

The mean value of the interferogram is the average of the maximum and minimum values, and is given by

$$\frac{I_{max} + I_{min}}{2} = I_{01} + I_{02} \exp\left[-4\pi n_{0I} \frac{d}{\lambda}\right]. \quad (7.13)$$

To get an absolute measurement of r, it is important to correctly understand what it is that the lock-in amplifier is measuring when the chopper is used. Recall that the lock-in amplifier measures the root-mean-square intensity $< I >$. Because the peak-to-peak amplitude is I (the light is chopped from fully on to fully off), the amplitude is $I/2$. A second complication is that the waveform is a square wave, which, based on how a lock-in amplifier works, adds a factor of $\sqrt{2}$. Thus, we find that the lock-in measures $< I >= I/2^{3/2}$ so that with the help of Equation (7.12), we get

$$\begin{aligned} \frac{I_{max} - I_{min}}{2} &= \frac{1}{2} \cdot 2^{3/2} \left(\langle I_{max}\rangle - \langle I_{min}\rangle\right) \\ &= 2\sqrt{I_{01} I_{02}} \cdot \exp\left[-2\pi n_{0I} \frac{d}{\lambda}\right]. \end{aligned} \quad (7.14)$$

Note that in this section, the brackets will denote the reading by the lock-in amplifier. Equation (7.13) can be similarly expressed in terms of the intensity measured by the lock-in,

$$\frac{I_{max}+I_{min}}{2} = \frac{1}{2}\cdot 2^{3/2}\left(\langle I_{max}\rangle + \langle I_{min}\rangle\right)$$
$$= I_{01} + I_{02}\exp\left[-4\pi n_{0I}\frac{d}{\lambda}\right]. \qquad (7.15)$$

While the lock-in amplifier reads the voltage of a detector, we will continue to express the reading as an intensity. In the final analysis, only intensity ratios are required to determine r, so as long as the detector is linear (the voltage is linearly proportional to the intensity), we can be sloppy with our our notation.

STEP TWO — Chopper OFF, Voltage ON

The chopper is turned off and a sinusoidal voltage at frequency Ω is applied to the sample. The lock-in amplifier is synchronized to this modulation frequency. According to Equation (7.8), a change in the real or imaginary parts of the refractive index will yield a change in the output intensity. Here we will make the assumption that the change in the complex refractive index leads to only a small phase change. Second, since we will be measuring the signal at the voltage source frequency Ω, we will only keep terms in the intensity that have a Fourier component at Ω. We apply these approximations to express the intensity as follows:

$$I = I_0 + \left(\left.\frac{\partial I}{\partial(\Delta n_R)}\right|_{\Delta n=0}\right)\Delta n_R + \left(\left.\frac{\partial I}{\partial(\Delta n_I)}\right|_{\Delta n=0}\right)\Delta n_I, \qquad (7.16)$$

where we have defined the intensity dependent refractive index as Δn, so $n = n_0 + \Delta n = n_0 + \Delta n_R + i\Delta n_I$ and I_0 is given by Equation (7.9). Note that because we are only interested in a change in the intensity that is due to a refractive index change, we will ignore the static intensity I_0.

Differentiating Equation (7.8), substituting the result into Equation (7.16), and ignoring I_0, we get

$$I(\Delta n) = -4\pi\frac{d}{\lambda}\cdot\exp\left[-4\pi n_{0I}\frac{d}{\lambda}\right]\Delta n_I I_{02}$$
$$-4\pi\frac{d}{\lambda}\sqrt{I_{01}I_{02}}\exp\left[-2\pi n_{0I}\frac{d}{\lambda}\right]\times\{\Delta n_I\cos\phi + \Delta n_R\sin\phi\}. \qquad (7.17)$$

We eventually need to solve for the refractive index change, Δn, in terms of the measured quantities. As such, we need rewrite Equation (7.17) in a more experimentally friendly form. First, we recognize that Equations (7.14) and (7.15) can be used to eliminate the two exponentials in Equation (7.17). Furthermore, we can express the term in brackets in the more convenient form by using a simple trigonometric identity:

$$\{\Delta n_I\cos\phi + \Delta n_R\sin\phi\} = |\Delta n|\cos\left(\phi - \tan^{-1}\left[\frac{\Delta n_R}{\Delta n_I}\right]\right). \qquad (7.18)$$

Substituting Equations (7.14), (7.15), and (7.18) into Equation (7.17), we get

$$I(\Delta n) = -4\pi \frac{d}{\lambda} \cdot (\sqrt{2}\,[\langle I_{max}\rangle + \langle I_{min}\rangle] - I_{01})\Delta n_I$$
$$-4\pi \frac{d}{\lambda}\left(\frac{\langle I_{max}\rangle - \langle I_{min}\rangle}{\sqrt{2}}\right) \times |\Delta n| \cos\left(\phi - \tan^{-1}\left[\frac{\Delta n_R}{\Delta n_I}\right]\right). \quad (7.19)$$

A static voltage applied to the sample results in a refractive index change, which leads to a change in the intensity leaving the interferometer, as given by Equation (7.19). With the applied voltage fixed, a scan of ϕ results in a sinusoidal signal whose amplitude is proportional to $|\Delta n|$ with a vertical offset proportional to the imaginary part of the refractive index change, Δn_I. In addition, the phase offset gives information about the ratio of the real to the imaginary part of the refractive index change.

Since the refractive index change is small (the resulting intensity change is much smaller than background light levels), a homodyne technique must be used. A sinusoidally varying applied voltage leads to a refractive index that is also sinusoidally varying. This in turn yields a sinusoidally varying intensity, which is read by a lock-in amplifier. Recall that $\Delta n = \Delta n_r + i\Delta n_I$, which for a sinusoidal voltage is given by Equation (7.7). Using Equation (7.1) and Equation (7.7), we get

$$\Delta n = \Delta n_r + i\Delta n_I = \frac{1}{2^{1/2}}\frac{V_{rms}}{d} n_0^2 \cos(\Omega t)\{r_R + i r_I\}$$
$$= \frac{1}{2^{1/2}}\frac{V_{rms}}{d} \cos(\Omega t)\left\{\left(\left[n_{0R}^2 - n_{0I}^2\right] r_R - 2n_{0R}n_{0R}\cdot r_I\right)\right.$$
$$\left. + \left(\left[n_{0R}^2 - n_{0I}^2\right] r_I + 2n_{0R}n_{0I}r_R\right) i\right\}, \quad (7.20)$$

where r_R and r_I are the real and imaginary parts of r; and where we have used $n_0^2 = n_{0R}^2 - n_{0I}^2 + 2i n_{0R} n_{0I}$. The idea is to substitute Equation (7.20) into Equation (7.19), then to calculate the signal at the frequency of the applied voltage, which is the quantity that is read by the lock-in. Since Equation (7.20) is quite messy, we will assume that the linear refractive index n_0 is mostly real — which is equivalent to assuming that the sample is reasonably transparent. It is straightforward to generalize the equations that follow to the complex n_0 case, and that is left to the reader should the need arise. The one wrinkle in this process is that to solve for the complex electrooptic coefficient, the real and imaginary parts of n_0 must be known *a priori*.

Under the assumption of real n_0, substituting Equation (7.20) into Equation (7.19), and taking the root-mean-square (rms) intensity at frequency Ω, we get

$$I_{rms}(\Omega) = -2\pi n_0^2 \frac{V_{rms}}{\lambda} \cdot (\sqrt{2}[\langle I_{max}\rangle + \langle I_{min}\rangle] - I_{01}) r_I$$
$$-2\pi n_0^2 \frac{V_{rms}}{\lambda}\left(\frac{\langle I_{max}\rangle - \langle I_{min}\rangle}{\sqrt{2}}\right)$$
$$\times \sqrt{r_R^2 + r_I^2} \cos\left(\phi - \tan^{-1}\left[\frac{r_R}{r_I}\right]\right). \quad (7.21)$$

It is interesting to note that Equation (7.21) is independent of sample thickness, d. The thickness cancels because the phase shift increases with increased d, but the field decreases as $1/d$. This result is convenient because it is not necessary to measure the sample thickness, which is a difficult task to do accurately when the sample is thin. Equations (7.21) and (7.11) are all that one needs to calculate the electrooptic coefficients.

7.1.1.2 Experimental Technique to Measure Electrooptic Coefficients

In this section, the reader is directed through the process of measuring an electrooptic coefficient of a thin film. After the interferometer is constructed as shown in Figure 7.2, the optics need to be adjusted to get a bull's-eye pattern at the detector. The most crucial component that needs to be aligned is the re-combining beam splitter. The first step is to direct the two beams that are to be interfered at the second beam splitter so that both beams cross at the interface of the beam splitter, as shown in the lefthand portion of Figure 7.3. The two mirrors in the interferometer are placed on tilt mounts to allow the beams to be steered. (Note that the two beams are shown in different shades and thicknesses so that the two can be easily differentiated.) One method for doing so is to first place a card at the approximate position of the beam splitter to make sure the two beams cross in space. Also, the two beams should be adjusted to be parallel to the optical table. Then, the beam splitter is placed on a translation stage and moved into place until its interface matches the position of the crossing point (the experienced researcher can do this by hand). At the detector, a set of bright lines, called interference fringes, can be observed. This is shown in the bottom part of the figure. The fringes are closer together when the interferometer is more out of alignment.

The beam splitter should be rotatable about an axis that is perpendicular to the two beams at the crossing point. The simplest method is to mount it on a rotation stage. The beam splitter can also be rotated by hand without a stage (the mount holding the beam splitter can be slid around the optics table surface and then mounted in place when the correct position and orientation is found). The beam splitter is rotated in a direction that brings the beams together, as shown in the righthand portion of Figure 7.3. As the interferometer gets closer to being tuned,

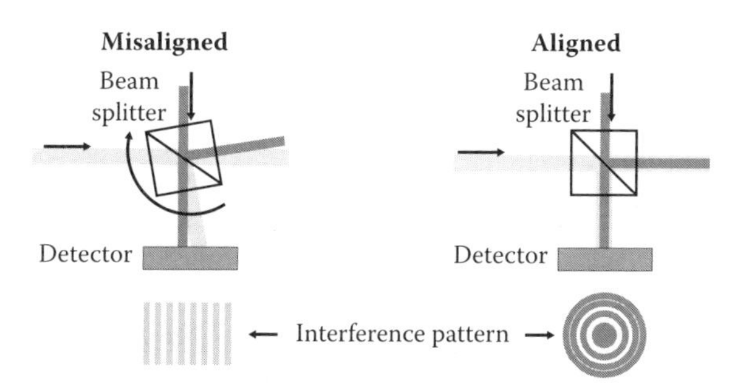

FIGURE 7.3 Aligning the second beam splitter in an interferometer.

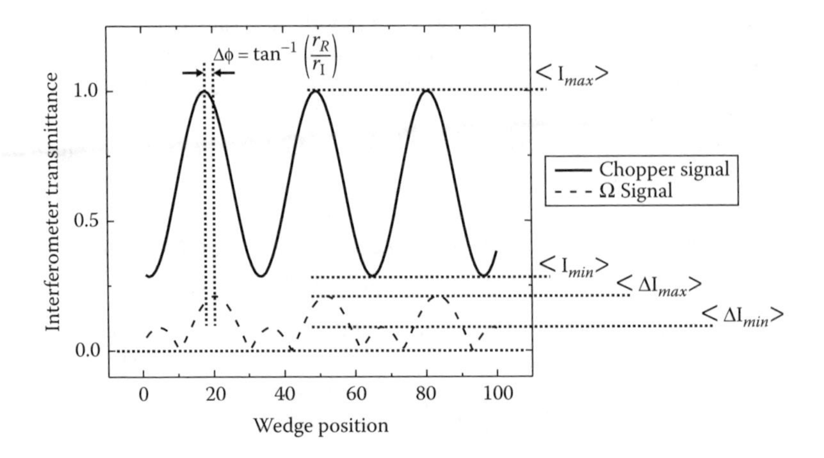

FIGURE 7.4 Theoretical prediction of the measured signal with the chopper on and no voltage (top) as a function of the phase; and the signal at the modulation voltage frequency as a function of the phase with the chopper off (bottom).

the interference lines get further apart and become curved. Finally, the bull's-eye pattern emerges. A pinhole, placed in front of the detector, can be used to ensure that the detector is only viewing the center of the bull's eye.

In the first step of the experiment, the chopper is turned on with no voltage applied to the sample. The lock-in is synched to the chopper and measures the intensity leaving the interferometer as a function of the phase, which is adjusted by translating the wedge. The top portion of Figure 7.4 shows a sketch of the typical data expected according to Equation (7.9). From this run, the maximum, $\langle I_{max} \rangle$, and minimum, $\langle I_{min} \rangle$, intensities are determined from the plot.

Next, the chopper is turned off and the voltage, at frequency Ω, is applied to the sample. The lock-in, which is synched to the applied voltage, measures the amplitude of the modulated intensity as a function of the wedge position. A sketch of typical data as predicted by Equation (7.21) appears in the bottom plot of Figure 7.4. In this plot, we assume that the lock-in is set to measure the *magnitude* of the signal. As such, since Equation (7.21) is a sinusoidal function with a vertical offset, the magnitude of the signal appears in a double humped structure. The heights of each peak, $\langle \Delta I_{min} \rangle$ and $\langle \Delta I_{max} \rangle$, can be determined from this plot.

When both runs are plotted on the same figure, the phase shift between the two curves can be determined from the horizontal offset between the peaks, $\Delta\phi$ — as shown by the vertical dotted lines in Figure 7.4. Note that to calculate the offset, we need to use the electrooptic modulation peak that is π out of phase with the applied voltage. This will correspond to the peak for which the lock-in reads a negative (or greater than $\pi/2$) phase. In our illustration, we have assumed that the larger peak has a relative π phase. In real experiments, the smaller peak may be the appropriate one for measuring the phase shift.

Finally, the *rms* voltage that is applied to the sample, V_{rms}, can be measured with a voltmeter provided that it is designed to operate in the frequency range of the voltage waveform. With these measurements in hand, we can determine r_R and r_I as follows.

The amplitude of the electrooptic modulation signal, I_{sig}, is calculated from the data according to

$$I_{sig} = \frac{\Delta I_{max} + \Delta I_{min}}{2}. \tag{7.22}$$

From Equation (7.21), the amplitude of the electrooptic modulation signal is of the form

$$I_{sig} = 2\pi n_0^2 \frac{V_{rms}}{\lambda} \left(\frac{\langle I_{max} \rangle - \langle I_{min} \rangle}{\sqrt{2}} \right) \times \sqrt{r_R^2 + r_I^2}. \tag{7.23}$$

With the help of Equation (7.22), we can solve Equation (7.23) for the magnitude of the electrooptic coefficient:

$$\boxed{|r| = \sqrt{r_R^2 + r_I^2} = \frac{\sqrt{2}\lambda}{4\pi n_0^2 V_{rms}} \cdot \frac{(\Delta I_{max} + \Delta I_{min})}{(\langle I_{max} \rangle - \langle I_{min} \rangle)}.} \tag{7.24}$$

Equation (7.24) expresses the magnitude of the electrooptic coefficient in terms of the heights of the peaks and troughs measured from the data, the applied voltage, and the wavelength of the light source.

Since the measured offset, $\Delta\phi$, gives the ratio of the real to imaginary parts of r, we can use this along with Equation (7.24) to get both the real and imaginary parts. Given that $\tan \Delta\phi = r_R / r_I$, we get

$$\boxed{r_R = |r| \sin \Delta\phi,} \tag{7.25}$$

and

$$\boxed{r_I = |r| \cos \Delta\phi.} \tag{7.26}$$

Note that when the imaginary part of the electrooptic coefficient vanishes ($r_I = 0$), the phase offset becomes $\Delta\phi = \pi/2$. If $r_R = 0$, the phase offset is $\Delta\phi = 0$.

It is interesting to note that the problem is overspecified since there are other ways to get the electrooptic coefficients. For example, the first term in Equation (7.21) (i.e., the first line of the equation) provides an offset to the sinusoidal function of the phase that is given by the remaining term. Notice that this offset is related to only the imaginary part of r. From Figure 7.4, we can see that the offset, δ, is simply given by

$$\delta = \frac{\Delta I_{max} - \Delta I_{min}}{2}. \tag{7.27}$$

Applying Equation (7.27) to the "offset" term in Equation (7.21) and solving for r_I, we get

$$r_I = -\frac{\lambda}{4\pi n_0^2 V_{rms}} \cdot \frac{\Delta I_{max} - \Delta I_{min}}{(\sqrt{2}\,[\langle I_{max} \rangle + \langle I_{min} \rangle] - I_{01})}. \tag{7.28}$$

All of the intensities in Equation (7.28) are measured from the interferograms (Figure 7.4) except for I_{01}. This quantity can be measured by blocking the sample arm with the chopper on and voltage off, and measuring the intensity at the detector. Thus, Equation (7.28) can also be used to get r_I and the result compared with Equation (7.26).

The above procedure works in a perfect world. In real experiments, electromagnetic (EM) pickup due to the high-voltage waveform generator is picked up by the cables and added to the signal. Since EM pickup is independent of phase as adjusted by the wedge, the measure of the magnitude of r using Equation (7.24) is the most reliable. However, since EM pickup leads to an offset, Equation (7.28) is the least reliable. In order to do a clean experiment, it is important for the experimenter to shield all of the cables. Even the most vigilant use of shielding never totally eliminates background. So, materials with large r result in the most accurate measurements. In the process of separating the real and imaginary parts of r from the phase offset, it is important that the researcher check that the phase measured by the lock-in amplifier is reliable. This can be a problem if EM pickup affects the phase of the electrooptic modulation signal. In these cases, a voltage dependent measurement can be used to isolate the problem. If r is found to be independent of applied voltage, then the experiment is more likely to be reliable.

We note that for the third-order process in which the change in the refractive index is proportional to the square of the applied voltage, a similar measurement and theoretical approach can be used to determine this higher-order coefficient. This calculation will not be presented here. A detailed discussion of third-order measurements as well as the microscopic mechanisms that lead to quadratic electrooptic effects in polymers can be found in the literature.[111] The intensity-dependent refractive index, also a third-order nonlinear-optical process, is discussed in Section 7.2 of this chapter.

7.1.1.3 Electrooptic Modulation in a Polymer Optical Fiber

A means for applying a voltage to a polymer optical fiber is required to make an electrooptic device. Since a large applied electric field is needed to get an appreciable change in refractive index, it is best if the electrodes are close together. So, the best way to apply a large field to the core of a fiber is to embed the electrodes in the fiber on either side of the core. Figure 7.5 shows a sketch of an electrooptic fiber that was made by Welker and coworkers.[112]

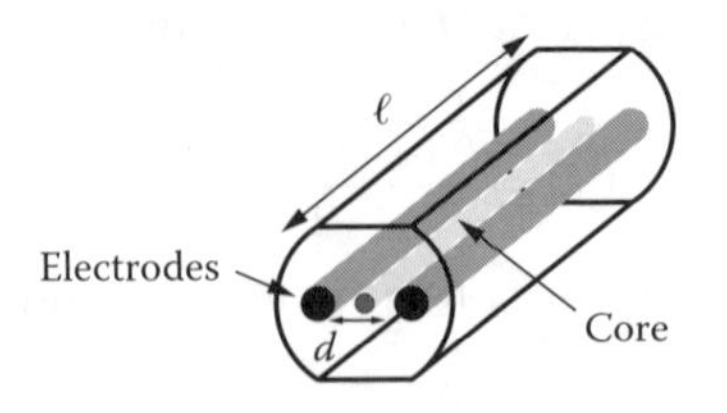

FIGURE 7.5 A schematic diagram of an electrooptic fiber.

In the previous section, we found that the signal was independent of the thickness of the sample. This was so because the electric field was applied along the propagation path of the light beam. In the electrooptic fiber, the phase shift will be larger by a factor of ℓ/d, where ℓ is the fiber length and d the spacing between electrodes. Since in a single-mode polymer optical fiber, d can be as small as 40μ (if the electrodes get too close to the core, the evanescent field of the light will interact with the electrodes, and increased optical loss will result), and the length of propagation can be made tens of centimeters, the amount of intensity modulation can be increased by a factor of over 1,000 over the longitudinal film geometry. This makes electrooptic fibers attractive components in an optical system. A channel waveguide device (as are commonly made on silicon wafer substrates) has a similar advantage (even though propagation lengths are somewhat shorter than in a fiber). The main advantage of an all-fiber device is that connecting it to another fiber is relatively easy. Planar devices require an expensive pigtailing operation. Furthermore, polymer fibers tend to have lower loss than planar devices due to the smoother surfaces that result from the fiber drawing process.

Since polymers are processed at low temperatures, an electrode with low melting temperature must be used. Welker and coworkers used an indium alloy. The preform was made with the dry process described in Chapter 3. Both the electrodes, which are $2mm$ in diameter, and the core, which is $800\mu m$ in diameter, are placed in semi-cylindrical groves that are milled into a PPMA half-round, and they are squeezed together at elevated temperature to fuse the preform. The diameter of the final preform is $12.7mm$ and the length is about $10cm$. Figure 7.6 shows the process and a photo of a preform that was made with a DR1 dye-doped PMMA core in a PMMA cladding. The flat surfaces are used for alignment, as described later.

The drawing parameters are affected by the presence of the indium: Heat conduction along the electrodes changes the temperature profile at the neck-down

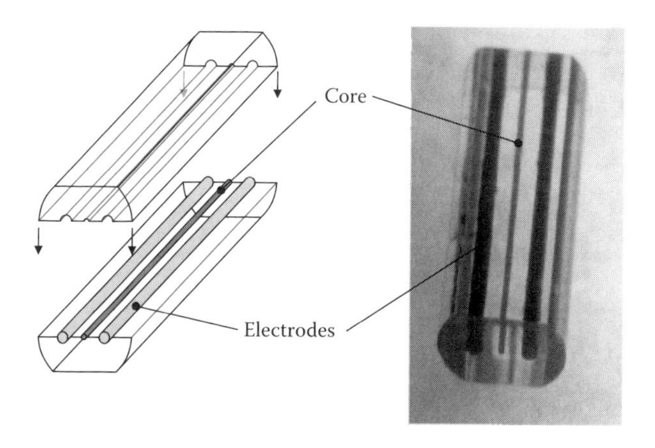

FIGURE 7.6 Fabrication of an electrooptic fiber preform (left) and photograph of a complete preform with a DR1-doped PMMA core (right). Adapted from Ref. [112] with permission of the Optical Society of America.

region. Increasing the temperature to compensate can cause the electrodes to flow out of the preform. Lower temperatures, however, make the indium stiff, resulting in delamination between the indium and the cladding. As such, getting the process to work well requires a tedious period of trial and error. In Welker's process, they estimated the preform temperature in the neck-down region to be about 230°C.

Since polymers are usually isotropic with respect to polar order, they need to be poled to align the dipole moments of the chromophores to impart a second-order susceptibility. For the dyes to freely orient in the presence of the applied electric field, the sample needs to be heated above the polymer's glass transition temperature, then cooled with the field remaining on to lock in the polarization as the polymer hardens. When a finished polymer fiber is poled at elevated temperature, the softened material can flow leading to deformation of the fiber. It is therefore best to pole the material while the fiber is being drawn. To do so, however, requires that leads be attached to the electrodes. Indium is a soft metal, so thin wires can be inserted into the end of the preform to make contact with the electrodes.

Figure 7.7a shows a schematic diagram of an electrooptic fiber being drawn. The electrodes are attached to a high-voltage power supply that poles the dye molecules in the core at a voltage V that is applied. A picoammeter is placed in series with the power supply to measure the current. (To protect the ammeter, a large resistance should also be placed in series with the sample. This resistor, however, should have an impedance, R, that is smaller than the fiber so that most of the voltage drop is across the fiber.) Since the resistivity of PMMA is high (for a thin film, the resistance is over $10^{10}\Omega$[113]), any decrease in impedance is a sign that there is a problem with the poling process. A catastrophic short yields a huge jump in the current, which is limited by the resistor to a current given by V/R. Figure 7.7 shows the current measured as a function of time as the fiber is being drawn. A voltage of $1 kV$ is applied to the electrodes for the first 2 minutes.

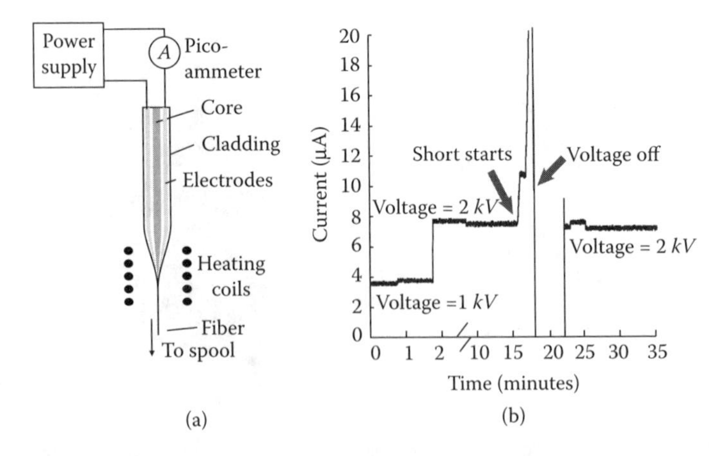

FIGURE 7.7 Drawing an electrooptic fiber preform. Adapted from Ref. [112] with permission of the Optical Society of America.

The voltage is then increased to $2V$, and the current increases by about a factor of two as would be expected for an ohmic material. At about 15 minutes, the current spikes, indicating a short between the electrodes. At this point, the voltage is turned off for about 5 minutes. At about 22 minutes into the drawing process, the voltage is turned back on and the current is comparable to its value before the short occurred. From this data, it appears that the short has repaired itself.

The current measured during the drawing process is most likely concentrated in the neck-down region where the polymer temperature is the highest. From the measured current and applied voltage, the electrooptic fiber acts like a resistor with $R = 2.5 \times 10^8$. If measured at the same temperature, we would expect that the resistance would be higher than for the thin film measurements reported by Zimmerman and coworkers [113] since the separation between the electrodes in the fiber is tens of microns in the neck-down region — compared with about $5\mu m$ for the thin films. However, we find that the resistance is much smaller. The resistivity of the polymer drops with increased temperature, so it is reasonable that the resistivity would drop by several orders of magnitude when the temperature is so far above the transition temperature — as it is in the neck-down region.

So, the current measurement is a direct probe of the neck-down region. The voltage spike found in Figure 7.7 is a clear indication of a short. While the voltage is applied, the resulting current flowing through the defect can result in further irreversible damage. When the voltage was turned off, the material seemed to be repaired by the time the voltage was turned on again. There are two explanations for why this happens while the voltage is off. First, because the material is well above the glass transition temperature, when the voltage is turned off, the material may flow and heal. Second, as the defect moves further and further away from the neck-down region, it cools and the resistivity goes up.

Once the preform is exhausted, about $1km$ of fiber has drawn onto a spool. The fiber is cut into 10 to $20cm$ pieces and the ends are polished. Attaching leads to the fiber, which are needed for connecting the device to an external source, is a challenging task since the electrodes are on the order of $100\mu m$ in diameter. As such, the fiber cladding must be machined down to the electrode. If the machining proceeds too far into the electrode, it can sever, leading to loss of electrical continuity. If the machining does not go deep enough, bad electrical contact to the leads results. The micro-machining process thus requires visual feedback.

To prepare the fiber for electrode attachment, it is placed in a groove that is machined in a polymer substrate and filled with epoxy. The flat fiber face ensures that the electrodes in the fiber are parallel to the substrate surface. Figure 7.8a shows the electrooptic fiber that is glued in place inside the groove.

Figure 7.8b shows the micro-milling process that exposes each electrode at two points. Because the fiber is securely attached to the substrate with epoxy, and the substrate is the same material as the fiber, the milling process does not harm the fiber. If the fiber is not rigidly attached, the mill can catch the fiber and break it. The milling process requires two steps. In the first step, the substrate is milled with a large bit until it reaches the surface of the fiber. Subsequently, a small bit running at high speed is slowly pushed into the side of the fiber while the operator

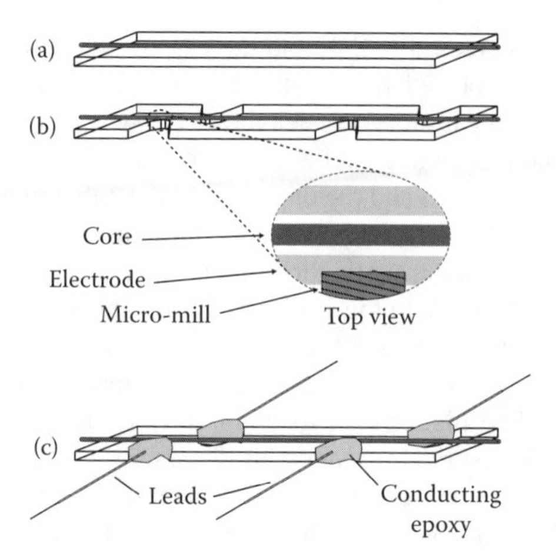

FIGURE 7.8 (a) An electrooptic fiber is held in groove in a polymer substrate with epoxy. (b) The fiber is micro-machined down to the electrodes at four contact points — two on each electrode. (c) Leads are attached to the electrodes with conducting epoxy.

observes the process from above through a microscope. Once the micro-mill is observed to reach the electrodes, the bit is retracted. Since four leads need to be attached, the milling process requires lots of skill and patience. Even then, the yield of this process is usually less than 25%. Clearly, if this process is taken beyond the research lab, an automated process could undoubtedly be developed to decrease the time to make a device as well as increase yield.

Figure 7.8c shows four leads that are attached to the electrodes with a conducting epoxy. The purpose of attaching two leads to each electrode is to test electrical continuity. Welker reports that the typical indium alloy electrodes of $40\mu m$ diameter have a resistance of about $0.1\ \Omega/cm$. The resistance between the two electrodes in electrooptic fiber devices with $20cm$ of fiber is about $50{,}000\ M\Omega$.

A photograph of an electrooptic device structure made by Welker and coworkers is shown in Figure 7.9. The device is about $1cm$ wide and $15cm$ long. It is

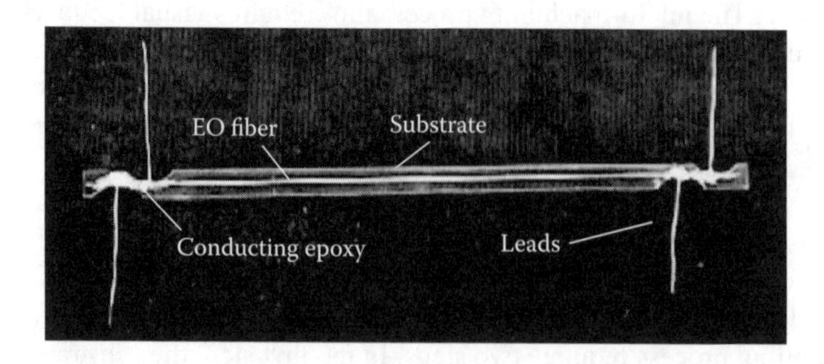

FIGURE 7.9 A photograph of an electrooptic fiber devices structure. Photograph provided courtesy of Sentel Technologies.

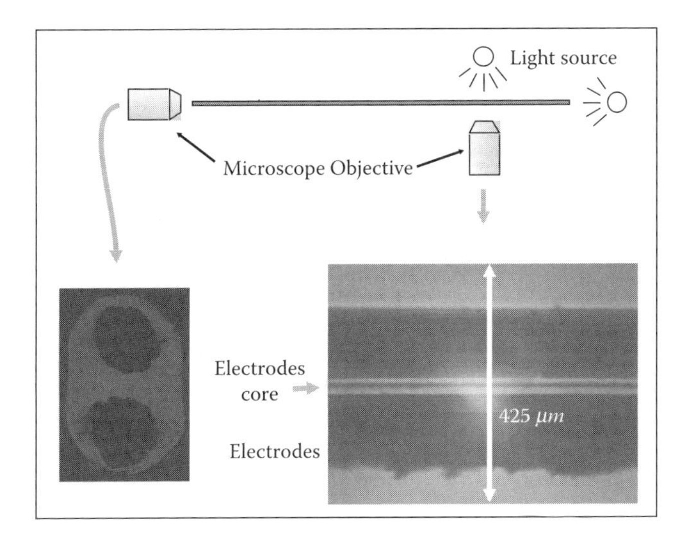

FIGURE 7.10 A microscopic view of an electrooptic fiber from the end and from the side under white-light illumination. Figure provided courtesy of Sentel Technologies.

possible to make structures with a large range of sizes. For commercial applications, such a structure can be packaged and hermetically sealed. Since the active region of the device is enclosed in a cladding material, packaging may only be necessary for mechanical protection and aesthetics.

Once the electrooptic fiber device is made, visual characterization provides information about the quality of the structure. Figure 7.10 shows the end view and side view of an electrooptic fiber under side illumination. From the side view, the core appears as a fine straight line while the electrodes have clearly diffused. Note that because of the shape of the cladding, it acts like a lens, magnifying the appearance of the size of the electrode. The end view shows a more accurate view of the dimensions of the electrodes, though the process of cutting the fiber with a razor blade has resulted in some deformation in both the shape of the cladding and the electrodes. The small finger-like protrudences are real. During the drawing process, the molten electrodes both flow and diffuse into the polymer. Even though the electrodes have diffused, there is still a solid core part of the electrodes that is pure metal and very similar in size to what would be expected from the drawing process if the polymer and metal were Newtonian liquids of similar viscosity.

Next, the waveguiding modes of the fiber core can be characterized by coupling light into one end of the fiber and viewing the other end with a microscope objective. Figure 7.11 shows the fiber end face for $1064nm$ light propagating in the core. It is interesting to note that the mode is elliptical. The ellipticity can be controlled by the drawing temperature and speed to make fibers span from high ellipticity cores to circular ones. The one shown in Figure 7.11 has large ellipticity. In such cores, the polarization of the mode along the long axis of the ellipse

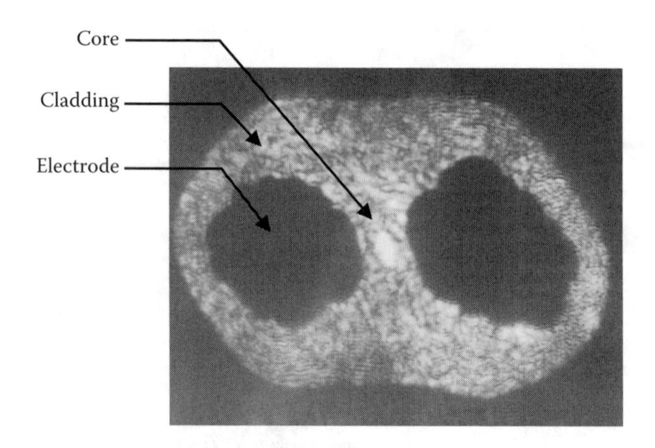

FIGURE 7.11 A microscopic view of the waveguiding mode of an electrooptic fiber. Photograph provided courtesy of Sentel Technologies.

propagates in signal mode while the perpendicular one does not. So, this type of fiber preserves polarization.

Welker and coworkers characterized the electrooptic properties of the device structure using a Mach-Zehnder interferometer with the homodyning technique. [112] The experiment is shown in Figure 7.12. The electrooptic device substrate is made short enough so that the length of fiber protruding from the ends is sufficient to be placed in a collet so that light can be coupled into the core.

The fiber is placed in one arm of the interferometer and the light is focused into the fiber with a microscope objective while an objective following the fiber is used to re-collimate the beam. The fiber and microscope objectives have two effects on the light. First, the beam width and divergence are changed. Second, the coupling loss into and out of the fiber, as well as the scattering and absorptive loss in the fiber, results in a much weaker beam than in the reference arm. To compensate for these two effects, an absorber and a lens pair are placed in the reference arm so that the two beams, when they recombine at the second beam splitter, have the same properties.

To evaluate the phase modulation properties of the fiber, a sinusoidal voltage is applied and the lock-in amplifier reads the magnitude of the voltage. The wedge is translated (Welker and coworkers used a rotating glass slide to adjust the phase), and the signal at the frequency of the applied voltage is measured as a function of the phase to produce an interferogram as described in Section 7.1.1. The peak value from this interferogram, called I_{sig}, is determined. The voltage is turned off and the intensity exiting the interferometer is measured as a function of wedge position (a chopper — not shown — and detector are used). The signal from the previous step is divided by the peak value determined from this interferogram to determine the signal. (This peak value is called I_{ref}.) Figure 7.13 shows a plot of the signal as a function of the applied voltage. For an electrooptic effect, this should be a straight line. Aside from small oscillations (which are probably

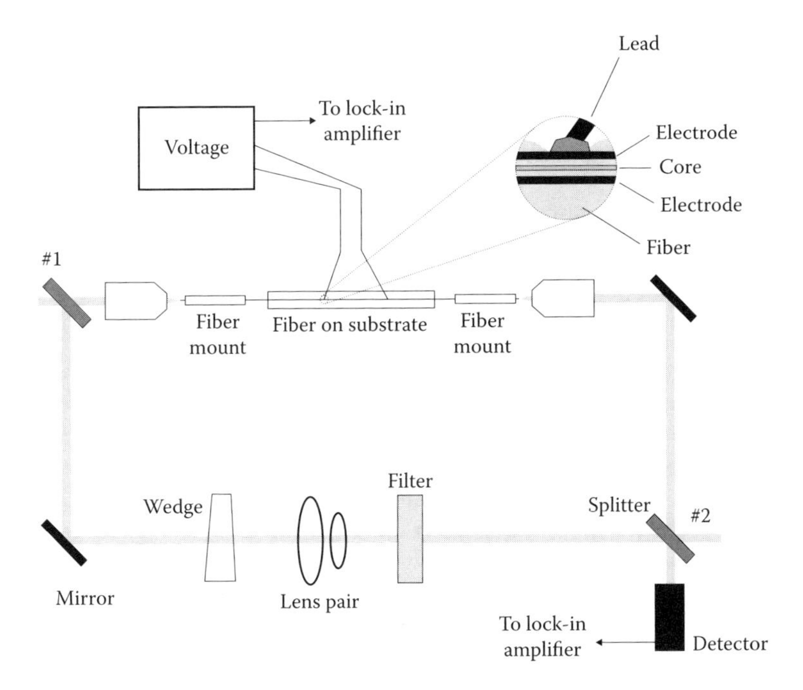

FIGURE 7.12 A Mach-Zehnder experiment used to measure the electrooptic coefficient of an electrooptic fiber. Adapted from Ref. [112] with permission of the Optical Society of America.

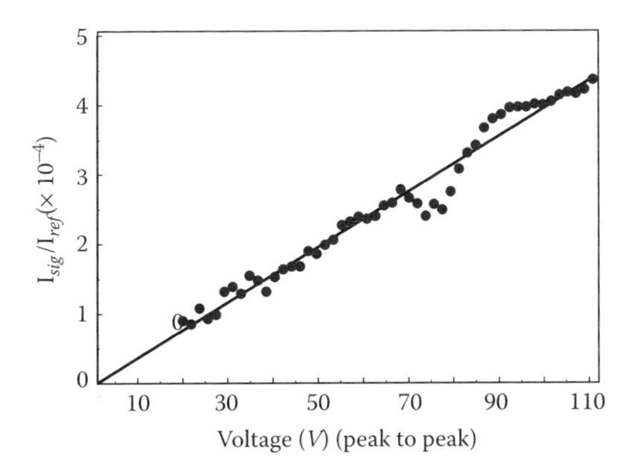

FIGURE 7.13 The electrooptic signal due to phase modulation in an electrooptic fiber as a function of applied voltage. Adapted from Ref. [112] with permission of the Optical Society of America.

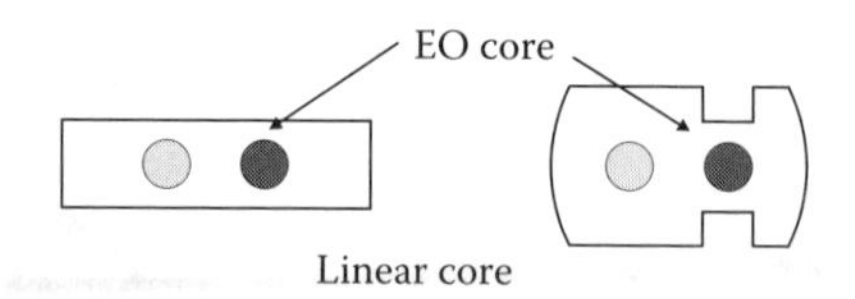

FIGURE 7.14 Two designs used by Peng and coworkers to make electrooptic fiber. Adapted from Ref. [114] with permission of SPIE.

due to drift in the interferometer), the data falls on a straight line that includes the origin.

One other point needs to be addressed. The signal measured in this experiment was small. The reason is that the researchers had depoled the sample by heating it, so the electrooptic coefficient of the fiber was small. If the fiber had remained poled, the ratio of signal to reference would have been near unity.

Peng and coworkers have taken a different approach for making electrooptic fiber.[114] The idea is that the polymer fiber is drawn first and the electrodes are later placed into the structure. Electric field poling is therefore the last step in the process. The advantage is that the fiber drawing process and the subsequent poling process is much simpler to implement. Furthermore, the preform is easier to make since metal is not involved. The disadvantage is that the structures that can be made are necessarily larger and the electrode spacing is greater. The net result is that for a given applied voltage, the electric field is smaller, leading to a smaller refractive index change.

Peng and coworkers have demonstrated their process with two difference structures, which includes a planar slab and an H-fiber with two waveguiding cores as shown in Figure 7.14. One of the cores is optically linear while the other is made of a material that when poled, renders it with an electrooptic response. The structure on the left is about $2mm$-$3mm$ thick and the cores are about $50\mu m$ thick. As such, the core cannot conveniently be made into a single-mode fiber, which typically requires cores of less than about $10\mu m$.

The electrooptic fiber (shown in the left part of Figure 7.14) is sandwiched between two metal plates, which act as electrodes. The whole fiber is placed in a hot oil batch as shown in Figure 7.15. The oil bath's temperature is raised to bring the sample above the glass transition temperature of the polymer so that it can be poled. The oil bath provides high thermal mass that keeps the temperature

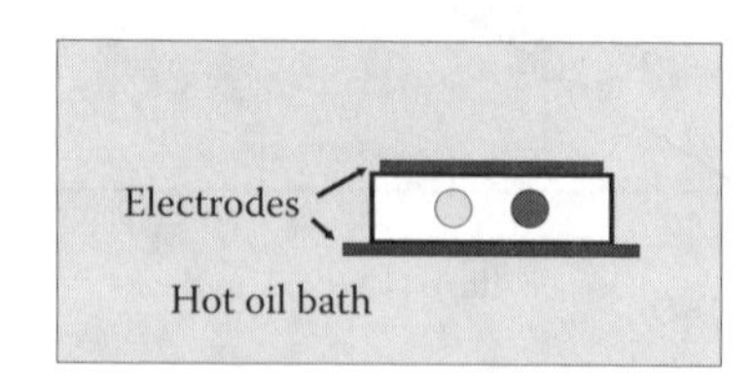

FIGURE 7.15 Electric field poling of an electrooptic fiber in a hot oil bath. Adapted from Ref. [114] with permission of SPIE.

stable over time. Second, the oil — being a dielectric — prevents electrical arcing between the places.

Because the electrodes are so far apart, the voltage applied is about $50kV$. A $30k\Omega$ resistor and ammeter are placed in series with the sample so that the poling current can be measured (the resister protects the ammeter should an electrical short develop). The temperature is lowered with the electric field applied until the fiber reaches ambient, well below the glass transition temperature of the polymer, at which point the voltage is turned off. Note that the electrooptic fiber with embedded electrodes that are $40\mu m$ apart requires a poling field of only $1kV$ to get the same electrooptic coefficient. Similarly, the embedded electrode device can run at $5V$ whereas the slab fiber requires $250V$ to obtain the same refractive index change.

The structure shown in the right-hand portion of Figure 7.14 is made into a preform and drawn into a fiber. After drawing, the grooves are about $80-100\mu m$ deep and $120-180\mu m$ wide. More importantly, the bottoms of the grooves are $100-150\mu m$ apart. After pulling, electrodes are placed into the grooves, so now they are 20 times closer together than in the case of the slab, so the required poling voltage and the operational voltage of the devices are lowered by a factor of 20.

The electrodes are formed by placing silver wires along the length of the grooves and filling the grooves with insulating epoxy or electrically conducting epoxy as shown in Figure 7.16. The silver wires extend beyond the end of the grooves and are used as leads. The fiber structure is heated to above the glass transition temperature so that it can be poled. Because the electrodes are closer together than in the slab fiber, the poling field used can be lower; $3.1kV$ is sufficient to pole the material.

When the insulating epoxy is used, the poling process is more reliable than when the conducting epoxy is used; but the silver wire's distance from the bottom of the groove is not known so that poling field cannot be accurately determined.

Once the fiber is poled, the electrooptic coefficient is measured. Instead of using a Mach-Zehnder measurement technique, the authors used polarization rotation. The advantage of the Mach-Zehnder interferometer is that the two tensor components of the electrooptic coefficient can be independently determined. The polarization-rotation technique, however, is intrinsically more stable, but measures

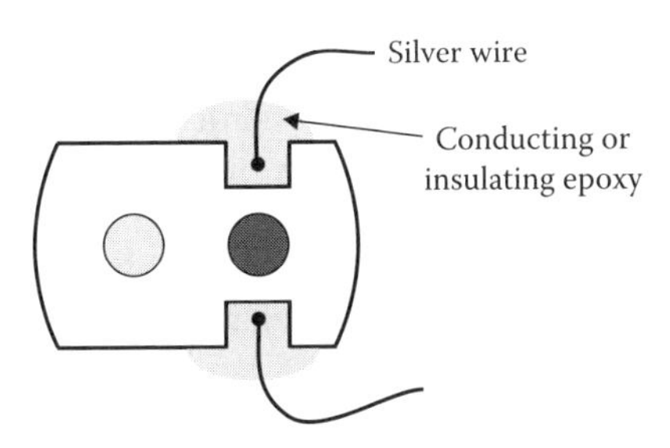

FIGURE 7.16 The electrodes are formed by placing silver wires along the length of the grooves and filling the grooves with electrically conducting epoxy. Adapted from Ref. [114] with permission of SPIE.

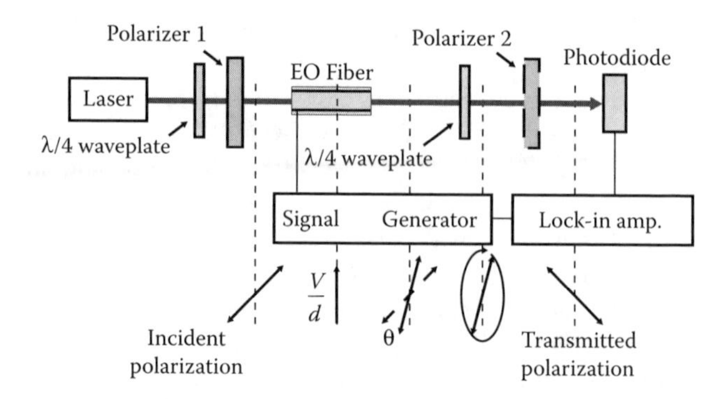

FIGURE 7.17 The top part of the figure shows the experimental layout of electrooptic-induced birefringence. The bottom part of the figure shows the polarization of the light beam and the electric field (E) from the applied voltage at various points along the beam.

a combination of tensor components of r. To determine one of the tensor components, an assumption about the material symmetry needs to be made. Even so, the polarization technique should still lead to a reasonably good estimate of r.

The experiment is shown in the top portion of Figure 7.17. Light is launched into a quarter wave plate to make it circularly polarized. Polarizer 1 can then be rotated to control the polarization of the light beam that enters the electrooptic fiber. The polarization is set to 45° with respect to the direction of the static electric field that is applied with a signal generator. When the voltage is applied, the induced birefringence in the sample causes the polarization to rotate by an angle θ (the polarization will also, in general, become elliptical, but this is not shown in the figure). The amount of rotation is related to the strength of the applied voltage and the electrooptic tensor. The next quarter wave plate is used to make a phase adjustment (described later). Finally, the second polarizer is set at an angle to optimize the signal as determined from the calculation that follows. Since the intensity of light that is transmitted by the second polarizer is related to the angle of rotation, an applied sinusoidal violate will lead to a sinusoidal intensity modulation at the detector. The amplitude of the modulated intensity is read by a lock-in amplifier.

In a poled polymer, the symmetry of the poling process for an initially isotropic material yields $3r_{33} = r_{13}$ (r_{33} and r_{13} is shorthand notation for r_{333} and r_{113}). Figure 7.18 shows r_{113} is being measured for the static field and the optical fields perpendicular (left) and r_{333} for the fields parallel (right).

We are now ready to calculate the intensity read by the detector. According to Figure 7.18, since the light is sent into the sample at 45°, the electric field can be written as

$$\vec{E} = \frac{E_0}{\sqrt{2}} \left(\exp\left[-\frac{2\pi i \ell}{\lambda} n_1 \right] \hat{1} + \exp\left[-\frac{2\pi i \ell}{\lambda} n_3 \right] \hat{3} \right), \tag{7.29}$$

where n_1 and n_3 are the refractive indices along $\hat{1}$ and $\hat{3}$; and ℓ is the length of the electrooptic fiber.

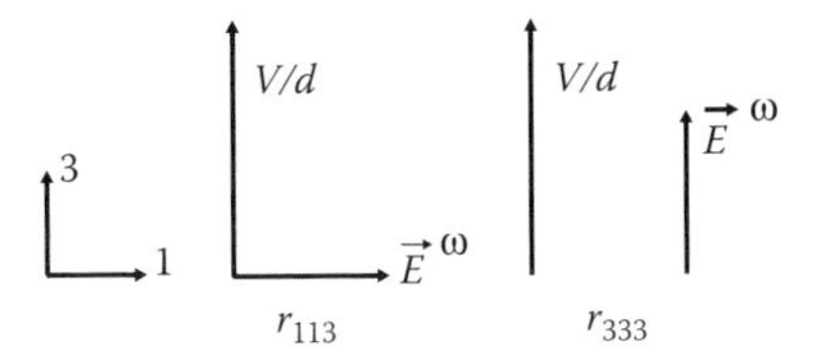

FIGURE 7.18 The tensor components measured when the static and electric field are perpendicular (left) and parallel (right).

We now make the following assumptions:

1. The refractive index is mostly real so that optical absorption is small. Note that the calculation can be generalized to include complex refractive index, but we do not do so here to simplify the expressions.
2. The electrooptic coefficient is mostly real so that the imaginary part will be ignored. This can be easily generalized to complex r but makes the problem more messy.
3. The refractive index of the material before the field is turned on is isotropic. While this is not true for a poled polymer, we can compensate by setting the angle of the second half-wave plate (see Figure 7.17).

If the half-wave plate induces a relative phase shift between the two polarizations of ϕ, and the polarizer is set to an angle α relative to the $\hat{3}$-axis (so that the polarization passed by the second polarizer is $\hat{p} = (\sin\alpha)\hat{1} + (\cos\alpha)\hat{3}$), the field passed by the second polarizer is given from Equation (7.29) by

$$\vec{E}_p = \vec{p}\cdot\vec{E} = \frac{E_0}{\sqrt{2}}\left(\exp\left[-\frac{2\pi i\ell}{\lambda}n_1\right]\sin\alpha + \exp\left[-\frac{2\pi i\ell}{\lambda}n_3 + \phi\right]\cos\alpha\right). \tag{7.30}$$

The intensity transmitted through the second polarizer is then given by the square of the magnitude of Equation (7.30),

$$I_p = \frac{I_0}{2}\left(1 + \sin(2\alpha)\cos\left[-\frac{2\pi i\ell}{\lambda}(n_3 - n_1) + \phi\right]\right), \tag{7.31}$$

where using Equation (7.2) and Equation (7.30),

$$n_3 - n_1 = \frac{n_0^2}{2}(r_{33} - r_{13})\frac{V}{d} = \frac{n_0^2}{2}\left(\frac{2}{3}r_{33}\right)\frac{V}{d}, \tag{7.32}$$

where V is the applied voltage and d is the separation between the electrodes. The last equality has used the fact that the thermodynamic model of a poled polymer yields $r_{33} = 3r_{13}$.

$n_3 - n_1$ is the electrooptic change in the refractive index and is usually small. Expanding Equation (7.31) to first order in $n_3 - n_1$, we get

$$I_p = \frac{I_0}{2}\left(1 + \sin(2\alpha)\left[\cos\phi + \frac{2\pi i\ell}{\lambda}(n_3 - n_1)\sin\phi\right]\right). \tag{7.33}$$

Clearly, the electrooptic signal (second term in brackets in Equation 7.33) is largest when $\phi = \pi/2$ and $\alpha = \pi/4$.

So, the electrooptic-induced birefringent experiment is tuned as follows: The second polarizer is set to $\alpha = \pi/4$. The sample is birefringent, so it will add to the phase shift ϕ. Since the sample birefringence is not known *a priori*, the second half-wave position needs to be determined empirically. With no voltage applied, the transmitted intensity for $\phi = \pi/2$ is $I_0/2$. So, the transmitted intensity as a function of the half-wave plate angle will yield a sinusoidal function. When the half-wave plate is rotated to the point where $I_p = I_0/2$, that corresponds to $\phi = \pi/2$. Note that this measurement determines I_0.

Once the polarizers and analyzers are tuned and I_0 determined, a sinusoidal voltage of amplitude V is applied to the sample and the lock-in amplifier measures the amplitude of the signal, I_s. Given these values, substituting Equation (7.32) into Equation (7.33) and solving for r_{33}, we get Equation (7.31) to first order in $n_3 - n_1$,

$$\boxed{r_{33} = \frac{3\lambda d I_s}{\pi n_0^2 V \ell I_0}}. \tag{7.34}$$

Recall that for the Mach-Zehnder interferometer, we had to scan the phase to get an accurate measurement of the optical bias point. Because the birefringence experiment is intrinsically more stable, the quarter-wave plate needs to be set only once.

Peng and coworkers used the induced birefringence technique to characterize several different dye-doped polymers made into electrooptic fiber. The chromophores they used are shown in Figure 7.19. Following the fabrication of 2.7*mm*-thick slab EO fibers, they were poled at a voltage of 35*kV* and at a temperature of 102.5°C for 4 hours followed by a 1.5 hour ramp-down process with the voltage applied. Figure 7.20a shows the signal as a function of the applied voltage for the CHAB molecule dopant. It is linear as expected by theory.

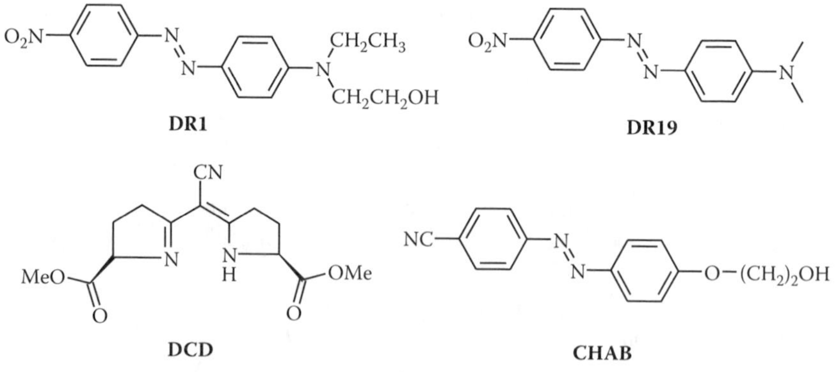

DR1 DR19

DCD CHAB

FIGURE 7.19 Dyes used by Peng and coworkers[114] as dopants to make EO fiber.

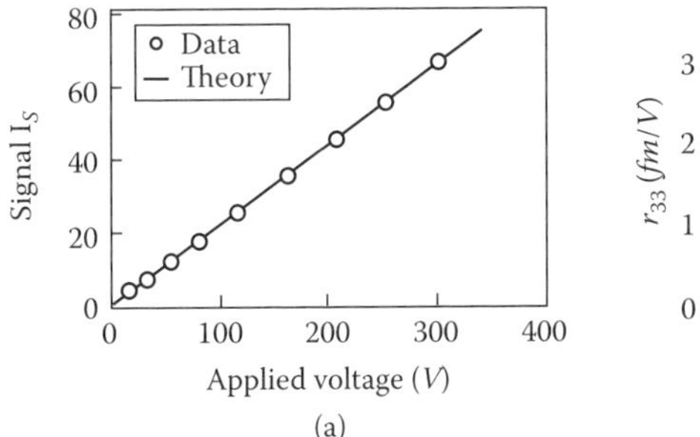

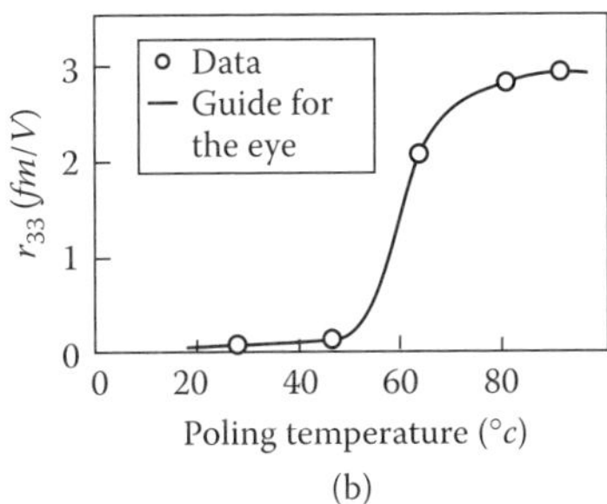

FIGURE 7.20 (a) EO modulation signal as a function of applied voltage for CHAB-doped EO fiber; and (b) the measured electrooptic coefficient as a function of temperature. Adapted from Ref. [114] with permission of SPIE.

Figure 7.20b shows the measured electrooptic coefficient as a function of poling temperature. The trend of the data is consistent with the thermodynamic model of poling — which predicts that the system should be poled near the glass transition temperature for maximum alignment. It is peculiar, however, that the magnitude of r_{33} is smaller than the value predicted by the thermodynamic model by two orders of magnitude. (The model has been shown to be accurate for thin films.) In Welker's work,[112] an r value that was two orders of magnitude smaller than the expected value was also observed — though the explanation was that the lead-attachment process required heating that had de-poled the sample. Clearly, more work is warranted to understand this issue.

An important property of any device material is long-term reliability. The decay of the electrooptic coefficient of dye-doped poled polymers is well documented in the literature, and many novel materials have been designed that show little long-term decay (though most materials show a small fast decay with a time constant on the order of days).[115] Thin-film-based devices made from such materials have been shown to be reliable even at elevated temperatures. Since many materials that are compatible with thin-film processing cannot be made into fiber, new materials will undoubtedly need to be developed to make reliable EO fibers.

Peng was the first to perform such reliability studies in EO fiber structures. Figure 7.21 shows a measurement of the electrooptic coefficient as a function of time for CHAB-doped EO fiber at room temperature and DCD-doped fiber at three different temperatures. (All values are normalized to r at $t = 0$.) Clearly, the CHAB chromophores are not as stable as DCD.

Measurements at elevated temperature accelerate the decay process. As such, they can be used to predict the stability of a material at longer times. The decay time for DCD at $T = 20°C$ and $T = 40°C$ is similar, but at $T = 60°C$, the decay process is accelerated. All of the measurements of DCD show the characteristic fast decay followed by a plateau. Extrapolating the elevated temperature measurements, the DCD-doped polymer should be stable on the order of years once the fast decay process is complete.

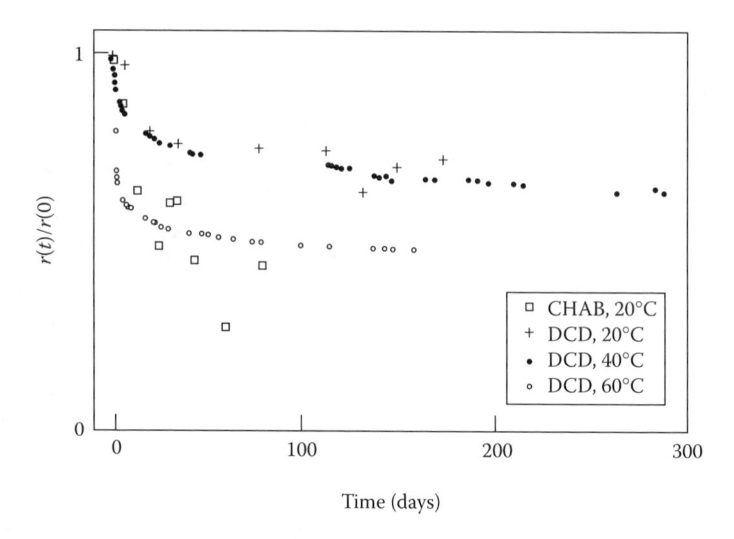

FIGURE 7.21 (a) EO coefficient as a function time for CHAB- and DCD-doped EO fiber at different temperatures. Adapted from Ref. [114] with permission of SPIE.

Clearly, electrooptic polymer optical fibers offer exciting potential for switching applications. Many technical challenges, however, need to be overcome before such fibers become a technology. Given the high intensities that are guiding in a single-mode core, and the large static electric fields that can be applied with the closely spaced electrodes, there is also undoubtedly room for new science that can take advantage of these properties. While electrooptic fiber is not yet a technology, this does not prevent us from considering the potential of this technology. The following section asks the practical questions, and shows how electrooptic fiber fares in meeting certain important device criteria.

7.1.2 Electrooptic Devices

If an electrooptic fiber technology were to be perfected, what are the types of devices that could be made and what would be the important issues? In this section, we speculate on the future potential of electrooptic devices based on extrapolations from current electrooptic fiber properties. We show that polymer fiber devices have the properties required to make fast and efficient devices.

One of the advantages of an electrooptic device is its potential speed. Since polymeric materials usually have an electrooptic response that originates in the electronic mechanism, the response is inherently fast. In fact, since the electronic mechanism's response time is on the order of a femtosecond, the inherent material response should never be an issue.

When a high-speed electrical signal is applied to the polymer through the electrodes, it can be treated as a traveling wave. The time it takes the electrical wave to penetrate the material is proportional to the square root of the dielectric constant, $\sqrt{\epsilon(\Omega)}$, where Ω is the dominant frequency (or carrier frequency) in the

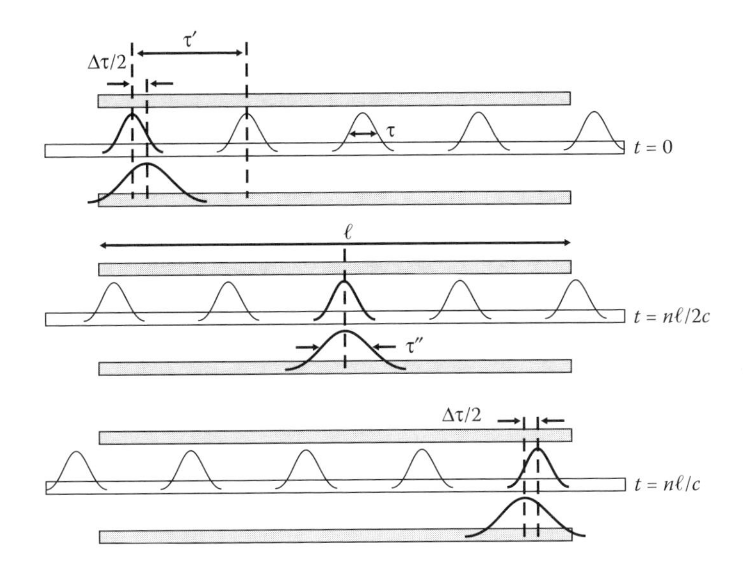

FIGURE 7.22 A diagram of optical and electrical pulse propagation in an electrooptic fiber at three different times.

electrical signal. If, for example, a pulse of light is traveling down the fiber that is to be electrooptically modulated by the voltage, the time it takes for this pulse to traverse the device is proportional to the refractive index of the polymer, n. If the full length of the device is to be used efficiently, the speed of the electrical wave and the optical pulse must not get significantly out of step with each other.

Consider an electrooptic fiber with an active waveguiding region of length ℓ that must modulate one pulse of temporal width τ in a stream of pulses that are separated by a time τ'. The width of the electrical pulse must therefore be on the order of the spacing between pulses, $\tau'' \simeq \tau$. For one light pulse to be efficiently modulated, the electrical pulse should completely overlap the optical pulse over the length of the device, so the best devices are those in which the electrooptic material's refractive index and square root of the dielectric constant are the same, or $n \approx \sqrt{\epsilon(\Omega)}$. This is the case for polymeric materials where they are within 15% of each other.

Figure 7.22 shows a diagram of an electrooptic fiber with a stream of optical pulses in the core and a voltage pulse in one of the electrodes for three different times. During the time that the modulated pulse is in the device ($t = n\ell/c$), the two pulses get out of step by an amount $\Delta\tau$, which is given by

$$\Delta\tau = \frac{\ell}{c}\left(\sqrt{\epsilon} - n\right). \tag{7.35}$$

To efficiently switch a pulse, this delay should be smaller than the width of the electrical pulse, or $\Delta\tau < \tau''$. On the other hand, to pack as much information on a fiber line as possible, the pulses should be as close together as possible, so ideally $\tau' = \tau$. In this limit, then, the electrical pulse needs to be the same width as the optical pulse, or $\tau'' = \tau$. But, in this high-bandwidth scenario, if the electrical pulse is the same width as the optical pulse, which is the same as the spacing between pulses, the relative delay between the electrical and optical pulses needs to be much less than the pulse width. Otherwise, the adjacent pulses may also be modulated.

As an example, let's consider a polymer with dielectric constant $\epsilon = 3.3$ and a refractive index of $n = 1.5$. This leads to $\tau = 1.05 \times 10^{-11} s \cdot cm$. This system would allow light pulses to be packed about $10ps$ apart for a device length of $1cm$ or $100ps$ for a $10cm$ device. These two cases correspond to a bandwidth of $100GHz$ and $10GHz$, respectively. So, shorter device lengths are desirable. However, the active device length required to modulate a pulse depends on the magnitude of the voltage that can be applied to the device as well as the electrooptic coefficient.

To determine the minimum required length of a device, let's assume that a π phase shift is needed. This is a middle-of-the-road value since most device designs require a phase shift of π within a factor of two. So, a device of length ℓ induces a phase shift,

$$\Delta\phi = \frac{2\pi}{\lambda} \cdot \Delta n \cdot \ell = \frac{2\pi}{\lambda} \cdot \frac{n_0^2}{2} r \frac{V}{d} \cdot \ell > \pi, \tag{7.36}$$

where d is the separation between electrodes (for this example, we assume planar electrodes) and where we have used Equation (7.1) with $E = V/d$. Solving for the required length, we get

$$\ell > \frac{\lambda d}{n_0^2 r V}. \tag{7.37}$$

So, for an electrooptic fiber with an electrode separation of $d = 40\mu m$ and electro-optic coefficient of $r = 50pm/V$ that operates at $\lambda = 1\mu m$ and has a $V = 5V$ power supply, for a refractive index of $n_0 = 1.5$, we require a device length of at least $\ell = 7.1cm$ to get a π phase shift. This length is a reasonable one for a polymer optical fiber; but this is getting long for a ribbed waveguide device on a planar substrate. Thus, the polymer fiber device has clear advantages.

Combining Equations (7.36) and (7.35), we can get an expression for the maximum bandwidth of an electrooptic device that requires a π phase shift:

$$\boxed{\text{bandwidth} = \frac{1}{\tau} = \frac{c n_0^2 r V}{\lambda d \left(\sqrt{\epsilon} - n_0\right)}}. \tag{7.38}$$

One final issue that needs to be addressed is the connection of the electrooptic device to electronics. For the electrical signal to reach the device without attenuation or distortion, the electrical impendence of the cable leading to the device needs to match the impedance of the device. We calculate the impedance of the electro-optic fiber by modeling it as two parallel cylindrical conductors of radius a that are separated by a distance d and placed in a dielectric of permittivity ϵ. We will work this problem in SI units, so we express $\epsilon = \kappa\epsilon_0$, where κ is the dielectric constant.

The top portion of Figure 7.23 shows a diagram of the two wires inside the fiber. Since the dielectric constant of the core and cladding are nearly equal, we can treat the dielectric as homogenous. Furthermore, if the electrodes are close

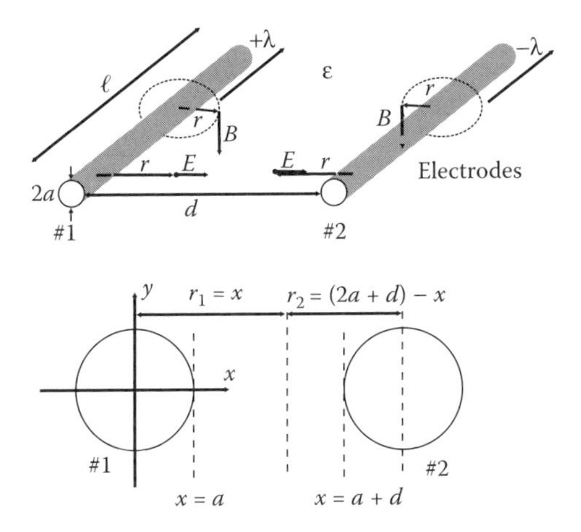

FIGURE 7.23 The electric and magnetic fields between two oppositely charged wires carrying opposite currents (top) and the coordinate system used to calculate the fields (bottom).

together compared with the size of the cladding, we can approximate the dielectric as filling all space. While the calculation of the impedance of parallel wires is well known, it is worthwhile to walk the reader through the calculation so that the role of the reactance in a transmission line is understood.

We begin by calculating the capacitance. If the electrodes are charged, the electric field outside is that of a line charge of linear charge density λ. If the electrodes have equal and opposite charge, as shown in Figure 7.23, in the plane containing the center of both cylindrical electrodes, the field from each one will point to the right. The total electric field in this plane is given by the superposition of the electric fields from each wire:

$$E = \frac{\lambda}{2\pi\kappa\epsilon_0}\left[\frac{1}{r_1} + \frac{1}{r_2}\right] = \frac{\lambda}{2\pi\kappa\epsilon_0}\left[\frac{1}{x} + \frac{1}{2a + d - x}\right], \qquad (7.39)$$

where the bottom part of Figure 7.23 defines the coordinate system. We have represented the electric field, E, as a scalar since it points in the same direction everywhere in the plane of the calculation.

The electric potential on a conductor is the same throughout the conductor. The magnitude of the potential difference is thus calculated by integrating over the electric field from the surface of one electrode (at $x = a$) to the other one (at $x = a + d$),

$$\begin{aligned}\Delta V &= \int_a^{a+d} E dx = \frac{\lambda}{2\pi\kappa\epsilon_0}\int_a^{a+d}\left[\frac{1}{x} + \frac{1}{2a + d - x}\right] dx \\ &= 2 \times \frac{\lambda}{2\pi\kappa\epsilon_0}\int_a^{a+d}\left[\frac{1}{x}\right] dx = \frac{\lambda}{\pi\kappa\epsilon_0}\ln\left(\frac{a + d}{a}\right),\end{aligned} \qquad (7.40)$$

where the first equality of the second line comes from symmetry; that is, the integral over the electric field due to one electrode (first term in square brackets on the first line of Equation 7.40) is the same as for the other one (second term in square brackets).

We can calculate the capacitance from Equation (7.40),

$$\boxed{C = \frac{Q}{\Delta V} = \frac{\ell \pi \kappa \epsilon_0}{\ln\left(\frac{a+d}{a}\right)},} \tag{7.41}$$

where we have used $Q = \lambda \ell$.

Next, we calculate the inductance. We begin by calculating the magnetic field due to a current-carrying cylinder. Since the magnetic field in the plane between the two conductors points downward, we can also represent it as a scalar. The magnetic field due to electrode #1 is

$$B = \frac{\mu_0}{2\pi} \cdot \frac{I}{r_1}, \tag{7.42}$$

where I is the current in the conductor. Again, we can use symmetry to simplify the problem. The magnetic flux between the two conductor surfaces is an integral over the area of the plane between them:

$$\Phi = 2 \times \int_a^{a+d} B \cdot \ell dx = \frac{\mu_0 I \ell}{\pi} \cdot \int_a^{a+d} \frac{1}{x} dx = \frac{\mu_0 I \ell}{\pi} \ln\left(\frac{a+d}{a}\right), \tag{7.43}$$

leading to an inductance,

$$\boxed{L = \frac{\Phi}{I} = \frac{\mu_0 \ell}{\pi} \ln\left(\frac{a+d}{a}\right).} \tag{7.44}$$

The two parallel wires have a resistance along the wires due to the indium alloy and between them due to the polymer. They also have an inductance and capacitance as calculated above. The equivalent circuit of the two-wire system is shown in Figure 7.24. The dashed box shows one of the infinitesimal slices of the equivalent circuit that represents a short segment of the transmission line (or it can represent the electrooptic fiber). In a transmission line, it can be shown that the

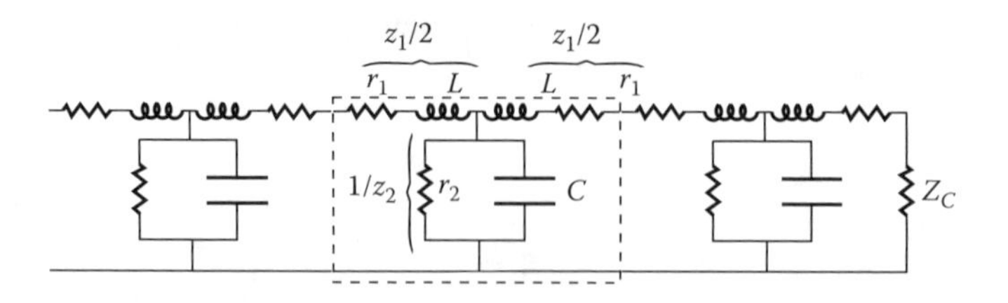

FIGURE 7.24 The equivalent circuit of two parallel wires in a dielectric.

signal is not distorted if the line is terminated with a characteristic impedance, Z_C, given by

$$Z_C = \sqrt{z_1 z_2} = \sqrt{\frac{r_1 + i\omega L}{1/r_2 + i\omega C}}, \tag{7.45}$$

where ω is the angular frequency of the signal.

Let's consider the electrooptic fiber as the last element in a transmission line. We saw in our previous calculations that the velocity mismatch between the optical wave and the electrical signal limits the device to about $100GHz$ bandwidth. (Note that we used conservative numbers, so higher bandwidths are attainable.) For the sake of argument, we use these numbers to determine the characteristic impedance. First let's consider the numerator in Equation (7.45). The resistance along an electrode is measured to be about $r_1 = 1\Omega$. Using Equation (7.44) with $\ell = 0.1m$ and $d/a = 2$, we get $L = 4.4 \times 10^{-8}H$. For $\omega = 10^{11} Rad/s$, the product $\omega L = 4,440$. This is clearly larger than the resistance of the conductor. Next, we evaluate the denominator. Using Equation (7.41) with $\kappa = 3$ and the same parameters as above, we get $C = 7.6 \times 10^{-12}F$. This gives $\omega C = 0.76$, which is clearly much bigger than $1/r_2$, for the measured resistance between electrodes of $r_2 \approx 10^{10}\Omega$.

Thus, for typical device parameters under a high-speed operation, the characteristic impedance can be approximated by

$$Z_C = \sqrt{\frac{L}{C}}. \tag{7.46}$$

We note that Equation (7.46) gives a real impedance, so the transmission line can be terminated with a resistor. We can calculate the characteristic impedance by substituting Equations (7.41) and (7.44) into Equation (7.46), yielding

$$\boxed{Z_C = \frac{1}{\pi}\sqrt{\frac{\mu_0}{\kappa\epsilon_0}} \ln\left(\frac{a+d}{a}\right) = \frac{120}{\sqrt{\kappa}} \ln\left(\frac{a+d}{a}\right) = \frac{276}{\sqrt{\kappa}} \log\left(\frac{a+d}{a}\right).} \tag{7.47}$$

The result given by Equation 7.47 can be easily evaluated for the electrooptic fiber. With $\kappa = 3$, $d/a = 2$, we get $Z_C = 76\Omega$. A common value of the characteristic impedance of commercially available coaxial cables is 72Ω. Clearly, by slightly tweaking the electrooptic fiber, we can make it match the impedance of standard cables without affecting its bandwidth.

Now that we have seen how electrooptic fibers have all the characteristics required for high-speed modulation, we can now consider how such fibers can be built into a system. The intensity modulator is the simplest example. Figure 7.25 shows one design for making such a device: an all-fiber Mach-Zehnder electrooptic modulator. In this design, the two modulators, when used together in a push-pull configuration (the phase in one arm decreases while the other one increases), require half the switching voltage of the single electrooptic fiber device. Another

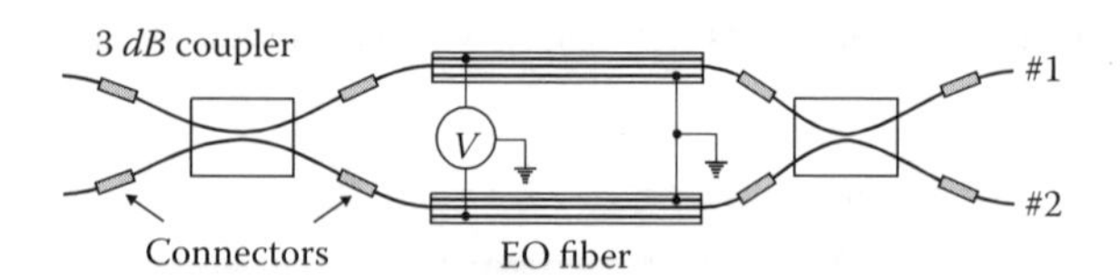

FIGURE 7.25 An all-fiber electrooptic modulator.

advantage of this design is that the two arms of the Mach-Zehnder are naturally balanced without the need for additional attenuators.

This device can act as both a modulator and a switch. When light is launched into one of the ports, and an AC voltage is applied, the light exiting the two ports will be modulated at that frequency, but π out of phase with each other. Alternatively, with no applied voltage, all the light will exit out of, let's say, port #1. When the voltage is applied, then all of the light will exit out of port #2. Thus, the applied voltage can be used to direct a signal carried on a beam of light between two fibers.

Aside from the electrooptic fiber, all of the components are commercially available. The 2x2 couplers (sometimes called 3*dB* couplers) made of glass fibers are universally available with connectors. If the core and cladding diameter of the electrooptic fiber matches the glass fibers in the 3*dB* coupler, similar connectors can be added to the polymer fibers. Then, the fibers can be snapped together. Having an electrooptic fiber component available would make it possible to build a system by just connecting together electrooptic fibers, couplers, and transmission fiber.

An issue of high importance for any system is the loss budget, which depends on the available amount of light and the amount of light that is required at the output end. Furthermore, a device must be stable under temperature fluctuations and must not send too much signal into the wrong port. Let's consider the components of the device above to see if it meets a set of reasonable requirements. (Note that the requirements that follow are an example from an Air Force call for proposals in 1998.)

Loss Budget of 1.2 *dB* (defined as total loss of the system): Several suppliers sell fiber connectors and couplers. Couplers can be purchased with an intrinsic loss of 0.32B (for example, AFOP manufactures the ultra model wideband single-mode fused coupler with this loss; see http://www.afop.com/) and fiber couplers (for example, FC, ST, and SC) with insertion losses of 0.2*dB*. The above device design has an input connector (0.2*dB*); 3*dB* coupler (0.3*dB*) connector to the active fiber(s) (0.2*dB*), connector to second 3*dB* coupler (0.3*dB*) and output connector (0.2*dB*). The maximum loss for this device is therefore less than 1.2*dB*. Lower losses are possible if fibers are fused together rather than using commercial connectors. Clearly, this spec can be met.

Return Loss of 55 *dB* (defined as total intensity reflected to the input): For the PC series connectors, the typical reflectance is -68*dB*. The 55*db* return loss specification is thus well within the sum of all component back reflections. This, of course, assumes that the polymer fiber connectors can operate with the same specs as glass fiber connections.

Optical Extinction of 55 *dB* (defined as off-to-on intensity ratio): The optical extinction of the AFOP 2x2 coupler is 65*dB*. Again, this is well within the optical extinction specification of 55*dB*. The two arms of the interferometer also need to be well balanced to reach these specs.

Bandwidth of less than 1μs (defined above): Almost a decade ago, polymer modulator devices in planar structures have been shown to perform at bandwidths of 100*GHz*.[116] Sub-microsecond switching time specification is therefore well within the capability of a polymer device. There is no intrinsic reason to prevent a polymer fiber device from operating at such speeds.

Temperature Stability: In the matched-arm design, both arms are affected by temperature changes in the same way, so the relative phase difference between them is small. As a result, such a device design should be stable.

In closing, it is important to stress that lots of device designs are made practical by having an electrooptic fiber technology that allows the electrooptic fiber to be easily placed into a fiber system. Figure 7.26 shows an elegant device design of an optical switch based on a Sagnac interferometer. Such interferometers are known for their stability, and details of how they work are discussed later in this chapter.

The interferometer is made of a loop of fiber and a coupler. The incident light is equally split into the clockwise and counterclockwise paths of the loop. Both of these pulses travel in the same fiber (except in opposite directions). At time t_1, an optical pulse is launched into the loop while the electrical pulse is sent to the electrooptic fiber, which is spliced into the loop. At time t_2, the clockwise pulse is coincident with the electrical pulse, thus inducing a phase shift. The voltage pulse has passed the EO fiber by the time the counterclockwise pulse reaches the EO fiber, so it experiences no phase shift. Thus with proper timing of the electrical pulse, the induced relative phase shift between the two counterpropagating pulses results in

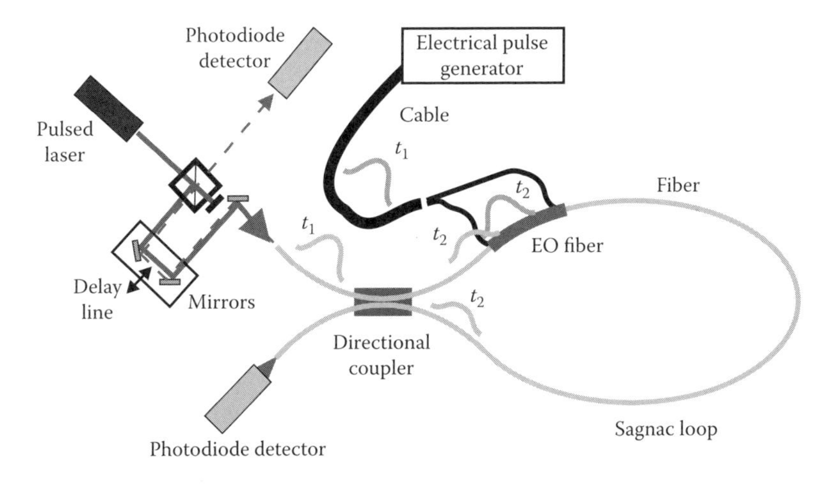

FIGURE 7.26 An all-fiber Sagnac optical switch.

the beam leaving the interferometer toward the detector. With the voltage off, the pulse goes back toward the laser (thin dashed ray) and into the other photodetector. The optical delay line is used to provide the correct timing between the electrical and optical pulses. This is an example of an electrooptical switch, where the light signal is directed between to two different output paths.

7.2 ALL-OPTICAL SWITCHING

7.2.1 INTRODUCTION

Normally, two beams of light do not interact with each other. This is the very property of light that makes it good for information transmission. However, light beams need to be able to interact with each other to affect their propagation properties if the information carried on the light is to be manipulated. For example, we may want to perform logic operations, where the output intensity depends on the input intensity. Or, multiplexed signals may need to be actively rerouted depending on their destinations. For such processes to occur, the material must mediate the interaction between light beams. Dye-doped polymers are well suited for this task since they can be made to possess a large third-order nonlinear optical susceptibility.

The intensity dependent refractive index is a third-order nonlinear optical process that leads to a change in the refractive index of a material,

$$n = n_0 + n_2 I, \tag{7.48}$$

where n_0 is the linear refractive index, I the intensity, and n_2 the intensity-dependent refractive index, which is proportional to the third-order susceptibility and is a property of the material. n_2 has units of inverse intensity. Experimentally, it is convenient to measure intensities in units of W/cm^2. The ultrafast intensity dependent refractive index, which arises from the electronic mechanism, is on the order of $10^{-12} cm^2/W$. n_2 is therefore sometimes expressed in units of cm^2/GW.

Figure 7.27 shows an example of the intensity-dependent refractive index of a prism that is made of a nonlinear optical material, such as a dye-doped polymer.

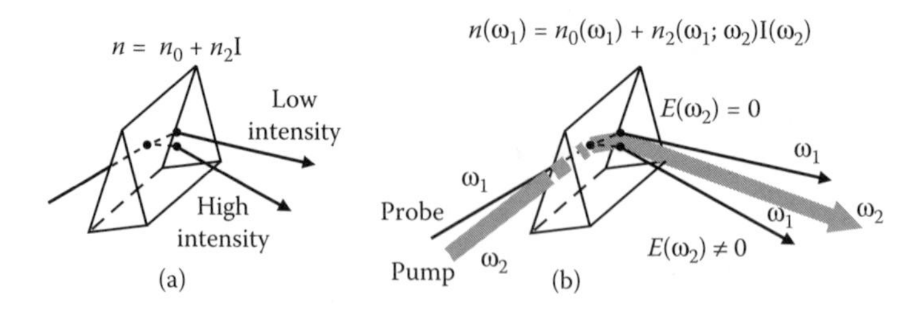

FIGURE 7.27 (a) A prism made of a material with an intensity-dependent refractive index will bend light differently at different intensities. (b) A bright beam of one color can change the amount of bending of another beam of a different color by using such a prism.

The amount that a single beam of light bends in such a prism depends on the intensity (Figure 7.27a). Alternatively, a strong beam of light, called a pump, can be used to deflect a weaker beam (Figure 7.27b). The strength of the effect, as quantified by $n_2(\omega_2; \omega_1)$, depends on the wavelength of the beams as well as their polarizations. For a single beam, the bending process is formally described by $n_2(\omega_1; \omega_1)$, but is expressed in shorthand as simply n_2. While the nonlinear bending process shown in Figure 7.27 would in principle make a good switching device (for example, as a function of pump intensity, the light coming out of one fiber can be switched between two fibers using a pump beam), the amount of deflection is too small to make such a practical. Interferometer designs for switching devices, which leverage the intensity-dependent effect, are more commonly used.

While most switching devices are based on the intensity-dependent refractive index, any process that leads to an intensity-depends phase can be used to make an active device. Since the phase is proportional to the refractive index and the length of the material, an intensity-dependent length change (called a photomechanical effect) can also be used to design a device. Chapter 9 provides a description of the mechanisms underlying the photomechanical effect.

7.2.2 Intensity-Dependent Phase Shift in Polymer Waveguide and Fibers

There are many experimental techniques that have been used to measure the n_2 of a material. Each of these has distinct advantages and shortcomings. Furthermore, since a given experiment can measure a combination of mechanisms, it is often difficult to interpret the data unambiguously. Experiments that are highly sensitive to the intensity-dependent refractive index are also sensitive to noise. The Sagnac interferometry, which is inherently stable against drift, has been found to be a good device for measuring the ultrafast n_2 without being sensitive to slowly varying background due to slower processes. For these reasons, the Sagnac interferometer is also well suited for making optical switching devices.

Gabriel and coworkers were the first to use the Sagnac interferometer to measure the intensity-dependent refractive index in polymer waveguides.[117] Later, Garvey and coworkers measured the intensity-dependent refractive index with a homodyne techniques using a Sagnac.[20] The basic Sagnac experiment is common to both experiments, and a schematic representation is shown in Figure 7.28.

The main part of a Sagnac interferometer is a loop that is made of a beam splitter and three mirrors. Ignoring all the other components for now, the beam splitter forms two beams that counterpropagate in the loop and recombine at the beam splitter. Since the path length is the same for these two paths, there is no phase difference between the beams at the splitter. If one of the mirrors were to move, causing only a change in the path length, the two counterpropagating waves would remain in phase at the splitter. This type of change in the interferometer is called a reciprocal phase change since both paths are equally affected. Any drift in the apparatus, including thermal changes in the components, will for the most part result in reciprocal changes.

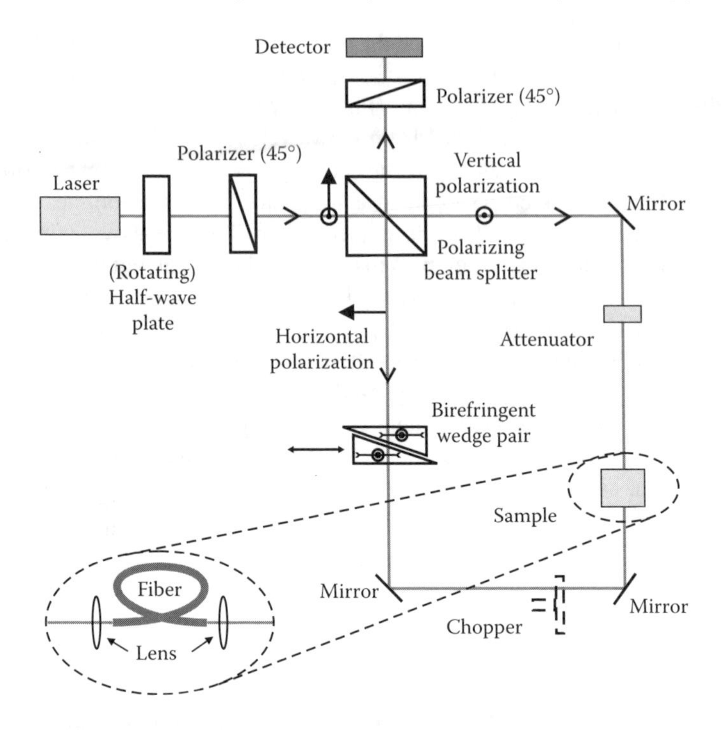

FIGURE 7.28 The Sagnac interferometer. A beam is split into two beans of orthogonal polarization that counterpropagate in the loop, recombine at the polarizing beam splitter, and interfere in the polarizer prior to entering the detector. Adapted from Ref. [20] with permission of the Optical Society of America.

In the design used by Gabriel, and later by Garvey, the beam splitter passes one polarization and reflects the orthogonal one. The half-wave plate following the laser source can be turned to adjust the intensity passing through the polarizer. The polarizer can be adjusted to control the polarization angle of the light entering the beam splitter. In the typical nonlinear measurement, the polarizer is set to 45° so that the counterpropagating beams are of equal intensity. In this design, then, a birefringent material placed in the beam within the interferometer will affect each of the beams differently since they are orthogonally polarized. A nonreciprocal phase shift results.

The nonreciprocal phase shift of the interferometer can be controlled using a birefringent wedge. As the thickness of material that is traversed by the beam is changed by moving the wedge into the beam, the phase difference between the two beams changes. For a single wedge, however, the beam is deflected. While the deflection angle is independent of the transverse wedge position, the beam will be shifted. To avoid this, a double wedge pair is used as shown in Figure 7.28.

Now let's consider the nonlinear-optical measurement. First, it is usually desirable to measure the ultrafast response, and to eliminate background from slower thermal effects (even on time scales of tens of nanoseconds, a signal from thermal heating can be comparable to the electronic mechanism). In most nonlinear-optical experiments, the signal is small and the background noise is large. As such, signal averaging over many laser pulses is required; but this leads to cumulative effects such as heating. Since the Sagnac interferometer is inherently not sensitive to slow thermal drifts, this problem is eliminated. However, if the repetition rate of the laser is high, the effects from one pulse may still be present when the next pulse reaches the sample. This effect is eliminated by interrogating the interferometer just prior to the measurement pulse's arrival.

The nonlinear measurement works as follows. An attenuator is placed in the loop near the sample, which is placed at an asymmetric point of the loop (in our drawing, this results in the clockwise-traveling pulse reaching the sample before the counterclockwise-traveling pulse). Thus, the dim pulse reaches the sample first and passes through without inducing an appreciable refractive index change due to its low intensity. Subsequently, the counter-propagating bright pulse passes through the sample, inducing a refractive index change that causes the light to experience an intensity-dependent phase shift. This bright pulse is then attenuated and recombines with the other pulse at the beam splitter. The relative phase shift between these two beams grows with increasing intensity, causing a change in the fraction of total light entering the detector.

The interferometer bias, which is defined as the phase difference between the two counterpropagating pulses in the absence of the nonlinear material, is adjusted using the birefringent wedge. Figure 7.29 shows the interferometer transmittance (the ratio of the output to input intensity) as a function of the bias for low (thin curve) and high (thick curve) intensities. The dashed curve shows the difference

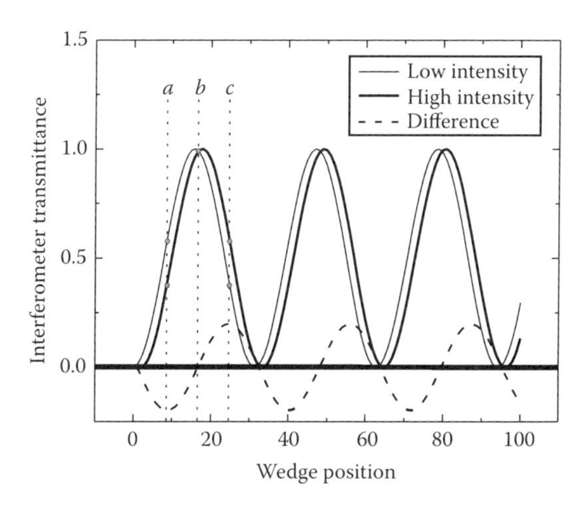

FIGURE 7.29 The transmitted intensity of a Sagnac interferometer as a function of the wedge position for two different intensities. The dashed curve shows the difference between the two curves.

between the two. Thus, if the interferometer bias is set to point a, as the intensity of the light is increased, the refractive index change causes a phase shift that leads to destructive interference; so the transmittance falls. At point b, the transmittance does not change with intensity; while at point c, it increases with intensity. The most sensitive point in the interferometer is thus at 50% transmittance; and the transmittance change is in phase with the light launched into the interferometer at point c and 180° out of phase at point a.

Gabriel and coworkers did their measurements by fixing the bias and and varying the intensity of the incident beam using the half-wave plate and measuring the transmittance. When the bias point is set at the equivalent of point a, the transmittance was found to decrease while at point c it was found to increase — as expected. For a large enough intensity range, the transmittance is a sinusoidal function of the incident intensity. Gabriel used the Sagnac interferometer to measure n_2 for a silica wafer, an inverted rib single-mode channel waveguide of PMMA polymer doped with the squaraine dye BSQ (see Figure 3.3 in Chapter 3), and a planar waveguide of poly-bis toluene sulfonate polydiacetylene. As shown in Figure 7.28, the sample can be both a bulk material or a fiber; though when measuring a waveguide, alignment is tricky. (Both counterpropagating beams must be coupled into the waveguide AND they must be co-linear everywhere in the interferometer.) Note that in these measurements, a chopper was used to modulate the signal so that a lock-in amplifier could be used to reduce stray signal from background light.

The main difference between Garvey's and Gabriel's methods is in the method of modulation. Garvey places a rotating half-wave plate directly in front of the laser. The light transmitted through the polarizer is thus modulated sinusoidally in time at frequency Ω. The repetition rate of the laser is much higher than the modulation frequency, leading to a time series of pulses whose amplitudes vary sinusoidally. The detector output is integrated with an RC circuit whose time constant is much shorter than the modulation frequency of the half-wave plate but slow compared with the repetition rate. This output is sent to a lock-in amplifier that is synched with the rotating half-wave plate. With no sample, the input to the lock-in amplifier is a sine wave at the modulating frequency. With the sample, the intensity-dependent refractive index leads to a small deformation in the waveform at twice the modulating frequency. Thus, the ratio of the signal at 2Ω to the signal at Ω is related to n_2.

One other complication is that the imaginary part of n_2 can also contribute. Whereas the real part results in an intensity-dependent shift as shown in Figure 7.29, the imaginary part results in an increase in the amplitude of this function. A measurement of the change in transmittance at one wavelength is therefore not sufficient to determine the real and imaginary parts. To understand how the two can be separated, consider a measurement performed at points a and c. If there is no imaginary part of n_2, the amount of signal decrease at point a is the same as the signal increase at point c. If there in an imaginary contribution, it will result in the same signal at both points (i.e., it will add to the signal at both points if the intensity-dependent absorption is negative and will subtract if positive). Thus, the

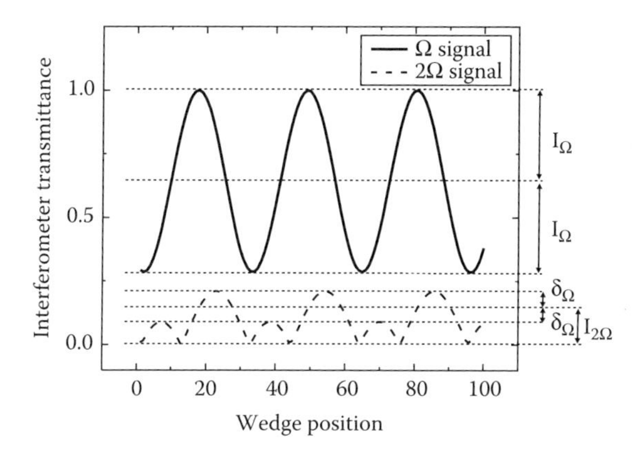

FIGURE 7.30 The theory of the transmitted signal at Ω and 2Ω for a Sagnac interferometer with a nonlinear material that has both real and imaginary n_2.

average of the change in signal at the two points yields the imaginary part and the difference gives twice the real part.

In Garvey's experiment, the magnitude of the signal read by the lock-in amplifier as a function of the bias is shown in Figure 7.30. Because of the homodyne nature of the experiment, it is possible to measure a signal at 2Ω that is three or four magnitudes lower than the signal at Ω. As such, the sensitivity of this apparatus is higher than that of Gabriel's, which requires that the signal be at least 5–15% of the background level. Note that the various amplitudes in Figure 7.30 are labeled as intensities, but experiments measure the voltage at the detector. Since n_2 will be calculated from ratios of intensities, as long as the voltage is proportional to the intensity, we will call the detector reading an intensity.

If the interferometer is well balanced (i.e., the intensity in each arm is the same) and if the mirrors are perfectly aligned so that all the beams are on top of each other, the pattern at the detector will be a bull's eye of concentric bright rings separated by dark rings. The intensity at the center of the ring, which is what the detector is set to read, will make an interferogram as depicted in Figure 7.30. Under perfect alignment, the Ω signal will be identically zero at the minima. In practice, this is never so.

The 2Ω signal has contributions from the real part of n_2, which is a sinusoidal function of a wedge position that is shifted by $\pi/2$ relative to the Ω signal, and an imaginary part, which is independent of phase. The two acting together leads to a signal with an offset, so that magnitude of the 2Ω signal forms a series of alternating peaks with cusps at the zeros, as shown in Figure 7.30. More specifically, the signal is of the form $|A + B \sin \phi|$, where ϕ is the phase shift due to wedge position.

The real and imaginary parts of the third-order susceptibility component of $\chi^{(3)}_{1111}$ are related to the real and imaginary parts of the intensity-dependent refractive index component $n^{(2)}_{1111}$ and to the measured quantities in Figure 7.30

according to [20]

$$Re\left[\chi_{1111}^{(3)}\right] = \frac{n_0^2 c}{12\pi^2} Re\left[n_{1111}^{(2)}\right] = \frac{\left(I_{2\Omega}^2 - f\delta_{2\Omega}^2\right)^{1/2}}{I_\Omega} \times \frac{c n_I^{(0)} n_0^2 f \alpha}{3\sqrt{2}\pi^2\left(\exp\left[4\pi n_I^{(0)}\ell/\lambda\right]\right)} \frac{1}{I_0^\Omega}, \tag{7.49}$$

and

$$Im\left[\chi_{1111}^{(3)}\right] = \frac{n_0^2 c}{12\pi^2} Im\left[n_{1111}^{(2)}\right] = \frac{c n_I^{(0)} n_0^2 f^{3/2} \alpha}{3\sqrt{2}\pi^2\left(\exp\left[4\pi n_I^{(0)}\ell/\lambda\right]\right)} \frac{1}{I_0^\Omega} \frac{\delta_{2\Omega}}{I_\Omega}, \tag{7.50}$$

where α is the fractional transmittance of the absorber, λ the wavelength, ℓ the sample thickness, c the speed of light, f the intensity ratio of the counterpropagating beams, and the other variables are as previously defined or in Figure 7.30. (See Sections 7.1.1 and 7.1.1 to acquire a taste for how these types of equations are derived.) Note that all intensities measured in the interferometer are rms values since that is what is normally returned by a lock-in amplifier. Also note that all intensities measured from Figure 7.30 appear as ratios of these parameters, so the formula holds for any units of intensity.

Table 7.1 shows measurement results of a 0.1%(by weight) ISQ dye in PMMA and 2% BSQ dye in PMMA in a channel waveguide. Both measurements were performed using a mode-locked Nd:YAG laser operating at a wavelength of $1064nm$.

In a dye-doped polymer, the intensity-dependent refractive index originates mostly in the dopant molecule when the host polymer is optically linear. This is the case with PMMA, where contributions from the polymer is negligible. When the molecules are noninteracting, the bulk susceptibility $\chi^{(3)}$ is proportional to the molecular susceptibility, γ and the number density of dopants, N. For a solid solution of randomly oriented one-dimensional dye molecules in a polymer host

TABLE 7.1
Intensity-dependent refractive index of polymer waveguides measured using a Sagnac interferometer

Material	Geometry	$n^{(2)}$ $(10^{-14} cm^2/W)$	$\chi_{1111}^{(3)}$ $(10^{-12} cm^3/erg)$	Ref.
0.1% ISQ in PMMA	SM fiber	2.1 (± 1.2)	1.2 (± 0.7)	[20]
2% BSQ in PMMA	Channel Waveguide	20 (± 10)	11.1 (± 5.5)	[117]

(which for the squaraine dyes is a good approximation), the bulk susceptibility is given by

$$\chi^{(3)}_{1111} = \frac{1}{5} N \gamma^*, \tag{7.51}$$

where γ^* is the dressed second hyperpolarizability (a property of the dopant molecule) of the only nonzero tensor component; and we have converted weight percent to number density as described in Chapter 4. Note that dressed values are related to vacuum ones through a multiplicative local field tensor, which in many cases can be approximated by a scaler.

From a bulk measurement, Equation (7.51) can be used to get the microscopic second hyperpolarizability. The values for BSQ (as measured by Gabriel and coworkers in a channel waveguide [117]) and for ISQ (as measured by Garvey and coworkers in a single-mode $25cm$ long polymer optical fiber with an ISQ/PPMA core [20]), yields $\gamma^* = 25 \pm 13$ and $\gamma^* = 38 \pm 21$, respectively. (The source of the large uncertainties originates in the difficulties associated with determining the optical loss in a given fiber. For the short pieces of fiber used, a large contribution to the measured loss is the coupling loss, which can only be roughly estimated.) These two values agree within experimental uncertainties. Though these molecules are not identical, their structures are similar and they are expected to have a similar value of γ^*.

7.2.3 Optical Devices in Polymer Fibers

7.2.3.1 Sagnac Device

Most optical devices require about a π phase shift to perform a useful operation. Consider a $10W$ peak power laser pulse that is launched into a single-mode polymer optical fiber with a 9μ diameter core. The peak intensity is $1.57 \times 10^7 W/cm^2$ (well below the damage threshold of $10^9 W/cm^2$). For 2% BSQ, this leads to a refractive index change of $3.1 \times 10^{-6}/cm$. For light with a wavelength of $1\mu m$, this yields a phase shift of $0.063\pi/cm$. A $16cm$ length of fiber would thus be long enough to make a switching device.

Let's first consider an all-optical switching device using a bulk Sagnac interferometer. Figure 7.31 shows the design. The light in the Sagnac loop, drawn in light gray, is different in color or polarization than the control beam, which is drawn in dark gray. The dichroic beam splitter passes the light inside the interferometer and reflects the control light. (This prism can be designed to operate on each of the beams differently depending on their color or polarization. In our example, it is the color that is different.) Without the control beam, the interferometer is balanced (the polarizations and intensities of both beams are the same), so there is no relative phase shift between the two counterpropagating pulses. The light destructively interferes in the direction of the detector, causing it to be reflected back toward the laser source. If the control pulse is launched into the interferometer coincident with the counterclockwise pulse in the interferometer, the two

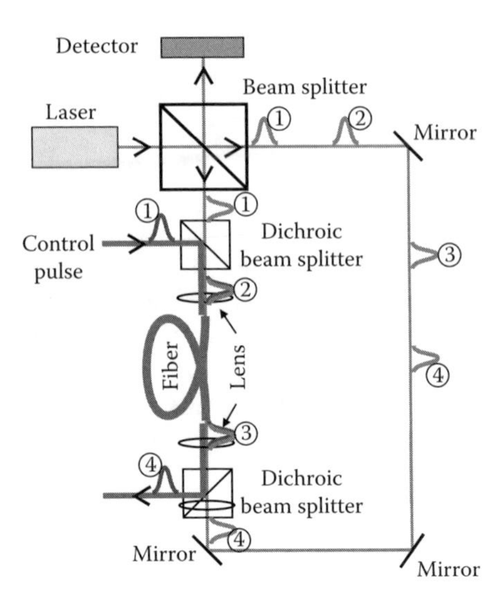

FIGURE 7.31 The Sagnac optical switch. Each number inside a circle represents a time index.

pulses will travel together in the fiber loop and are then removed by the second dichroic splitter. The counterclockwise pulse remains in the loop. The number in each circle labels the time sequence. Clearly, if the active fiber is placed asymmetrically within the loop as shown, the clockwise pulse will have traveled through the fiber before the clockwise pulse arrives. Consequently, the counterclockwise pulse will be phase shifted through the intensity-dependent refractive index relative to the other unaffected pulse. The light pulse will then emerge in the direction of the detector.

This device is a simple AND gate. Both pulses have to be launched into the device for there to be an output pulse at the detector. If we also consider the light that heads back toward the laser, it is on only when a light pulse is launched into the loop with no control pulse. These combinations of outputs can be cascaded together to form a variety of logic operations. If all the pulses are of equal intensity, then any of the outputs (control, detector, and laser) can be used as the laser input into the next devices. Additionally, the output from the loop can be launched back into another port (such as the control port, for example) to give feedback.

The bulk interferometer is a particularly clumsy one. It is difficult to align and requires lots of components. The all-fiber Sagnac interferometer is much simple to build and alignment of the components is automatic. Figure 7.32 shows an electrooptic and all-optical all-fiber Sagnac switch.

Clearly, the bottom part of Figure 7.32 is analogous to the bulk device shown in Figure 7.31. The directional coupler is an all-fiber component with four fiber pigtails. Thus, a Sagnac loop is made using a one-directional coupler and one loop of fiber. To make the switch, a fiber with an intensity-dependent refractive index is

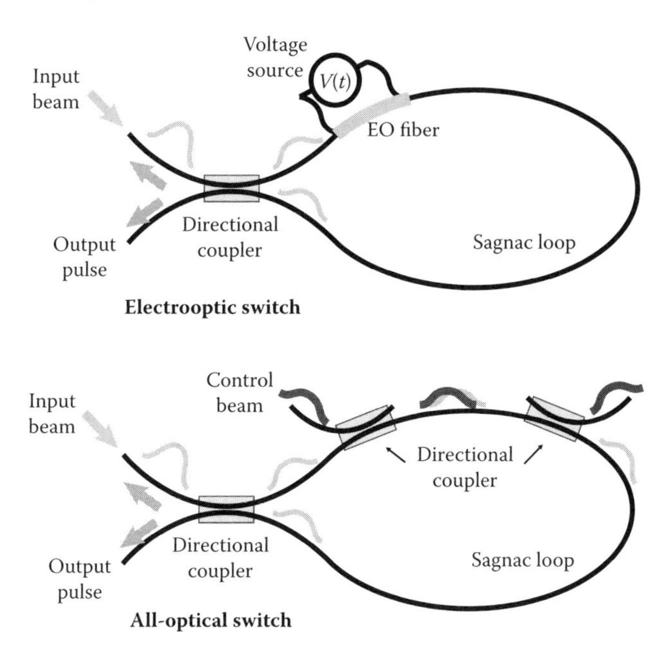

FIGURE 7.32 Electrooptic Sagnac switch (top) and all-optical Sagnac switch (bottom).

spliced into the loop with two other directional couplers. These two couplers are designed to pass the light in the input beam but to reflect the light in the control beam. Thus, the control beam propagates with the clockwise propagating pulse. Switching occurs in the same way as in the bulk device. The beauty of the polymer fiber component is clear. It can be made with a connector on each end, making it easy to snap into the system without the need for alignment. Second, the fiber provides a large intensity-dependent refractive index change, so that only a short length is required.

In the electrooptic switch, an electrooptic fiber is placed in the loop so that an applied voltage leads to a phase shift. If the applied voltage is timed to be coincident with one of the counterpropagating pulses, the light can be switched between the two ports.

Silica glass fibers have been used to make such devices, [118–120] but by virtue of the long lengths required (i.e., kilometers), the devices suffer from a long latency time; that is, the switched light leaves the loop at a time that is delayed relative to the incident light by an amount ℓ/c, where ℓ is the loop length and c the speed of light in the fiber (i.e., the mode velocity). The latency times are on the order of tens of microseconds. Polymer optical fibers with nonlinear-optical cores therefore offer some unique advantages for making better switching devices. In all fairness, we note, however, that there are some applications where the long latency times are not a problem. [118–120]

7.2.3.2 Dual-Core Polymer Fiber Switch

Xiong and coworkers made a twin core fiber with β-carotene-doped polymer cores to demonstrate all-optical switching.[121] When two cores are placed near enough to each other so that the evanescent field of light propagating in one of the cores overlaps the other guide, energy will be transferred between them. If all of the light is initially launched into one of the cores, the energy will transfer back and forth between the cores as the light travels down the length of the fiber. The distance over which all the light transfers from one core to the other one is called the coupling length or the beat length.

The beat length depends on the refractive index difference between each core and the surrounding cladding, the distance between the waveguides, and their shapes. If the cores are moved closer together, the evanescent field in the other waveguide gets larger, resulting in greater coupling between the waveguides, leading to a shorter beat length. Similarly, if the refractive index of each core is lowered, the tail of the evanescent field gets longer and the beat length decreases. Through the intensity-dependent refractive index, the beat length will depend on the intensity. So, for a fixed length of fiber, the light output can be made to alternate between the two cores simply by changing the intensity of the beam launched into one of the cores.

The end of Chapter 2 discusses the propagation of light in a twin core fiber with nonlinear waveguides. To understand the results of Xiong and coworkers, it is useful to calculate how light propagates in a waveguide in their specific geometry and to fill in some of the details, which help build an intuition to this complex behavior. To keep the calculation simple, we can make some reasonable assumptions. First, we assume that the modes in each waveguide are the same as the modes of an isolated waveguide. For this to be so, the interaction between the guides must be weak. Second, we will assume that both waveguides are identical in shape and composition. Finally, we assume that each waveguide supports one and only one mode, i.e., they are single-mode waveguides.

The electric field inside waveguide i can be expressed as

$$\vec{E}_i = a_{0i}\vec{e}_i(x, y)\exp[-i\beta z + i\omega t], \tag{7.52}$$

where a_{0i} is the amplitude of the field, $\vec{e}_i(x, y)$ is a normalized function that gives the transverse profile, and β is the mode's wavevector — so, the mode velocity is ω/β. Since the two cores are identical, Equation (7.52) describes the fields in both waveguides. The only difference between the two will be the amplitude a_0.

When there is an interaction between the waveguides, the amplitude a_0 will change as the beam travels down the fiber, so a_{0i} will be a function of z. It is convenient to combine all of the z-dependence in one function (this form is particulary useful when the waveguides are made of a material with an intensity-dependent refractive index), so we define

$$a_i(z) = a_{0i}\exp[-i\beta z]. \tag{7.53}$$

When the two waveguides interact, the equation for the fields is derived from Maxwell's wave equation. For the z-dependence of the field in guide 1, a_1, in terms of the field in guide 2, a_2, we get

$$\frac{da_1}{dz} + i\beta a_1 = ia_2 C, \tag{7.54}$$

where C is a constant that is proportional to the strength of energy exchange from guide 2 to guide 1 and depends on the geometry of the waveguides. C is given by the amount of mode overlap,[122]

$$C = \frac{\omega}{2} \int_{A_\infty} n^2(x, y) \left[\vec{e}_1(x, y) \cdot \vec{e}_2{}^*(x, y)\right] dxdy, \tag{7.55}$$

where the integration is over the whole infinite plane (A_∞) that contains the cross-section of both cores in the fiber at z. (Because the transverse intensity profile of an eigenmode of a waveguide propagates without changing, this integral is independent of z.) $\vec{e}_i(x, y)$ is the electric field in guide i, so, if the two waveguides are offset from each other by the displacement (x_0, y_0), and if both modes are the same, then $\vec{e}_1(x, y) = \vec{e}_2(x - x_0, y - y_0)$. The equation for the field in core 2 is given by Equation (7.54) with the subscripts 1 and 2 interchanged.

Before proceeding to the nonlinear case, we begin by solving the linear one. First, we note that with no coupling between waveguides ($C = 0$), the solution to Equation (7.54) is simply Equation (7.53), as we would expect. Now to solve the coupled equations for the field in guide 1, we eliminate a_2 from the two first-order differential equations (given by Equation 7.54 and Equation 7.55 with 1 and 2 interchanged) to get the second-order differential equation:

$$\frac{d^2 a_1}{dz^2} + 2i\beta \frac{da_1}{dz} + \left(C^2 - \beta^2\right) a_1 = 0, \tag{7.56}$$

with general solutions,

$$a_1(z) = A \exp\left[i(-\beta + C)z\right] + B \exp\left[i(-\beta - C)z\right], \tag{7.57}$$

where A and B are integration constants. If $a_1 = 0$ and $z = 0$, and the maximum amplitude of the field is unity, this boundary condition yields $A = -B = 1$, so the final result is

$$a_1 = \exp[-i\beta z] \sin(Cz). \tag{7.58}$$

The electric field in the second guide can be calculated using the same approach. Alternatively, we can use Equation (7.58) and energy conservation to get the intensity $|a_2|^2$,

$$|a_2|^2 = 1 - |a_1|^2 = \cos^2(Cz). \tag{7.59}$$

So, energy is transferred between the two guides sinusoidally with a beat length of $\pi/2C$.

Now we are ready to tackle the problem of the nonlinear waveguide, where the refractive index depends on the intensity. To simplify the discussion, let's ignore the constants that depend on units, and define the intensity dependent refractive index in the form $n = n_0 + n_2(\vec{E}^* \cdot \vec{E})$. In Equation (7.54), both β and C depend on the refractive index, so both terms need to be considered. We treat each separately.

Let's first consider the overlap integral given by Equation (7.55). The square of the refractive index profile in the nonlinear case is given by,

$$n^2(x, y) = (n_0 + n_2\vec{E}(x, y)^* \cdot \vec{E}(x, y))^2 \approx n_0^2 + 2n_0n_2\vec{E}(x, y)^* \cdot \vec{E}(x, y), \quad (7.60)$$

where we have used the fact that the refractive index change is usually small compared to the refractive index. Using Equation (7.52), and the fact that the total electric field is the sum of the fields from each core ($\vec{E} = \vec{E}_1 + \vec{E}_2$), we get

$$\begin{aligned}\vec{E}^*(x, y) \cdot \vec{E}(x, y) &= |a_1\vec{e}_1(x, y) + a_2\vec{e}_2(x, y)|^2 \\ &= |a_1|^2\, \vec{e}_1^*(x, y) \cdot \vec{e}_1(x, y) + |a_2|^2\, \vec{e}_2^*(x, y) \cdot \vec{e}_2(x, y) \\ &+ a_1a_2^*\vec{e}_1(x, y) \cdot \vec{e}_2^*(x, y) + a_1^*a_2\vec{e}_1^*(x, y) \cdot \vec{e}_2(x, y)\end{aligned} \quad (7.61)$$

Substituting Equation (7.61) into Equation (7.60) and then substituting the result into Equation (7.55), the nonlinear refractive index contribution to the overlap integrals becomes

$$C_{ij}^{NL} = \omega \int_{A_\infty} n(x, y)n_2(x, y)[\vec{e}_1(x, y) \cdot \vec{e}_2{}^*(x, y)][\vec{e}_i(x, y) \cdot \vec{e}_j{}^*(x, y)]dxdy, \quad (7.62)$$

where, for our particular case, $n_2(x, y)$ is the constant n_2 inside the cores and zero in the cladding. The overlap integral thus becomes

$$C \rightarrow C + \sum_{i,j=1}^{2} C_{ij}^{NL} a_i a_j^*. \quad (7.63)$$

Next, we need to consider the term $i\beta a_1$ in Equation (7.54). The intensity-dependent refractive index results in an intensity-dependent β in waveguide i of the form [123]

$$\delta\beta_i = k \int_{A_\infty} \delta n(x, y, z)\vec{E}_i^*(x, y) \cdot \vec{E}_i(x, y)dxdy, \quad (7.64)$$

where k is the free-space wavevector and $\delta n(x, y, z)$ is the change in the refractive index as induced by the total electric field in all of space,

$$\delta n(x, y, z) = n_2(x, y)(\vec{E}_1^*(x, y) + \vec{E}_2^*(x, y)) \cdot (\vec{E}_1(x, y) + \vec{E}_2(x, y)). \quad (7.65)$$

The dot product of the fields in Equation (7.65) is the same as given by Equation (7.61). Defining β_{ijk}^{NL} as

$$\beta_{ijk}^{NL} = k \int_{A_\infty} n_2(x, y) \left[\vec{e}_k(x, y) \cdot \vec{e}_k{}^*(x, y)\right] \left[\vec{e}_i(x, y) \cdot \vec{e}_j{}^*(x, y)\right] dxdy, \quad (7.66)$$

so, β_k for core k can be written as

$$\beta_k \rightarrow \beta + \sum_{i,j=1}^{2} \beta_{ijk}^{NL} a_i a_j^*. \tag{7.67}$$

Substituting Equation (7.67) and Equation (7.62) into Equation (7.54), we get the final result for nonlinear coupling:

$$\frac{da_1}{dz} + i\left(\beta + \sum_{i,j=1}^{2} \beta_{ij1}^{NL} a_i a_j^*\right) a_1 = ia_2\left(C + \sum_{i,j=1}^{2} C_{ij}^{NL} a_i a_j^*\right) - \alpha a_1, \tag{7.68}$$

where we have added a loss term, $-\alpha a_1$, which accounts for energy being lost from the system due to absorption or scattering. In glass fibers, loss is usually ignored, but in polymer fibers, this term can be important. For waveguide 2, we get the same expression with 1 and 2 interchanged. The coupled equations derived from Equation (7.68) are too complex to solve analytically, so numerical techniques must be used.

There are two ways to express Equation (7.68). We have done so by defining the a_i according to Equation (7.53), so it represents the total dependence of the field on z. Other researchers choose to write Equation (7.68) in terms of what we call a_{0i}, the slowly varying envelope of the electric field. With this choice,

$$\frac{da_i(z)}{dz} + i\beta a_1 \rightarrow \frac{da_{0i}(z)}{dz} \exp[-i\beta z]. \tag{7.69}$$

Now we are ready to proceed with the experiments. The dual-core preform was fabricated by Xiong by polymerizing a rod around two parallel Teflon wires, removing the wires, then filling them with a mixture of initiator, monomer, and the dopant β-carotene at $250ppm$. The researchers were careful to polymerize the cores in a dark oxygen-free environment to prevent the highly sensitive chromophores from decomposing. After polymerizing the cores, the preform was drawn into a fiber of $3.0\mu m$ diameter cores with a center-to-center separation of $10.5\mu m$. The refractive index difference between core and cladding was $\Delta n = 0.0056$ yielding a cut-off wavelength of $790nm$. At $810nm$, the attenuation coefficient was $\alpha = 1.13/m$.

As shown in Figure 7.33, $810nm$ light from a diode source is launched into one of the cores (labeled 1) of the $1.3m$ long twin-core fiber through a single-mode

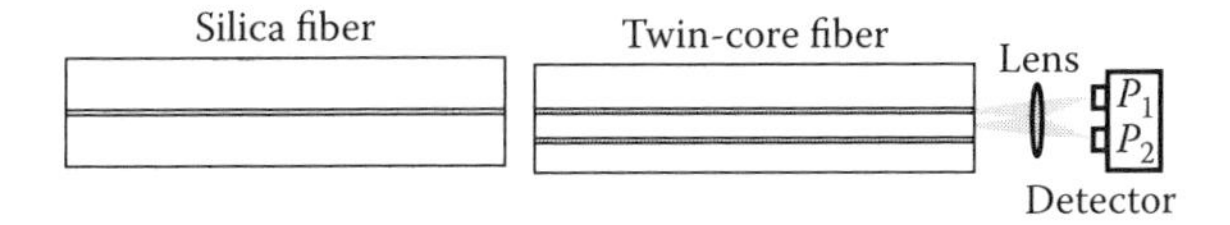

FIGURE 7.33 A schematic diagram of the twin-core nonlinear coupling experiment (the fibers are not drawn to scale).

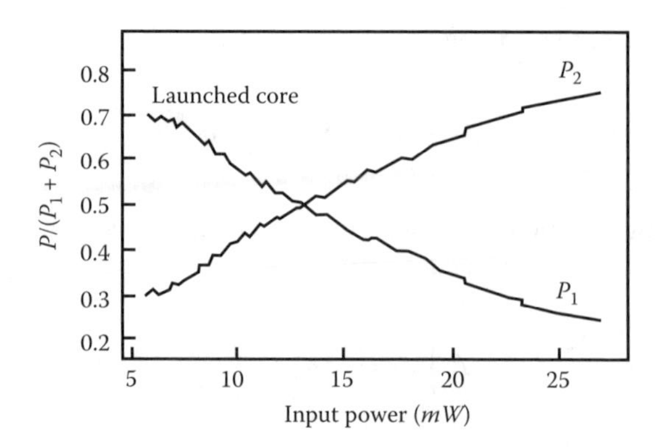

FIGURE 7.34 The normalized output power from each core as a function of the input power. Adapted from Ref. [121] with permission of SPIE.

silica optical fiber. A 100× objective imaged the light on the output side to two parallel detectors — each of which measures the light intensity exiting one of the cores. The laser intensity is a triangular waveform with $4ms$ full width and a maximum peak power of $27mW$. The input power is coupled into one of the cores and the output of each core is recorded as a function of time.

Figure 7.34 shows the normalized power in each core as a function of the input power. (At each input power, the output power plotted is divided by the sum of the powers in each core.) The output power of core 1 (the core into which the light is originally launched) starts high and drops with input power while the light in the other core rises. So, with $27mW$ of input power, the nonlinear interaction causes most of the light from one core to switch into the other one. Based on experiments that varied the rep-rate of the pulses, Xiong determined that the response time of the material had to be shorter than $300ns$.

The theory that Xiong and coworkers use to analyze the data is of the form given by Equation (7.68), but using the convention that $a_i(z)$ is the slowly varying envelope of the electric field. Specifically, their coupled equations are

$$\frac{da_1}{dz} = iC_\ell a_2 \exp[2i\delta z] + iC_{n\ell} a_1 \left|a_1\right|^2 - \alpha a_1, \tag{7.70}$$

for core 1 and the same equation for core 2, with 1 and 2 interchanged. Furthermore, $\delta = (\beta_2 - \beta_1)/2$,

$$C_\ell = \frac{kc\epsilon_0}{4} \int\int n^2(x, y) E_1(x, y) E_2^*(x, y) dxdy, \tag{7.71}$$

and

$$C_{n\ell} = \frac{kc^2\epsilon_0^2}{4} \int\int n^2(x, y) n_2(x, y) \left|E_{1,2}(x, y)\right|^2 dxdy, \tag{7.72}$$

where from the constants, it is clear that they have used SI units.

Solving the nonlinear coupled equations numerically, the data shown in Figure 7.34 could be reproduced with reasonable values of the parameters. For example, the data was used to determine that the change of the refractive index induced with a 10mW beam was $\Delta n = 9.23 \times 10^{-7}$.

While the authors did not discuss the material implications of their measurement, this is a topic that deserves some discussion. The intensity-dependent refractive index can be estimated from the refractive index change, Δn, the cross-sectional area of the beam, A, and the intensity of light, I, according to $n_2 = \Delta n / I$. If we assume that the beam uniformly illuminates the $3\mu m$-diameter core, we get $n_2 = 6.6 \times 10^{-13} cm^2/W$. This value is in the range of what is typically measured for an electronic or reorientation response of an organic liquid. Given the low concentration of the β-carotene molecules, however, the values seem a bit large.

One of the biggest nuisances in an all-optical device is photothermal heating, which leads to a thermal intensity-dependent refractive index. Thermal effects tend to be large on second to millisecond time scales (the time it takes the material to absorb energy and heat the sample), though thermal effects can still be large in the nanosecond regime. Given the large amount of light absorbed by the fiber, even on millisecond time scales, the temperature increase is enough to change the refractive index many orders of magnitude larger than measured.

The geometry of the device is an important consideration to understand why the thermal contribution is smaller than expected. Recall that the cores are close together, so when heating occurs in one core, the heat can flow to the other core. Since the cores are so close together, this can happen quickly. Since this nonlocal heating effect is not taken into account by the theory (i.e., the process where the intensity in one of the cores affects the refractive index in the other one), this could lead to a balancing effect that decreases the effects of heating. So, it is likely that the mechanism that is largely responsible is the thermal mechanism. Whatever the mechanism, this demonstration shows promise for future applications.

7.2.4 Optical Bistability and Multistability

An optically bistable device is one that can be found in one of two states (characterized by two output intensities) for a single input intensity. This is equivalent to saying that the output versus input function is double valued for some range of input intensities. Figure 7.35 shows an example of such a device. As the input intensity is turned on, the output rises, as shown by the bottom arrow. At the input intensity I_{on}, the output jumps to the upper branch. With further increase of the intensity, the function remains single valued. As the intensity is decreased, the same path is traced out until the input intensity falls below I_{on}, at which point the upper part of the curve is traced out. When the input intensity falls below I_{off}, the output intensity drops abruptly to the lower branch.

Such a device can be used as an optical memory, where an output intensity of I_{out}^{off} characterizes the binary zero level and an output intensity of I_{out}^{on} characterizes the binary one state. This device needs to be powered with a bias intensity I_{bias} that falls halfway between I_{off} and I_{on}. (The bias intensity enters the device at the

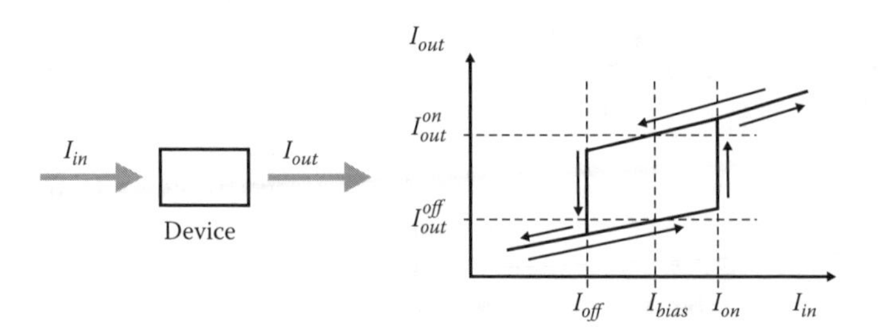

FIGURE 7.35 The output versus input plot of an optically bistable device is a double-valued function.

input along with the input intensity.) So, if the device is in its off (or zero) state, it can be turned on with an input pulse whose intensity exceeds $I_{on} - I_{bias}$. When in the on state, the device can be turned off by providing a transient at the input that drops the bias intensity below I_{off}.

The material from which the device is made changes its state to cause the output intensity of the device to change. The most common demonstration of optical bistability uses a Fabry-Perot interferometer, which is made of a material with an intensity-dependent refractive index that is placed between parallel reflectors. In this case, it is the refractive index of the material between two mirrors that is a double-valued function of the input intensity. If the intensity of the light beam causes a change in the separation between the reflectors, then the different states of the device are characterized by different interferometer lengths.

Bistable devices are a subset of the more general class of multistable devices. Most systems are in fact multistable, where for some range of input intensity, the output intensity can take on many values, depending on the time sequence in which the light beam is turned on.

Multistable devices operate on feedback, where the output light is fed back into the material and interacts through the optical nonlinearity with the incident light. The feedback can be extrinsic, where the output light is redirected back into the material, or intrinsic, where the feedback is built into the device. In the Fabry-Perot interferometer, the light can be viewed as bouncing back and forth between partially transmitting reflectors. Upon each bounce at the output side, some of the light exits and the remaining light is reflected back into the material where it interacts with the incident light.

In the following sections, the output versus input function is calculated for a Fabry-Perot interferometer, which is made of a fiber between two partially transmitting mirrors, and for a bare fiber with no mirrors. Optical feedback in the latter device is due to fresnel reflections from the ends.

7.2.4.1 Multistability of a Fabry-Perot Interferometer with End Reflectors

Figure 7.36 shows a fiber Fabry-Perot interferometer that is made of a polymer fiber and a reflector on both ends. The bottom part of the figure shows the complex

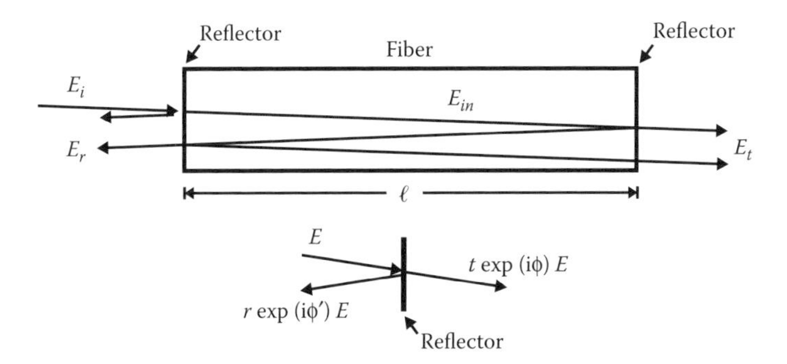

FIGURE 7.36 An optical fiber with a reflector on each end makes a Fabry-Perot interferometer (top). The reflected and transmitted field from the reflector is shown on the bottom.

transmittance and reflectance of an electric field from the reflector. At this point, we ignore the waveguiding modes, so there are no reflections off the sides of the fiber. Under these assumptions, the transmitted electric field is related to the incident field according to

$$E_t = E_i\left(t\exp\left[i\phi\right]\right)^2\left[1+\left(r\exp\left[i\phi'\right]\exp\left[2\pi in\ell/\lambda\right]\right)^2+()^4+\cdots\right],$$
$$= \frac{t^2\exp\left[2i\phi\right]}{1-r^2\exp\left[2i\left(\frac{2\pi nl}{\lambda}+\phi'\right)\right]}E_i \qquad (7.73)$$

where the coefficient $(t\exp[i\phi])^2$ represents the fact that all rays that exit on the right are transmitted through front and back reflectors once, the first term in the brackets represents the ray that goes straight through, the second term, $\left(r\exp\left[i\phi'\right]\exp\left[2\pi in\ell/\lambda\right]\right)^2$, a ray that has reflected once from the back and front reflector before exiting, the next term for two round trips, etc. The factor $2\pi in\ell/\lambda$ is the accumulated phase for a wave that travels from one reflector to the other one. ℓ is the length of the fiber, n the refractive index; and ϕ and ϕ' the phase upon transmission and reflection from the reflector.

To take into account waveguiding in a large-diameter multimode fiber, the ray picture is sufficient. Figure 7.37 shows one of the rays as it undergoes total internal reflection. The rays, as they exit either on the right or the left, will interfere as before, but the path length ℓ is replaced with the longer path length given by $\ell/\sin\theta$. The fact that the phase changes with each bounce of the light as it reflects from the surface of the fiber due to total internal reflection also needs to be taken into account. The number of bounces, N_b, for each single pass of the fiber length is

$$N_b = \frac{\ell}{D\sin\theta}, \qquad (7.74)$$

where D is the fiber diameter. Thus, the phase shift upon one traversal of the fiber, $\Delta\phi$ is

$$\Delta\phi = \frac{\delta(\theta)\ell}{D\sin\theta}, \qquad (7.75)$$

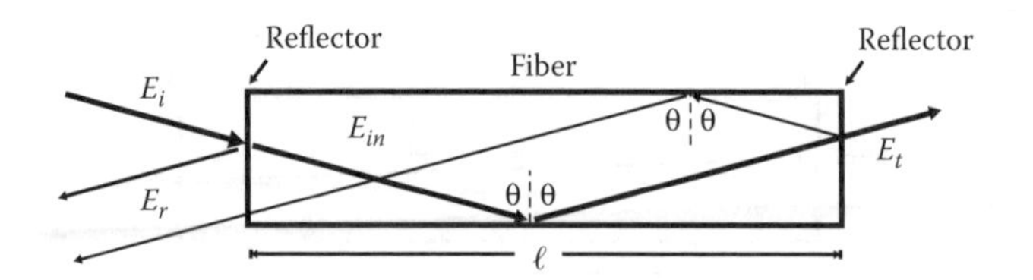

FIGURE 7.37 A multimode optical fiber with a reflector on each end.

where $\delta(\theta)$ is the phase shift upon one bounce. Note that $\delta(\theta)$ depends on the angle of the ray and the polarization of the light beam with respect to the plane of incidence of that bounce.

Taking both the phase shift and the ray propagation distance into account, Equation (7.73), with the help of Equation (7.74), becomes,

$$E_t = E_i \left(t \exp\left[i(\phi + \Delta\phi)\right]\right)^2 \left[1 + \left(r \exp[i(\phi' + \Delta\phi)] \exp\left[\frac{2\pi i n \ell}{\lambda \sin\theta}\right]\right)^2 + \cdots\right]. \tag{7.76}$$

In the fiber, then, the energy travels along its length at a speed of $c \sin\theta / n$. We can call $n/\sin\theta$ the mode index.

It must be pointed out that this result is for a ray that falls in a plane that contains the axis of the fiber. In this case, both the reflected and transmitted rays (labeled as E_r in Figure 7.37 and E_t, respectively) can either emerge parallel to the incident ray or parallel to the *first* reflected ray from the left surface. As such, there are two possible rays that can emerge in the plane defined by E_i and E_r. If the light is launched into the fiber, not in the plane containing the fiber axis (called a skew ray), then the output light will fall on the surface of a cone that contains the two rays found in the planar case. In both cases, the transmitted field can be calculated using Equation (7.74). One other complication needs to be taken into account. If the beams are polarized either in the plane of incidence or parallel to it, the polarization of the light beam will remain unchanged after each reflection. Since the reflection coefficient depends on the polarization, the polarization of a light beam that is not along one of these two directions will rotate. A rigorous calculation must take this effect into account. Clearly, skew rays are more difficult to treat and will not be further considered in this chapter.

In the limit where there are only a few waveguiding modes, where the waveguide region's transverse dimension is on the order of $\lambda/\Delta n$ (Δn is the refractive index between the core and the cladding), the wave nature of the electric field must be used to calculate the interference condition. The simplest case is the one in which the fiber only supports one waveguiding mode, or if the modes do not couple to each other. In this case, the energy travels at a speed given by the speed of light divided by the mode index, and the intensity profile of the beam remains unchanged. Thus Equation (7.73) is generalized by replacing the refractive index with the mode index. The transverse intensity profile remains the same throughout the length of the fiber. If more than one mode is generated, each mode travels at a different mode index so that the transverse intensity profile changes as the light propagates. Thus, the intensity at the output plane of the fiber will vary differently with fiber length depending on where in the output plane the detector is placed.

To calculate the transmitted intensity of the Fabry-Perot fiber, we return to the ray picture given by Figure 7.36 and the fields given by Equation (7.73). Recalling that the intensity is proportional to the square of the magnitude of the electric field and the refractive index of the material, Equation (7.73) yields the transmitted intensity,

$$\boxed{I_t = \frac{I_i}{1 + \left(\frac{2r}{1-r^2}\right)^2 \sin^2\left(\frac{2\pi n\ell}{\lambda} + \phi'\right)}}, \tag{7.77}$$

where we have used the fact that $1 - r^2 = t^2$; and the infinite series were summed using $\sum_{n=0}^{\infty} \epsilon^2 = 1/(1 - \epsilon^2)$.

Equation (7.77) gives the transmittance (I_t/I_i), which is plotted as a function of the optical path length, $2\pi \ell n/\lambda$, in Figure 7.38. The period of this function depends on the spacing between the mirrors and the wavelength of the light. Indeed, Fabry-Perot interferometers are used to accurately determine the wavelength spectrum of light. As the reflectance of the mirrors increases, the fringes become sharper and the contrast increases. Fabry-Perot interferometers with high-reflectivity ends are often used to resolve small wavelength differences in light spectra.

To get multistability, we need a device where the output is fed back into the device. Equivalently, this will happen when the optical path length depends on the intensity of light inside the Fabry-Perot interferometer. Such an intensity dependence leads to an internal feedback — that is, the light does not leave the device before providing feedback. The most common demonstrations of optical multistability are for systems where the refractive index depends on the intensity.

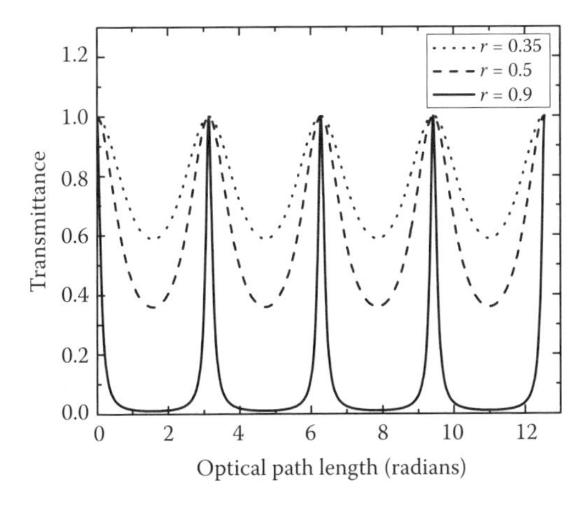

FIGURE 7.38 Transmittance of a Fabry-Perot interferometer as a function of optical path length between the reflectors.

Interestingly, the results will be the same if the length of the material depends on the intensity. Since the intensity-dependent length change is not as commonly treated, we will focus on this process.

To determine the length of the device, we need to calculate the intensity inside the interferometer. Summing over all the rays inside the material, the electric inside is given by

$$\begin{aligned} E_{in} = {} & E_i t \exp[i\phi](\exp[ikx][1 + (r\exp[i\phi']\exp[ik\ell])^2 + ()^4 + \cdots], \\ & + r\exp[-i\phi']\exp[2ik\ell]\exp[-ikx][1 + (r\exp[i\phi']\exp[ik\ell])^2 + \cdots]), \end{aligned} \tag{7.78}$$

where x is the distance inside the interferometer from the left reflector and $k = 2\pi n/\lambda$.

We can understand this complex expression by associating a ray with each term. Starting with the pre-factor, t appears once since all of the light inside the interferometer is transmitted through a reflector only once (to get inside from the left). The first line in Equation (7.78) represents the right-going rays and the second line the left-going ones. The factor $\exp[ikx]$ is clearly a right-going wave (the term $-i\omega t$ in the exponent is omitted because it will eventually be replaced with a time average when the intensity is calculated from the electric field). The first term in the sum (the "1") represents the transmitted ray. The next term in the sum represents a ray that has bounced off the right reflector, then the left one; hence, the term goes as the square of the reflectance. The round-trip phase upon this double bounce is $2ik\ell$.

The left-going rays (second line in Equation 7.78) suffer an extra reflection, hence the extra reflectance pre-factor. The first term in the brackets (the "1" term), corresponds to the ray that has reflected from the right reflector once, the next term corresponds to two bounces, etc. The pre-factor $\exp[2ik\ell]\exp[-ikx]$ represents the accumulated phase of a ray at position x after one bounce inside the interferometer, which has traveled a distance ℓ to the right and a distance $\ell - x$ to the left — leading to a total phase of $k(2\ell - x)$.

Summing the infinite series and combining terms, we get the inside electric field:

$$\begin{aligned} \frac{E_{in}}{E_i} = {} & \frac{t\exp[i\phi]}{1 - (r\exp[i\phi']\exp[ik\ell])^2} \\ & \times [\exp[ikx] + r\exp[i\phi']\exp[2ik\ell]\exp[-ikx]]. \end{aligned} \tag{7.79}$$

Taking the square of the modulus of Equation (7.79) and multiplying by the refractive index, we get the intensity inside:

$$\boxed{I_{in} = nt^{-2}I_t\left[(1+r)^2 - 4r\sin^2\left(\frac{2\pi n}{\lambda}(x-\ell) - \frac{\phi'}{2}\right)\right]}, \tag{7.80}$$

where we have explicitly displayed $k = 2\pi n/\lambda$. Also, we have used Equation (7.77) to eliminate I_i.

The first term in brackets is a constant while the second term results in intensity modulation. The mean intensity inside the interferometer is given by

$$\bar{I}_{in} = \frac{1}{\ell}\int_0^\ell I_{in}dx = \frac{nI_t}{\ell t^2}\left[\left(1+r^2\right)x - \frac{\lambda r}{2\pi}\sin\left(\frac{4\pi n}{\lambda}(x-\ell)+\phi'\right)\right]\Bigg|_0^\ell$$
$$= \frac{nI_t}{t^2}\left[\left(1+r^2\right) + \frac{r}{2\pi}\left(\frac{\lambda}{\ell}\right)\left\{\sin\phi' - \sin\left(\frac{4\pi n}{\lambda}+\phi'\right)\right\}\right], \tag{7.81}$$

where we have used the trigonometric identity $\sin\theta = (1-\cos\theta)/2$. When the fiber is many wavelengths long ($\ell >> \lambda$), the second term in Equation (7.81) will be much smaller than unity, so the mean intensity is then

$$\boxed{\bar{I}_{in} = \frac{n}{t^2}I_t(1+r^2) = n\left(\frac{1+r^2}{1-r^2}\right)I_t}\,. \tag{7.82}$$

Note the limiting cases of Equation (7.82): when $r = 0$, the mean intensity inside is the same as the transmitted intensity; but when the reflectivity approaches unity, the intensity inside diverges. When $r \to 1$, the intensity builds up as the rays bounce to and fro, leading to the divergence. A Fabry-Perot cavity is therefore a good light intensifier, which makes it useful as part of a device whose response depends on the intensity.

Now we focus on a material whose length depends on the intensity (i.e., a photomechanical effect) and consider what happens when it is placed in a sinusoidally varying spatial intensity, of the form

$$I(x) = \bar{I} + A\sin(kx), \tag{7.83}$$

where A is of the same order as $\bar{I}$. We assume that the stress is linearly proportional to the intensity, i.e.,

$$\delta\ell(x) = \xi\ell I(x), \tag{7.84}$$

δ is the photomechanical constant that depends on the material's properties.

If we break up a fiber of length ℓ_0 into small pieces dx, the total length of the fiber is given by

$$\ell = \int_0^{\ell_0}\left(dx + dx\,\xi I(x)\right), \tag{7.85}$$

where the first term is the length of the dark fiber and the second term represents the photomechanical length change at position x in the fiber as given by Equation (7.84). Upon integrating Equation (7.85), and using the modulated intensity given by Equation (7.83), we get

$$\ell = \ell_0 + \xi\ell\bar{I} + \frac{\lambda\xi A}{2\pi}\left(1-\cos(k\ell_0)\right). \tag{7.86}$$

When the length of the fiber is long compared with the spatial period of modulation ($\ell >> \lambda$), the second term in Equation (7.86) vanishes. For our case of the fiber Fabry-Perot with a photomechanical effect, we have

$$\boxed{\ell = \ell_0 + \xi\ell\bar{I}_{in}}\,. \tag{7.87}$$

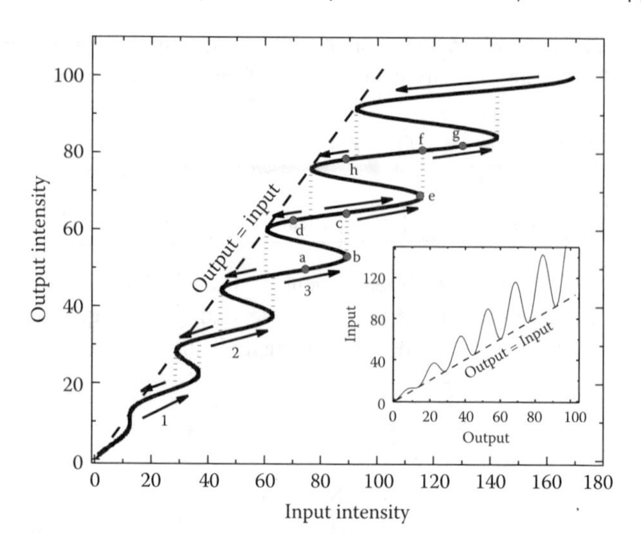

FIGURE 7.39 The transmitted intensity as a function of incident intensity for a multistable device. The inset shows the incident intensity as a function of the transmitted intensity.

As such, the mean fiber intensity inside the fiber determines the length change.

Now we are ready to include the photomechanical effect as the source of feedback in a fiber Fabry-Perot fiber. Substituting Equation (7.82) into Equation (7.87) and using the result in Equation (7.77), we get

$$\boxed{I_t = \frac{I_i}{1 + \left(\frac{2r}{1-r^2}\right)^2 \sin^2\left(\frac{2\pi n}{\lambda}\left[\ell_0 + \xi n\left(\frac{1+r^2}{1-r^2}\right) I_t\right] + \phi'\right)}}. \tag{7.88}$$

We can easily solve Equation (7.88) for I_i as a function of I_t, and the result is a single-valued function. However, it is not possible to solve for I_t as a function of I_i.

There are two ways to attack Equation (7.88). The simplest is to numerically evaluate I_i as a function of I_t, then plot the values of I_t as a function of I_i. The second method is graphical. We start with the former and defer the latter until Chapter 9. The inset in Figure 7.39 shows the plot of I_i as a function of I_t and the main figure shows a plot of the inverse. Clearly, the transmittance is a multivalued function of the incident intensity.

As an aside, one can ask whether it would be possible to write the length change of the fiber as a function of the incident intensity. Then, Equation (7.88) would not have to be inverted. But, because the output is a multivalued function of the incident intensity, this is not possible. The length state of the fiber is only uniquely determined by the transmitted intensity.

Now, we consider in some detail how the transmitted intensity depends on the incident intensity. If we start at zero intensity and turn up the incident intensity, the transmitted intensity will follow the arrow labeled 1 in Figure 7.88, then make a vertical jump along the dotted line, then follow the curve above arrow 2, then make another vertical jump, etc. As long as the incident power is being *increased*, the output will go through a series of steps. Now consider the output to be determined by point a. If the incident intensity is increased to just beyond point b, the output will jump to point c. If now the intensity is *decreased*, the output will follow the curve from points c to d. If the intensity is continually decreased from this point, the output intensity will make a series of steps that follow the upper arrows. Clearly, this type of response leads to a series of loops as shown in Figure 7.35.

While the very first loop (near arrow 1) resembles Figure 7.35, the higher loops are connected. For example, above point b, we find points c and h, so in this region, we observe tristability. As we go to higher input intensities, it is possible to have more and more length states of the system for a single incident intensity. This general case is referred to as multistability.

Another important point that needs to be made is that even if a system appears to be bistable, it may be multistable. For example, if we start with the system in a state given by point f, if we move between h and g we will see only that single curve. If we lower the intensity below h, the system jumps down to the d-c-e branch. As long as we keep the input intensity between d and g, we will only map out that one loop.

When the multistability loops are due to the intensity-dependent refractive index, it is not possible to access the inside of the loop. So, for example, it would not be possible to get the system to a state corresponding to being on the curve between points a and d or points c and h. One interesting implication of the length change mechanism, however, is the fact that it may be possible to mechanically shock the system and bring it to one of the inner curves.

7.2.4.2 Multistability of a Fabry-Perot Interferometer due to Fresnel Reflections

In Section 7.2.4, we talked about a Fabry-Perot interferometer with reflectors on the ends. A polymer optical fiber with polished ends will experience a small amount of reflection. While the amount of reflection is only about 4%, it is still enough to observe multistability. All of the nuances associated with a Fabry-Perot waveguide, as discussed in the beginning of Section 7.2.4, will also be important here. Since these issues have already been discussed, we will jump right into the calculation. The geometry is similar to Figure 7.36 except now we treat the light as a plane wave. Figure 7.40 shows the geometry.

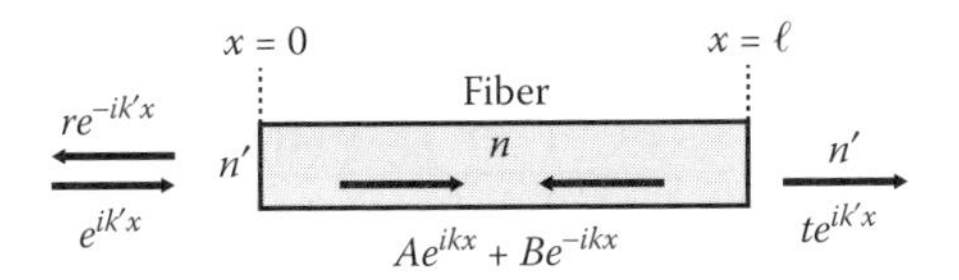

FIGURE 7.40 The electric fields inside a fiber of refractiveness n for an incident plane wave.

The incident wave travels to the right and we will arbitrarily define its amplitude to be unity since all the other fields will be referenced relative to it. Here, the wavevector is defined as $k = 2\pi n'/\lambda$, with n'=1. (Note that we will keep the problem general, by carrying the symbol n'.) Inside the fiber, the wavevector depends on the refractive index at the wavelength of the plane wave, and is given by $k = 2\pi n(\lambda)/\lambda$. The complex reflectance, r, represents the field amplitude of the light traveling to the left. Similarly, the transmitted complex amplitude is given by the transmittance t. Inside the fiber, the light travels both to the right (with amplitude A) and to the left (with amplitude B).

We are now ready to apply the boundary conditions. To simplify the problem, we will assume that the fiber is isotropic so that the electric polarization vector remains in the same plane. This is usually the case for most short lengths of fiber. Maxwell's equations demand that the electric field component parallel to the interface (called the tangential component, E_{tan}) must be continuous. Also, the tangential part of the cross product $\vec{k} \times \vec{E}$ must be continuous at each interface. This yields the following:

E_{tan} continuous at $x = 0$:

$$1 + r = A + B. \tag{7.89}$$

E_{tan} continuous at $x = \ell$:

$$A \exp[ik'\ell] + B \exp[-ik'\ell] = t \exp[ik\ell]. \tag{7.90}$$

$\vec{k} \times \vec{E}$ continuous at $x = 0$:

$$k'(1 - r) = k(A - B). \tag{7.91}$$

$\vec{k} \times \vec{E}$ continuous at $x = \ell$:

$$k\,(A \exp[ik\ell] - B \exp[-ik\ell]) = tk' \exp[ik'\ell]. \tag{7.92}$$

Equations (7.89), (7.90), (7.91) and (7.92) are four equations in terms of four unknowns. We can thus solve for all four unknowns (A, B, r, and t). The results are as follows:

$$A = \frac{2n'\left(n' + n\right)}{(n' + n)^2 - (n' - n)^2 \exp\left[2ik\ell\right]}, \tag{7.93}$$

$$B = \frac{2n'\left(n - n'\right) \exp\left[2ik\ell\right]}{(n' + n)^2 - (n' - n)^2 \exp\left[2ik\ell\right]}, \tag{7.94}$$

$$r = \frac{\left(n'^2 - n^2\right)\left(1 - \exp\left[2ik\ell\right]\right)}{(n' + n)^2 - (n' - n)^2 \exp\left[2ik\ell\right]}, \tag{7.95}$$

$$t = \frac{4n'n \exp\left[2i(k-k')\ell\right]}{(n'+n)^2 - (n'-n)^2 \exp\left[2ik\ell\right]}. \tag{7.96}$$

Armed with the solutions for the four unknowns, we can determine the intensities inside and outside of the fiber. First, we calculate the transmittance, T, for a fiber of refractive index n that is surrounded by vacuum ($n' = 1$). Taking the square of the magnitude of Equation (7.96), after some algebra we get

$$T = |t|^2 = \frac{1}{1 + \frac{\left(n^2-1\right)^2}{4n^2} \sin^2\left(\frac{2\pi\ell}{\lambda} n\right)}. \tag{7.97}$$

Aside from unit-dependent multiplicative constants, the intensity inside the fiber is given by

$$I_{in} = n \left|A \exp\left[ikx\right] + B \exp\left[-ikx\right]\right|^2. \tag{7.98}$$

We calculate the intensity inside the interferometer by substituting Equation (7.93) and Equation (7.94) into Equation (7.98). As we did in Section 7.2.4.1, however, we are interested in the average intensity inside. Using similar arguments as in Section 7.97, we get

$$\bar{I}_{in} = \frac{n^2+1}{2n} T = \frac{n^2+1}{2n} I_i, \tag{7.99}$$

where we have used Equation (7.97) to simplify the expression (analogous to how this was done in Section 7.2.4.1). The last equality stems from the definition of the transmittance for an incident wave of unit amplitude (that is, $T = I_t/I_i$, which for unit incident intensity yields $T = I_t$).

Substituting Equation (7.99) and Equation (7.87) into Equation (7.97) yields

$$\boxed{I_t = \frac{I_i}{1 + \left(\frac{(n^2-1)^2}{4n^2}\right)^2 \sin^2\left(\frac{2\pi n}{\lambda}\left[\ell_0 + \xi\left(\frac{n^2+1}{2n}\right) I_t\right] + \phi'\right)}}. \tag{7.100}$$

Clearly, this has the same form as Equation (7.88), except Equation (7.100) is represented in terms of refractive indices rather than reflectivities. This result is not surprising since the reflectivity of the bare fiber end depends on the refractive index of the fiber. As such, the Fresnel calculation yields a more complicated dependence of the transmitted intensity on the refractive index. However, most of the refractive-index's effect on the variation of the transmittance arises from the $2\pi n$ factor in the sinusoidal term. For small refractive index changes, Equation (7.88) and Equation (7.100) have the same functional form of the transmittance as a function of refractive index.

7.2.4.3 Comparison of the Fresnel and Fabry-Perot Results

It is an interesting exercise to reconcile Equations (7.88) and (7.100). The former is calculated by considering a geometric ray bouncing back and forth between two reflectors. The later method uses a wave approach with boundary conditions. In this section, we will determine the reflectivity and transmittance of a single

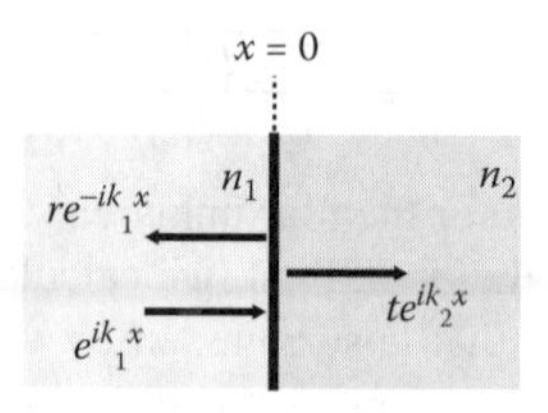

FIGURE 7.41 Reflection and transmission from an interface.

interface; then, we will apply the ray tracing method to see if we get the Fresnel reflection results. In essence, we will show the equivalence of Equation (7.88) and Equation (7.100).

Let's consider a single interface as shown in Figure 7.41 with a normal incident ray. The continuity of E_{tan} yields

$$1 + r = t, \tag{7.101}$$

and the continuity of the tangential component of $\vec{k} \times \vec{E}$ gives

$$k_1(1 - r) = tk_2. \tag{7.102}$$

These two equations can be solved for the two unknowns (t and r):

$$\boxed{r = \frac{n_1 - n_2}{n_1 + n_2}}, \tag{7.103}$$

and

$$\boxed{r = \frac{2n_2}{n_1 + n_2}}. \tag{7.104}$$

We can also invert Equation (7.103) and Equation (7.104) to express the refractive index of the two materials on either side of the interface in terms of r and t:

$$\boxed{\frac{n_1}{n_2} = \frac{1 + r}{1 - r},} \tag{7.105}$$

and

$$\boxed{\frac{n_1}{n_2} = \frac{t}{2 - t}.} \tag{7.106}$$

Now we consider a fiber of refractive index n in air, where the refractive index of air is about unity. In this case, the reflectivity of the ends for light that is incident on the interface from the inside of the fiber is given by

$$r' = \frac{n - 1}{n + 1}. \tag{7.107}$$

The transmittance from the outside into the fiber is

$$t = \frac{2}{n + 1}, \tag{7.108}$$

and the transmittance from inside the fiber to the outside is

$$t' = \frac{2n}{n+1}. \tag{7.109}$$

We can now use these values in a ray-tracing calculation similar to Section 7.2.4. In analogy to the derivation leading to Equation (7.77), the intensity is given by

$$I_t = \frac{I_i t^2 t'^2}{(1-r'^2)^2 + 4r'^2 \sin^2\left(\frac{2\pi n\ell}{\lambda} + \phi'\right)}, \tag{7.110}$$

where we have taken into account that the transmittance is different at the front (entrance) end and the back (exit) end of the fiber. Substituting Equations (7.107), (7.108), and (7.109) into Equation (7.110), we get

$$T = \frac{1}{1 + \frac{(n^2-1)^2}{4n^2}\sin^2\left(\frac{2\pi\ell}{\lambda}n\right)}. \tag{7.111}$$

This is identical to the Fresnel result given by Equation (7.97). So, when the Fresnel reflections and transmissions of the fiber end are substituted into the Fabry-Perot ray-tracing calculation, the results that are obtained are the same as the wave equation result for the two-boundary problem. Similarly, if we substitute the reflectivity and transmittances given by Equations (7.107), (7.108), and (7.109) into the ray-tracing method to calculate the intensity inside the interferometer, we get Equation (7.99). Going through this calculation serves no purpose other than to illustrate the equivalence of two seemingly different approaches. Suffice it to say that all of the results obtained using the ray-tracing method can be reproduced by solving the boundary value problem with Maxwell's wave equation.

7.2.4.4 Graphical Solution to Transcendental Equations

Solving an equation "by hand" with paper and pencil often leads to a deeper understanding than plugging equations into a computer for numerical solution (as we have admittedly done in generating Figure 7.39). We would like to solve Equation (7.88) with graphical methods. This question is of the form

$$\frac{I_t}{I_i} = f(I_t), \tag{7.112}$$

where $f(I_t)$ is a periodic function of I_t. The idea behind the graphical solution is to plot the left-hand and right-hand sides of Equation (7.112) as a function of I_t. The intersections between these two curves yields the solution.

Figure 7.42 shows a schematic representation of the periodic function $f(I_t)$ and I_t/I_i for several values of the incident intensity I_i. At low intensity, the slope of the line given by I_t/I_i versus I_t is large and the oscillating curve intersects this line only in one place. In this region, the solution, corresponding to the point of intersection, is single-valued. For a higher intensity (the line with intermediate

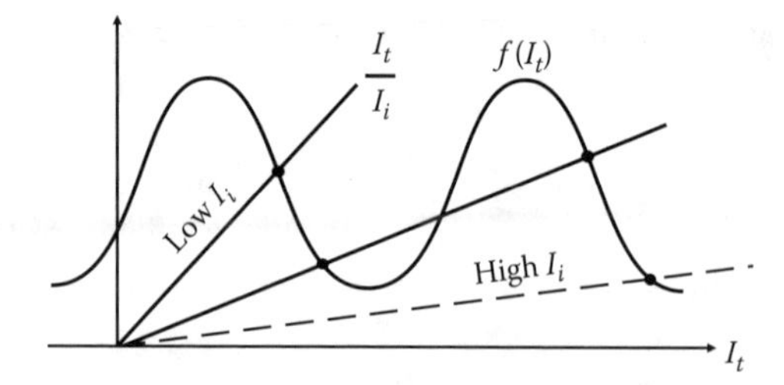

FIGURE 7.42 $f(I_t)$ and I_t/I_i as a function of I_i for three different intensities.

slope), there are three intersections of the line with $f(I_t)$. Two of these (labeled with heavy points) yield solutions that are physically attainable, while the third is found to be unstable and not accessible. As such, for this input intensity the system is bistable. Note that the third unattainable solution corresponds to an interior point in the hysteresis curve that has a negative slope (see Figure 7.39).

As the intensity is increased further, the number of intersections increases and the system becomes multistable. The graphical method can be used to obtain a sketch of the hysteresis loop, as will be discussed in more detail in Chapter 9.

8 Structured Fibers and Specialty Applications

In this chapter, we discuss specialty fibers that are designed for a specific application. For example, a Bragg grating is used to pass certain wavelengths of light while reflecting others. As such, they can be used to make filters in fibers — which have applications in wavelength division multiplexing or can be used to make a distributed fiber laser. Photonic crystal fibers, on the other hand, are made with a series of parallel cylindrical holes, which can be used to manipulate the propagation of light while photorefractive fiber can be used to store large amounts of information in the form of holograms or for image correction.

We begin this chapter with the theory of Bragg gratings and discuss applications in a polymer fiber. Then, the concept of a photonic crystal fiber is explained in terms of simple quantum mechanical arguments. The analogy between quantum mechanics, which describes electronic crystals, and electromagnetism originally motivated researchers to consider periodic photonic structures. We end this chapter with a discussion of photorefractive fibers and fiber chemical sensors.

We note that each of the areas we discuss in this chapter are topics of whole books, so our intention is to provide an overview of the basic physics rather than to provide an exhaustive study of all recent developments.

8.1 BRAGG GRATINGS

Polymer optical fibers can be doped with photosensitive materials that change refractive index upon exposure to bright light, usually in the ultraviolet range of the spectrum. Figure 8.1 shows how such a grating can be made.

There are many mechanisms that result in a refractive index change in a polymer, which include

- Ablation, where the light damages the material through the accumulation of heat in a small volume of space
- Bond breaking, where the light breaks a polymer chain, making the polymer molecular weight lower and yielding a lowered refractive index
- Photopolymerization, where the light initiates polymerization in unreacted monomer
- Cross-linking, where the light causes nearby chains to form chemical bonds between each other
- Photoisomerization, often a reversible process where light causes dopant molecules to change shape and orientation

Independent of the mechanism, as long as the material is not saturated (i.e., the change in refractive index is proportional to the time of exposure and to the

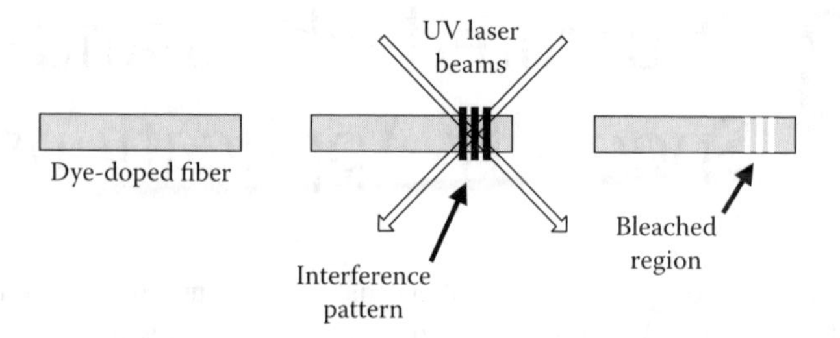

FIGURE 8.1 A Bragg fiber is made by burning a grating into a polymer fiber.

intensity), a sinusoidal refractive index profile will result at the intersection of two crossed beams provided that the light of each beam at the point of intersection can be approximated by a plane wave.

8.1.1 Theory

In this section, we calculate the reflectance and transmittance of a Bragg grating. First, we note that when two beams are launched from outside a fiber as shown in Figure 8.1, the fiber will focus the light, making the plane-wave approximation inaccurate. The light entering the fiber can be made into a plane wave if the fiber is emersed in a rectangular cell with an index-matching liquid. Under these conditions, the reflected intensities that we shall calculate in this section hold. While index-matching liquid is not always used, we shall only consider the plane-wave approximation, which still correctly predicts the wavelengths of reflection (i.e., Bragg Condition) and qualitatively predicts the exchange of energy between incident and reflected waves in a cylindrical fiber.

When two plane waves cross, as shown in Figure 8.1, the refractive index profile is sinusoidal. Before considering the reflected intensity from a sinusoidal refractive index profile, we first consider the Bragg reflection condition from two partially reflecting planes that are embedded in a material of refractive index n as shown in Figure 8.2.

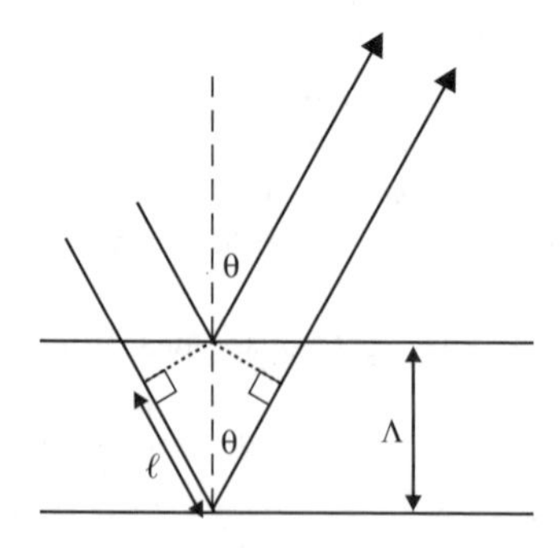

FIGURE 8.2 Interference between two parts of a beam that are partially reflected from parallel planes.

The constructive interference condition results when the extra path followed by the longer ray, 2ℓ, is an integer, N, times the wavelength of the light in the material, λ/n. From this simple geometry, we get the Bragg condition,

$$2\ell = 2\Lambda\cos\theta = N\left(\frac{\lambda}{n}\right). \tag{8.1}$$

Sometimes, it is convenient to express the Bragg condition in terms of the wavevector (defined as $k = 2n\pi/\lambda$),

$$\Lambda\cos\theta = N\left(\frac{\pi}{k}\right). \tag{8.2}$$

Clearly, if we add a series of reflecting planes, where each two consecutive ones are separated by Λ, the Bragg condition remains the same — as can be seen by considering any two adjacent planes. For a fixed-light wavelength λ and grating periodicity Λ, a bright spot will be observed at a distinct angle for a given value of the integer N. The angle corresponding to integer N is called the Bragg angle of order N.

In an optical fiber, the Bragg grating is usually (but not always) perpendicular to the fiber axis as shown in Figure 8.1. In this case, if a white-light spectrum is launched down such a fiber, the reflected wavelengths are $\lambda_N = 2\Lambda n/N$, where in the fiber, the refractive index n is replaced by the mode index. So, the transmitted spectrum will show a series of dark lines at λ_N.

The above calculation gives the wavelengths of reflection but not the intensity. To solve for the intensity, we must use coupled mode theory. We will assume that the grating planes are perpendicular to the wavevector as in the fiber; but we confine ourselves to a free-space plane wave, which can be generalized to the guided mode solutions in a straightforward albeit messy manner. Since the refractive index profile is sinusoidal, we express it as

$$n(z) = n_0[1 + \epsilon\cos(Kz)], \tag{8.3}$$

where $K = 2\pi/\Lambda$ and the wavevector, β, is along the $\hat{z}$-direction and takes the form

$$\beta(z) = n_0\frac{\omega}{c}(1 + \epsilon\cos(Kz)) \equiv \beta_0(1 + \epsilon\cos(Kz)). \tag{8.4}$$

Note that in a weekly guiding mode, Equation (8.4) is a good approximation. We will also make the assumption that the depth of modulation of the refractive index is small, so $\epsilon \ll 1$.

We express the electric field in the grating as a sum of forward- and backward-traveling plane waves of amplitude $A_f(z)$ and $A_b(z)$, respectively. These amplitudes depend on z to reflect the fact that energy is exchanged between them. Figure 8.3 shows the geometry of the problem. Since the electric field polarization is the same for the incident and reflected waves, we can write it in scalar form:

$$E = A_f(z)\exp\left[i\left(\omega t - \beta z\right)\right] + A_r(z)\exp\left[i\left(\omega t + \beta z\right)\right], \tag{8.5}$$

where β is the wavevector and ω the frequency.

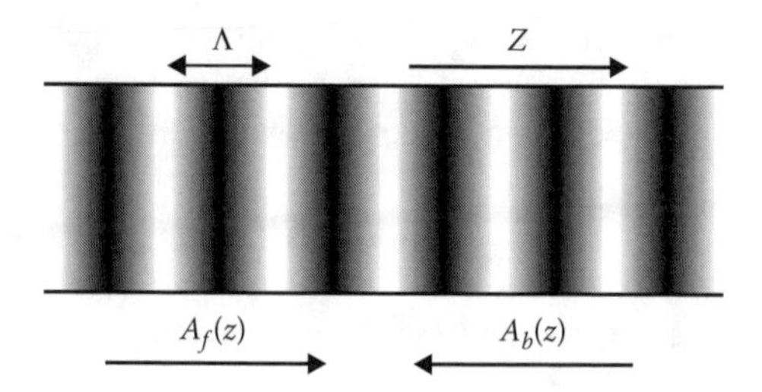

FIGURE 8.3 The forward (incident) and backward (reflected) waves from a Bragg grating with sinusoidal refractive index profile.

Substituting Equation (8.5) into Maxwell's Wave Equation,

$$\left(\frac{\partial^2}{\partial z^2} + \frac{\omega^2}{c^2} n^2\right) E = 0, \tag{8.6}$$

we get

$$\begin{aligned}
&\left[\left(A_b \epsilon \beta_0 (-i K^2 z \cos(Kz) + 2K(z\beta_0 - i)\sin(Kz)) + 2i\beta_0 \frac{\partial A_b}{\partial z}\right)\right] \\
&\quad \times \exp\left[+i\beta_0 z\right] \\
&+ \left[\left(A_f \epsilon \beta_0 (i K^2 z \cos(Kz) + 2K(z\beta_0 + i)\sin(Kz)) - 2i\beta_0 \frac{\partial A_f}{\partial z}\right)\right] \\
&\quad \times \exp\left[-i\beta_0 z\right] \\
&= 0,
\end{aligned} \tag{8.7}$$

where we have only kept terms up to order ϵ, the common factor $\exp[i\omega t]$ cancels, we assume that the conversion efficiency is small ($\partial^2 A_i/\partial z^2 \ll \beta\, \partial A_i/\partial z$ where i is f or b), and we use the definition

$$\beta_0 = n_0 \frac{\omega}{c}. \tag{8.8}$$

Details of the derivation leading to Equation (8.7) are given in the appendix at the end of this chapter.

To decouple Equation (8.7) into two differential equations, we first multiply it by $\exp[+i\beta z]$ and integrate over one optical cycle using the assumption that neither the amplitudes A_f or A_r, nor their derivatives change over one optical cycle. In Section 8.1.2, we illustrate how Equation (8.7) is decoupled at the Bragg condition.

8.1.2 The Bragg Condition, $K = 2\beta$

We start with Equation (8.7) under the Bragg condition $K = 2\beta$. To decouple the backward wave from Equation (8.7), we multiply both sides by $\exp[-i\beta_0 z]$

and integrate over one optical cycle under the assumption that A and $\partial A/\partial z$ are approximately constant. Any term that is sinusoidal with a spacial wavelength that is an integer multiple of the optical wavelength vanishes. As such, Equation (8.7) takes the form,

$$\begin{aligned}&\int_0^{2\pi/\beta_0}\left(A_b\epsilon\beta_0(-4i\beta_0^2 z\cos(2\beta_0 z)+4\beta_0^2 z\sin(2\beta_0 z))+2i\beta_0\frac{\partial A_b}{\partial z}\right)dz\\&+\int_0^{2\pi/\beta_0}\left[A_f 4\epsilon\beta_0^2(i\beta_0 z\cos(2\beta_0 z)+(z\beta_0+i)\sin(2\beta_0 z))\right]\exp\left[-2i\beta_0 z\right]dz\\&=0.\end{aligned} \tag{8.9}$$

Evaluating the integrals, we get

$$\frac{\partial A_b}{\partial z}+i\epsilon\beta_0 A_b-\frac{i}{2}\epsilon\beta_0 A_f=0. \tag{8.10}$$

A similar procedure can be used to get an equation for A_f, which yields

$$\frac{\partial A_f}{\partial z}-i\epsilon\beta_0 A_f+\frac{i}{2}\epsilon\beta_0 A_b=0. \tag{8.11}$$

Equation (8.10) and (8.11) can be decoupled into two equations that each only depend on one field amplitude, A_f or A_b. To get an equation for A_f, we solve Equation (8.11) for A_b and substitute the result into Equation (8.10), which yields

$$\frac{-\frac{\partial^2 A_f}{\partial z^2}+i\epsilon\beta_0\frac{\partial A_f}{\partial z}+i\epsilon\beta_0\left(-\frac{\partial A_f}{\partial z}+i\epsilon\beta_0 A_f\right)}{i\epsilon\beta_0/2}-\frac{i}{2}\epsilon\beta_0 A_f=0. \tag{8.12}$$

Combining terms, Equation (8.12) reduces to

$$\frac{\partial^2 A_f}{\partial z^2}=-\frac{3}{4}\epsilon^2\beta_0^2 A_f. \tag{8.13}$$

Similarly, for A_b we get

$$\frac{\partial^2 A_b}{\partial z^2}=-\frac{3}{4}\epsilon^2\beta_0^2 A_b. \tag{8.14}$$

The solutions to Equation (8.13) and (8.14) are sinusoidal. Let's consider the boundary condition that the light is launched into the grating at $z=0$, then $A_f(0)=A_0$ and $A_r(0)=0$. Under these conditions, the solutions are

$$A_b(z)=A_0\sin\left(\frac{\sqrt{3}}{2}\epsilon\beta_0 z\right);\qquad I_b(z)=I_0\sin^2\left(\frac{\sqrt{3}}{2}\epsilon\beta_0 z\right) \tag{8.15}$$

and

$$A_f(z)=A_0\cos\left(\frac{\sqrt{3}}{2}\epsilon\beta_0 z\right);\qquad I_f(z)=I_0\cos^2\left(\frac{\sqrt{3}}{2}\epsilon\beta_0 z\right) \tag{8.16}$$

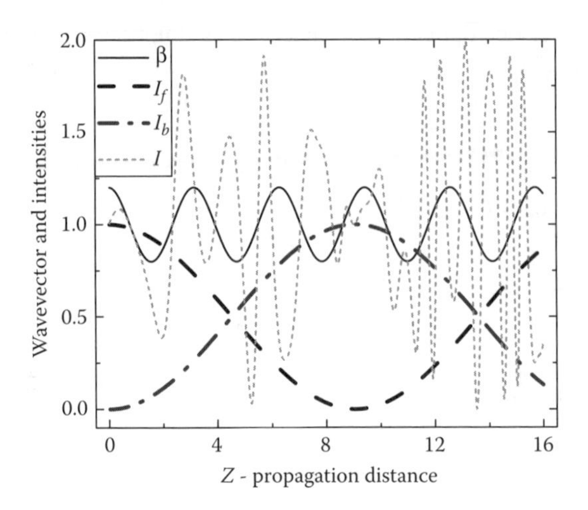

FIGURE 8.4 Wavevector, forward intensity, backward intensity, and total intensity for a beam at the Bragg condition when $\epsilon = 0.2$ and $\beta_0 = 1$.

where the intensity is proportional to the square of the electric field. I_0 is the intensity launched into the grating, which is fully reflected into the backward wave after propagating a distance $\ell = \pi/(\sqrt{3}\epsilon\beta_0)$. Since ϵ is small, ℓ is much longer than both the wavelength of the light and the grating period.

Figure 8.4 shows a plot of the wavevector β (with $\beta_0 = 1$ and $\epsilon = 0.2$) as a function of propagation distance for a forward wave that enters the grating at $z = 0$. Also plotted are the intensities of the forward and backward waves as well as the total intensity. This graph illustrates the conversion of energy from the forward wave to the backward wave and back again. Also note that the total intensity is a complicated function of position due to interference between the two waves and the oscillatory behavior of β.

8.1.3 Bragg Gratings in Polymer Fibers

As we saw in the previous sections, a Bragg grating acts as a filter that reflects light at the Bragg condition, which depends on the grating period. To make a tunable Bragg grating requires the ability to change the grating period. Since polymers have both a larger thermal expansion coefficient and are less stiff than glass fibers, it's possible to make temperature- or stress-tuned Bragg fibers with a large tuning range.

Figure 8.5 shows the transmittance of an optical fiber while a grating is being recorded with UV light as reported by Chu and coworkers.[124] The UV light is passed through a phase grating and reflected by prisms to produce an interference pattern in the fiber which writes the grating. An ASE white light source is coupled

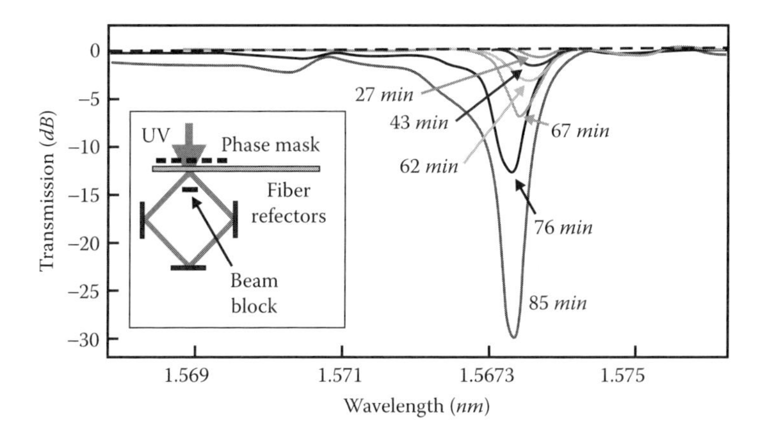

FIGURE 8.5 The transmittance of a fiber Bragg grating as a function of exposure time. The inset shows the setup for writing the grating. White light from an ASE source is coupled into one end of the fiber and the transmitted light is measured with a spectrometer. Adapted from Ref. [124] with permission of the Optical Society of America.

into one end of the fiber and a spectrometer measures the transmittance at the other end.

There are two types of gratings, called Type I and Type II. In a Type I grating, the refractive index modulation parameter, ϵ, grows linearly with time while a Type II grating shows nonlinear behavior. In the Bragg gratings measured in Figure 8.5, Chu and coworkers reported that during the first hour of UV exposure, the gratings are of Type I. Gratings formed after 1 hour of exposure are found to be Type II. Note that at the 85 minute mark, the transmittance minimum is down to noise levels.

At longer exposure times, the UV light begins to erase the grating and the transmittance dip disappears. However, if the UV source is removed, the grating is observed to recover over a period of 8 hours, at which point the reflectance from the grating levels off to about 62%. While the mechanism of the recovery process is unknown, Chu argues that during exposure, there is competition between photothermal stress and UV photoreaction, which are of opposite signs. When the grating light is turned off, the thermal stress relaxes while the results of photoreaction are irreversible and therefore permanent.

The wavelength of maximum reflectance of a Bragg polymer fiber can be tuned by stretching it.[124] For a 1.3% strain, the wavelength shift is about 20*nm* and returns without hysteresis to its original state of reflectivity when the strain is released. As such, straining a fiber provides a suitable means for making a tunable filter. When the fiber is stretched until it breaks, a shift of over 70*nm* is observed.

A polymer Bragg fiber can also be tuned by heating. For a temperature increase of 50°C from ambient (20°C), the peak reflectance can be blue-shifted by 18*nm* and returns back to its original value without hysteresis when the temperature comes back to ambient. Because the range of tunability in polymer fibers is much higher than for silica fibers, they are better suited for many applications.

8.1.4 Applications

In this section, we discuss two applications of Bragg gratings that require the separation of wavelengths: the optical fiber spectrometer and a fiber-optics system for wavelength division multiplexing and demultiplexing.

8.1.4.1 Spectrometer

Figure 8.6a shows an example of a spectrometer that is formed from a tunable Bragg fiber grating and circulator. The circulator is a standard nonreciprocal optical fiber component for which light entering into port a leaves through port b and all light entering port b exits through port c. So, if the grating is tuned to reflect light at wavelength λ_1, a light beam with many wavelengths will pass from port a to b and only the light of wavelength λ_1 is reflected from the grating and exits out of port c. As such, to measure the spectrum of the incident light, the power exiting port c is measured as a function of the center wavelength of the Bragg grating, which is either temperature or stress tuned. The polymer's larger coefficient of thermal expansion makes such a device tunable over a broader spectral range than the equivalent glass fiber Bragg grating.

8.1.4.2 Wavelength Division Multiplexing and Demultiplexing

Wavelength multiplexing allows one to pack information onto a fiber by encoding different packages of data into different colors. Figure 8.6b shows an example of a wavelength division multiplexing system. Signals at distinct wavelengths enter the device in the three input ports labeled 1, 2, and 3. Note that such a device can be generalized for an arbitrarily large number of ports by cascading the circulator/Bragg grating units together.

Let's first consider the inputs at wavelengths λ_2 and λ_3. The light at λ_3 enters the circulator and passes through to the Bragg grating where it is reflected. The light at wavelength λ_2 passes through the grating unchanged, so that the two colors are deflected downward into the next Bragg grating, which is set to reflect only light at λ_1. So, the light entering port 1 passes through the circulator and reflects from the Bragg grating, combining with the other two colors. All three colors are then launched into a transmission fiber. This stage acts as the multiplexer.

When the light arrives at the demultiplexer input, all three colors pass the first circular and the first Bragg grating except for the light at λ_1, which is reflected by the Bragg grating and exits the demultiplexer through port 4. Similarly, at each subsequent Bragg grating, one particular wavelength is reflected and exits through a port. Given that all the Bragg gratings can be made to be tunable, the various colors of light can be redirected in many ways. So, such a system is reconfigurable on the fly to reroute signals.

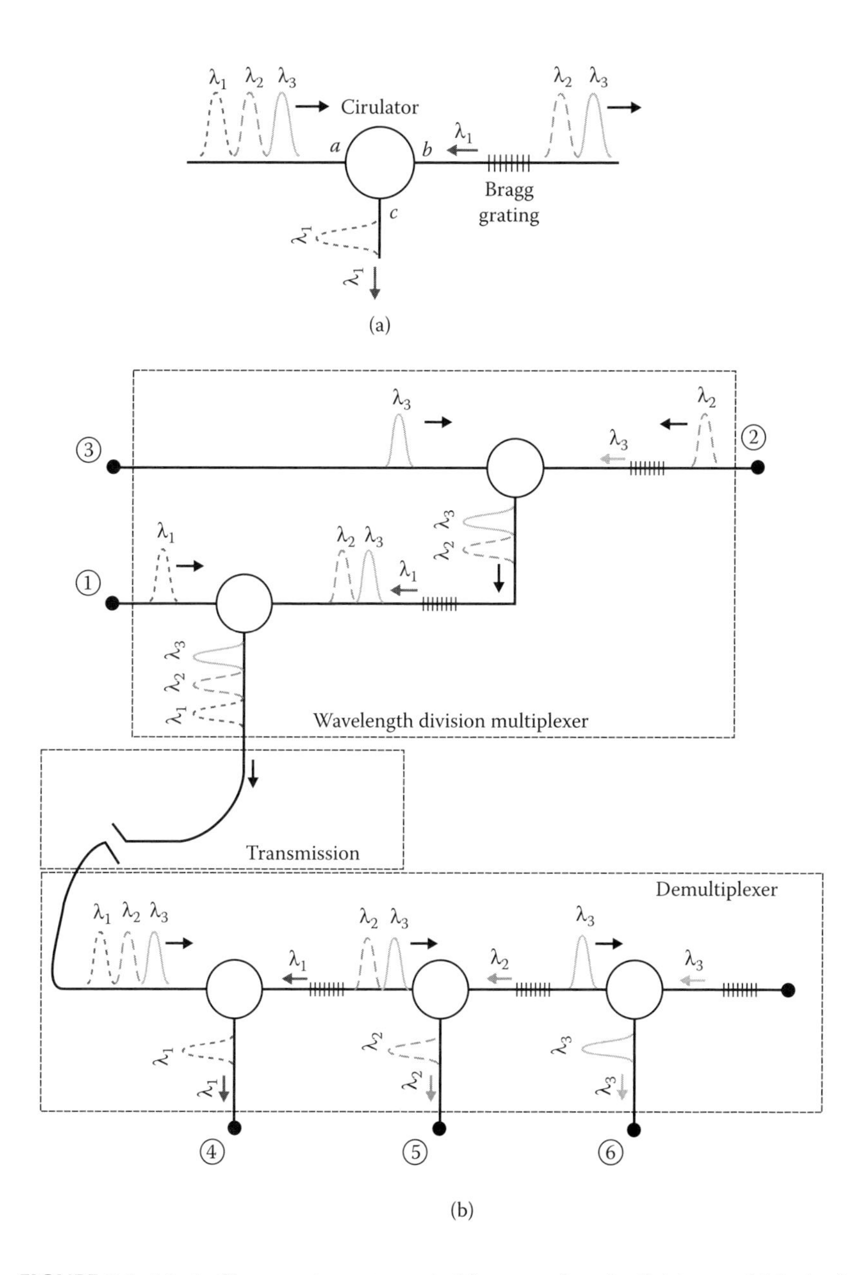

FIGURE 8.6 (a) A fiber spectrometer and (b) a wavelength division multiplexing/demultiplexing system.

8.2 ADVANCED STRUCTURED FIBERS

A structured fiber is one that can have a combination of many cores, holes, electrodes, etc. The electrooptic fiber, discussed in Section 7.1.1, is one example of a simple structured fiber. In this section, we focus on fibers with large numbers of structures.

8.2.1 Imaging Fibers

An imaging fiber is made of multiple noninteracting cores so that the intensity profile entering one end exits, unchanged, at the other end. Such a fiber is typically made by bundling together many smaller fibers, as shown in Figure 8.7. An imaging fiber thus extends the focal plane from one place to another. Applications include endoscopes that are used in medical applications to directly image interior organs.

To be useful, an imaging fiber needs to be made of small cores to maximize the resolution, high numerical aperture so that light is effectively collected, and the material should be flexible to permit the imaging fiber to be easily snaked into tight places. Medical applications require the smallest possible diameter fiber bundle to both allow for comfort of the patient as well as to allow access to small spaces. As such, the individual fibers that make an imaging bundle must be made from high-refractive index cores and low-refractive index cladding. Furthermore, the cladding must be made as thin as possible to allow for the highest possible packing density of the waveguides.

Redrawing is a common method for making imaging fiber. First, a fiber is pulled with the required aspect ratio and refractive index difference between core and cladding. The fiber is then cut into small sections, which are bundled together in a honeycomb pattern to make a preform, which is pulled into the imaging fiber. Since this is the second time the fibers in the bundle are pulled, the result is called

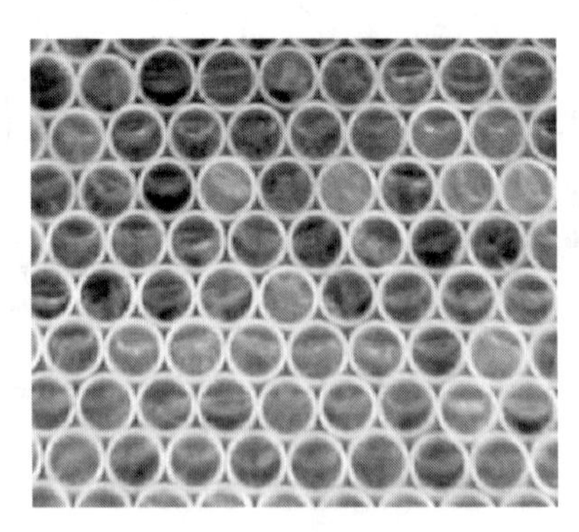

FIGURE 8.7 A closeup of an imaging fiber. Photograph courtesy of and with permission from Paradigm Optics, Inc.

FIGURE 8.8 An image inverter made by twisting an imaging fiber bundle. Photograph courtesy of and with permission from Paradigm Optics.

a second-generation fiber. (This process is described in Section 3.2.4.) In turn, the imaging fiber that results can be cut into pieces to form a preform that is drawn again. Each generation leads to smaller waveguides. However, errors introduced in the process of making preforms leads to more imperfections in later generation fiber. Usually, the benefits of the reduced core size are marginal relative to the increased number of misregistrations after three generations.

Because of the flexibility of a polymer, it is possible to deform an imaging fiber to manipulate the image. For example, one can imagine an adiabatic taper, where the diameter of the fiber slowly decreases along its length. This results in a smaller image at the output of the fiber. Similarly, the fiber bundle can be twisted at an elevated temperature to form an image inverter. Figure 8.8 shows an example of an image inverter made by Paradigm Optics.

The top portion of Figure 8.8 shows two different 180° image inverters, where the effects of the twist can be observed on the surface. Any arbitrary twist angle can be easily made. Since the image inverters are lightweight and thin, they are useful in imaging applications where the size and weight of the system are critical.

8.2.2 Capillary Tubes

A capillary tube is simply a small-diameter hole in a cylinder that draws a liquid due to interactions between the liquid and walls of the hole. Since the weight of a column of liquid scales as the volume and the force between the liquid and walls scales as the surface area, a small-diameter cylindrical hole provides the largest surface-to-volume ratio, thus yielding the most efficient drawing force. This effect is called capillary action.

FIGURE 8.9 Three views of a capillary tube array. Photograph courtesy of and with permission from Paradigm Optics, Inc.

There are many applications that use capillary action. Recently, the human genome has been mapped and the next major push is to understand the half million proteins that are encoded by the less than 30,000 genes. Such studies are called proteomics. Proteomics is to proteins what genomics is to genes. An important laboratory tool in proteomics is capillary action. Since a single capillary tube is used to run a test, having the availability of large arrays of small-diameter capillary tubes greatly increases the throughput of these experiments. Polymer capillary tubes have been found to have advantages over glass in proteomics applications, so polymer capillary arrays may be ideal components in such applications.

Section 3.2.4 describes the process for making preforms that can be drawn into arrays of holes. Figure 8.9 shows a polymer capillary tube array from three different perspectives.

Polymer fibers can be tailored for custom applications. For example, one can imagine making an endoscope with a fiber bundle for imaging, electrodes (inside the fiber) to provide current for cauterization, and a capillary tube, to inject medications or retrieve body fluid samples. Similarly, electrodes might be placed around capillary tubes so that an electric field can be applied to the liquid in the tube.

Figure 8.10 shows an example of a small-diameter capillary between two circular indium electrodes. Note that in the drawing process, the cross-sections of

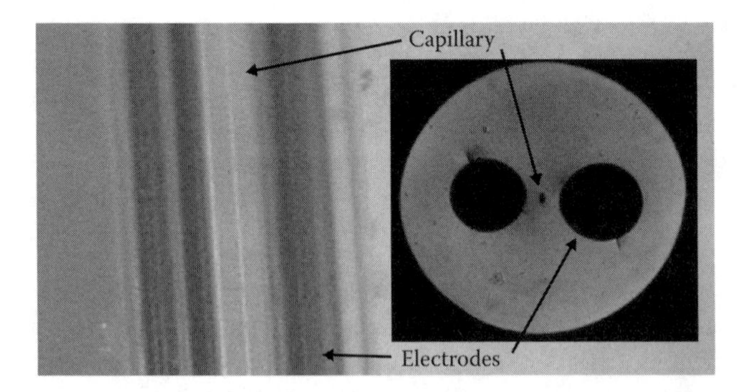

FIGURE 8.10 A capillary tube straddled by two electrodes in side view (left) and end view (right). Photograph courtesy of and with permission from Paradigm Optics.

the final electrodes and capillary are in the same proportions as originally in the preform before drawing. The improvement in the electrode quality in this picture, compared to previous-generation electrooptic fibers (see Figure 7.10) is obvious. Such improvements in the quality of polymer optical fiber structures requires meticulous and tedious trial-and-error processing studies.

8.2.3 Photonic Crystal Fibers

A photonic crystal is a periodic structure with a period on the order of a wavelength of light. Light interacts with a photonic crystal in the same way that electrons interact with an atomic crystal. In order to understand the principle of light interaction with a periodic structure, we will consider two separate cases: light guiding in parallel waveguides and light propagating perpendicular to a periodic refractive index profile. We'll see that these two cases are intimately related to each other.

8.2.3.1 Two Parallel Waveguides

For simplicity, let's start with the simplest possible case: two identical non-interacting parallel waveguides. When light is launched into one waveguide, it remains in that waveguide. Furthermore, we assume that only the lowest-order mode is excited and that all the modes propagate in the same direction. In Dirac notation, when a mode of unit power propagates in a particular waveguide, let's say in waveguide #1, it is explicitly represented as a column vector of the form

$$|1\rangle = \begin{pmatrix} 1 \\ 0 \end{pmatrix} \exp[i(\beta z - \omega t)], \tag{8.17}$$

where β is the wavevector of the mode and ω its angular frequency. So, in Dirac notation, $|1\rangle$ literally represents a mode that propagates in waveguide #1. Similarly, for light propagating in the second waveguide, we can express it as

$$|2\rangle = \begin{pmatrix} 0 \\ 1 \end{pmatrix} \exp[i(\beta z - \omega t)]. \tag{8.18}$$

The two column vectors are specifically chosen to be orthogonal since the parallel waveguides do not interact.

As an example of how this notation is used, let's first consider how one would calculate the power if all of the light is confined to waveguide #1. The power is calculated by simply taking the inner product of the state vector $|1\rangle$ (called a *ket*) and $\langle 1|$ (called a *bra*):

$$\langle 1|1\rangle \equiv \langle 1| \cdot |1\rangle = ((1 \quad 0)[-i(\beta z - \omega t)]) \left(\begin{pmatrix} 1 \\ 0 \end{pmatrix} \exp[i(\beta z - \omega t)] \right), \tag{8.19}$$

where a *bra* is derived from a *ket* by turning the vector (column matrix) into a row matrix and then taking the complex conjugate. Clearly, this leads to $\langle 1|1\rangle = 1$, or unit power. The most general state of the two-waveguide system is given by

$$|E\rangle = \frac{1}{\sqrt{|a|^2 + |b|^2}} \begin{pmatrix} a \\ b \end{pmatrix} \exp[i(\beta z - \omega t)], \tag{8.20}$$

where a and b are complex constants and the factor $\sqrt{|a|^2 + |b|^2}$ ensures normalization. It is simple to show that $P = \langle E|E\rangle = 1$.

Next, let's calculate the power in each waveguide. The formula for the power in waveguide n when the field is in a general state given by Equation (8.20) is

$$P_n = |\langle n|E\rangle|^2 . \tag{8.21}$$

Thus, it is straightforward to verify that $P_1 = |a|^2 / \left(|a|^2 + |b|^2\right)$ and $P_2 = |b|^2 / \left(|a|^2 + |b|^2\right)$. Note that this correctly yields that the combined power is $P_1 + P_2 = 1$. The process of multiplying the general state vector of the light by the *bra* representing one of the waveguides simply yields the electric field in that waveguide, the square of which yields the power.

Next, we define the K-matrix as follows:

$$K = \begin{pmatrix} \beta & 0 \\ 0 & \beta \end{pmatrix} . \tag{8.22}$$

This matrix, when operating on the state vector, yields an eigenvalue times the state vector where the eigenvalue is just the wavevector of the mode. Since both waveguides are identical, the wavevector is the same for each waveguide i and the K-matrix effects a ket as

$$K|i\rangle = \beta|i\rangle. \tag{8.23}$$

Thus, if light is propagating down a waveguide in a stationary state (i.e., the power distribution between waveguides does not change as it propagates), then operating with K on the state vector will always yield an eigenvalue times the state vector. Note that if we operate with K on the general state vector given by Equation (8.20), this will also yield β times the state vector. Since the waveguides are noninteracting, independent of the original distribution of power between the guides, the power distribution remains the same at all points along each waveguide. As we will see below, if the waveguides interact, only certain power distributions will remain invariant as the light propagates. These modes are called the eigenmodes of the system. Linear combinations of these modes, when operating on by K will yield a different state vector. In these mixed modes, power will transfer between modes as the light propagates in the waveguide.

An interaction between the waveguides is quantified by the length of propagation over which the light is transferred from state $|1\rangle$ to $|2\rangle$ and from state $|2\rangle$ to $|1\rangle$. If the two waveguides are identical, the rate of power transfer is the same in both directions. For a transfer length proportional to $1/\Delta$, the K-matrix is of the form

$$K = \begin{pmatrix} \beta & \Delta \\ \Delta & \beta \end{pmatrix} . \tag{8.24}$$

The eigenvalues of this matrix are λ_+ and λ_-,

$$\lambda_\pm = \beta \mp \Delta \tag{8.25}$$

with normalized (to unit power) eigenvectors $|+\rangle$ and $|-\rangle$ given by

$$|\pm\rangle = \frac{1}{\sqrt{2}}\begin{pmatrix} 1 \\ \pm 1 \end{pmatrix} \exp\left[i\left([\beta \mp \Delta]\, x - \omega t\right)\right]. \tag{8.26}$$

This analysis shows that an interaction between the waveguides results in a change of the wavevectors ($\beta \rightarrow \beta \mp \Delta$), and the state vectors correspond to an equal power distribution between the two waveguides. However, the eigenstate $|+\rangle$ corresponds to the electric fields in both waveguides being in phase while the eigenstate $|-\rangle$ corresponds to the fields being 180° out of phase.

Next, we consider how the light propagates if it is coupled into the first waveguide. Thus, at $t = 0$ and $x = 0$, let's assume that the state vector is given by $|1\rangle$ which yields as the initial condition $|1\rangle = (|+\rangle + |-\rangle)/\sqrt{2}$. The evolution of the state vector as it propagates along x is thus given by

$$|E\rangle = \frac{1}{2}\begin{pmatrix} \exp\left[i\left(\beta - \Delta\right)x\right] + \exp\left[i\left(\beta + \Delta\right)x\right] \\ \exp\left[i\left(\beta - \Delta\right)x\right] - \exp\left[i\left(\beta + \Delta\right)x\right] \end{pmatrix} \exp\left[-i\omega t\right], \tag{8.27}$$

where clearly, $|E\rangle$ at $x = 0$ and $t = 0$ is just $|1\rangle$. The power in waveguide 1 is calculated using Equation (8.21), which yields

$$P_1 = |\langle 1|\, E\rangle|^2 = \cos^2\left(x\Delta\right). \tag{8.28}$$

Therefore, all of the power is transferred from waveguide #1 to waveguide #2 over a distance $x = \pi/2\Delta$. Similarly, the energy of the second waveguide is $P_2 = \sin^2\left(x\Delta\right)$, so $P_1 + P_2 = 1$ and energy is conserved as expected.

We see from this approach that we can derive the mode coupling conditions without the need for knowing the mode's intensity profile. This kind of general approach is useful in building a physical intuition of how modes couple and can be easily generalized to a multiple-waveguide system. For a specific system of coupled waveguides, to calculate the exchange of energy between modes as a function of position one must calculate Δ. This task, however, can be quite complex depending on the geometry and can be determined by using the methods described in Chapter 2.

8.2.3.2 Infinite Number of Parallel Waveguides

We now generalize the calculations from Section 8.2.3 to an infinite number of waveguides of periodicity δ, as shown in Figure 8.11. Light propagating in waveguide n is represented by the state vector $|n\rangle$ and three guided modes are shown (as before, we assume that only the lowest-order mode can be excited in each waveguide). Note that if the light is launched perpendicular to the waveguides, this geometry is identical to the case of the Bragg grating. We will see that our formalism of coupled state vectors can be used to treat both these cases. We start with the waveguide calculation in analogy to Section 8.2.3.1.

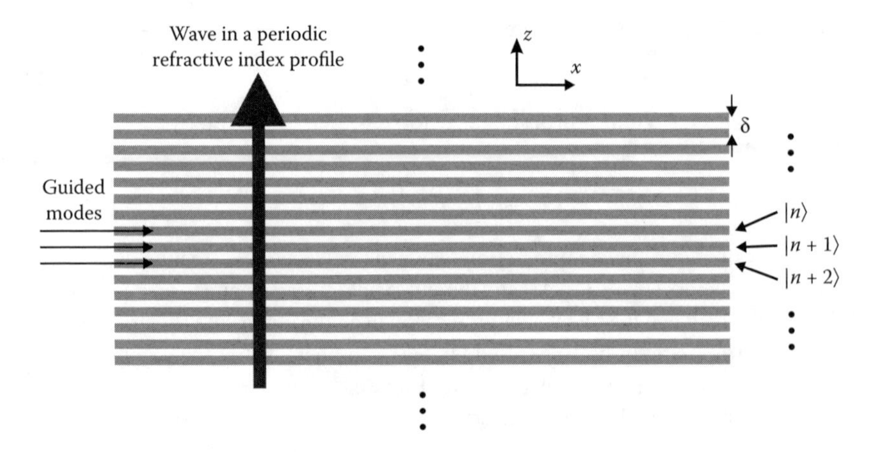

FIGURE 8.11 An infinite number of parallel waveguides of periodicity δ.

At first, this problem may seem intractable due to the fact that there are an infinite number of waveguides so that the most general state vector has an infinite number of components. However, we can use translational invariance to greatly simplify the problem. Let's define the translation operator $\tau(\delta)$, which shifts the waveguides up by one period. If we represent the refractive index profile as $n(x)$, the translation operator has the effect

$$\tau(\delta)n(x) = n(x+\delta) = n(x), \tag{8.29}$$

and

$$\tau(\delta)\,|n\rangle = |n+1\rangle\,. \tag{8.30}$$

When the waveguides are noninteracting, the infinite K-matrix is diagonal with β on the diagonals, so we get $K\,|n\rangle = \beta\,|n\rangle$.

Since $|n\rangle$ is an eigenstate of the K-matrix with eigenvalue β, the state vector $|\theta\rangle$ is defined by

$$|\theta\rangle = \sum_{n=-\infty}^{+\infty} e^{in\theta}\,|n\rangle\,. \tag{8.31}$$

The state vector $|\theta\rangle$ has the additional property that it is also an eigenstate of $\tau(\delta)$,

$$\begin{aligned}\tau(\delta)\,|\theta\rangle &= \sum_{n=-\infty}^{+\infty} e^{in\theta}\tau(\delta)\,|n\rangle = \sum_{n=-\infty}^{+\infty} e^{in\theta}\,|n+1\rangle = \sum_{n=-\infty}^{+\infty} e^{i(n-1)\theta}\,|n\rangle \\ &= e^{-i\theta}\sum_{n=-\infty}^{+\infty} e^{in\theta}\,|n\rangle = e^{-i\theta}\,|\theta\rangle\end{aligned} \tag{8.32}$$

with eigenvalue $e^{-i\theta}$. Thus, $|\theta\rangle$ is a special state vector that is an eigenstate of both K and $\tau(\delta)$.

It is important to note that the state of the system is infinitely degenerated with respect to β. That is, it doesn't matter which waveguide carries a mode, the wavevector is the same. Since light can be launched into an infinite number of waveguides, there are an infinite number of states with wavevector β. However, the state vector in the form given by Equation (8.31) represents an infinite number of wavefunctions, distinguished by the value of the continuous parameter θ. As such, any wavefunction of this system of an infinite number of waveguides can be expressed as a superposition of modes with different values of θ.

Next, let's allow the modes to interact with each other. If we allow for interactions between nearest waveguides, the K-matrix, in component form, becomes

$$K_{nm} = \delta_{n,m}\beta + \delta_{n,m\pm 1}\Delta, \tag{8.33}$$

where $\delta_{n,m}$ is the Kronecker delta function. Δ thus represents the rate of energy transfer between nearest waveguides, i.e., between state $|n\rangle$ and $|n \pm 1\rangle$. Let's now calculate the K-matrix's effect on $|\theta\rangle$:

$$\begin{aligned} K \sum_{n=-\infty}^{+\infty} e^{in\theta} |n\rangle &= \beta \sum_{n=-\infty}^{+\infty} e^{in\theta} |n\rangle + \Delta \sum_{n=-\infty}^{+\infty} e^{in\theta} \left(|n+1\rangle + |n-1\rangle\right) \\ &= \sum_{n=-\infty}^{+\infty} \left(\beta + \delta\left[e^{i\theta} + e^{-i\theta}\right]\right) e^{in\theta} |n\rangle \\ &= (\beta + 2\Delta\cos\theta) \sum_{n=-\infty}^{+\infty} e^{in\theta} |n\rangle . \end{aligned} \tag{8.34}$$

Therefore, we get the important result that each eigenstate is a mode whose power is spread evenly between all of the waveguides and only the phase of the electric field varies in a well-defined way from one waveguide to the next. In particular, for the mode $|\theta\rangle$[1] the phase difference between adjacent waveguide modes is simply θ. So, we find that the mode wavevector is given by

$$\beta(\theta) = \beta + 2\Delta\cos\theta, \tag{8.36}$$

so its range when $0 \leq \theta \leq \pi$ is

$$\beta - 2\Delta \leq \beta(\theta) \leq \beta + 2\Delta. \tag{8.37}$$

Thus, the range of allowable mode wavevectors is reminiscent of a band in a crystal, where Equation (8.36) is analogous to the dispersion relationship of an electron in a crystal.

---

[1] In this treatment, we did not normalize the modes to unit power. If we do so, the modes are of the form

$$|\theta\rangle = \frac{\sum e^{in\theta} |n\rangle}{\sum \left|e^{in\theta}\right|^2}. \tag{8.35}$$

Clearly, since there are an infinite number of waveguides, each carries an infinitesimal power with mode wavevector $\beta + 2\Delta\cos\theta$.

The result given by Equation (8.36), recall, was derived under the assumption that only the lowest-order mode is excited in each waveguide. If higher-order modes are excited, each of these will lead to a similar band. Thus, we will get a series of bands in which modes propagate. In the gaps between the bands, that particular range of modes will not propagate, yielding a band gap.

As we mentioned at the start of this section, the calculations presented here also hold if a light beam propagates perpendicular to the waveguides. In this instance, consider the waveguides as being slabs of alternating refractive indexes. If the refractive index contrast is very high, the light will be trapped in the regions of high refractive index, and these high refractive index regions will only interact weakly. As such, the state vectors, instead of representing guided modes, represent standing waves. So, in both cases, light is confined to a particular guide or slab, so the mathematics behind the calculations is identical. Thus, the resulting band structure we found in Equation (8.36) will apply to the case of light propagating through a periodic refractive index.

One other important issue needs to be mentioned. If one of the waveguides (or slabs) is removed, the symmetry is broken, and a discrete mode of propagation appears in the middle of a gap. This is analogous to an impurity in a crystal that shows discrete states in the band gap.

8.2.3.3 Holey Fiber

Holey fiber is an example of a two-dimensional periodic structure of holes parallel to the axis of a fiber with a missing hole in the center of the fiber that acts as a defect along which light will propagate. The first holey fiber was made in 1996 in silica glass and reported by a group at the University of Southampton.[125] Such structures in polymer fibers were first demonstrated in 2001 by a group at the University of Sydney.[126] Figures 8.12a and 8.12b show a preform made of DR1-doped PMMA polymer with six concentric hexagonal layers of holes. The preform was made by drilling a DR1-doped PMMA cylinder. Figure 8.12c shows a holey fiber made in PMMA polymer with three layers of holes. The guided mode in the center is clearly observed visually using a microscope.

Holey fibers have several advantages over traditional waveguides that are made with a region of elevated refractive index. First, as observed by Knight and coworkers,[125] holey fibers will meet the single-mode condition over a broad range of wavelengths. So, they are suitable for applications that require waveguiding of multiple colors such as parametric frequency generators and spectrometers. For example, note that the light guiding in the fiber shown in Figure 8.12b appears white because of the broad band of colors that are present. Second, holey fibers with a hexagonal array are observed to act as polarization-maintaining fibers for two orthogonal linear polarizations.[125] Even when the fiber is bent or twisted, the polarization remains linear. For any other polarization, elliptical light results.

Several other interesting features appear in Figure 8.12c. First, note that the hexagonal pattern of holes and uniformity of hole diameter is far from perfect.

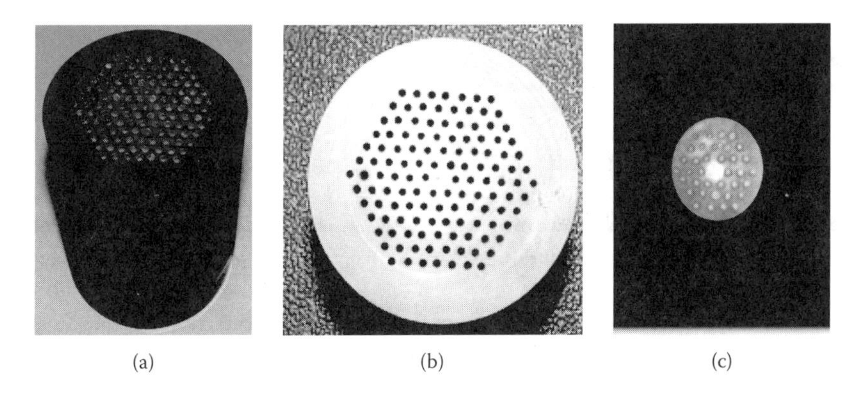

FIGURE 8.12 (a) Six layers of a hexagonal array of holes in a DR1-doped PMMA preform, (b) an end view of the same preform, and (c) a guided mode in a holey fiber with three layers of holes. Structures made and photographed by Juefei Zhou at Washington State University.

Even with these imperfections and the fact that this fiber only has three layers, the waveguiding mode is well confined. However, there is clear evidence that some of the field leaks away from the center where the holes deviate most from hexagonality, such as at the upper-left portion of the fiber.

Such fibers can also act as stress sensors. When the fiber is squeezed from the side, the polymer becomes birefringent and some of the light scatters out of the core. Since the blue light scatters more than the red, more of the blue light is cut off. Figure 8.13 shows a plot of the transmission spectra as a function of pressure. Note that the sharp cutoff makes this a sensitive sensor since small changes in the

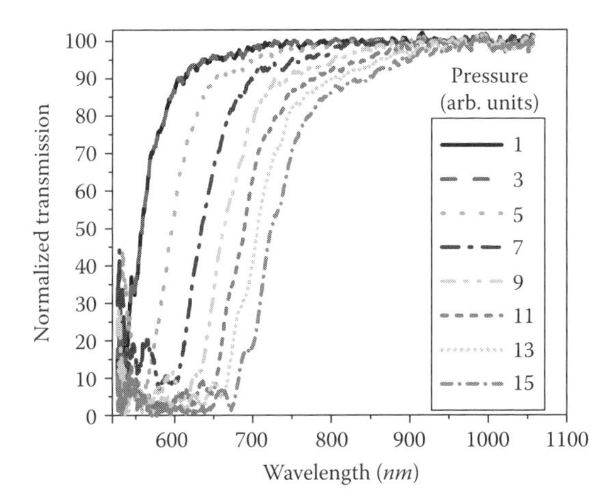

FIGURE 8.13 The transmission spectrum of light exiting a holey fiber as shown in Figure 8.12c as a function of applied pressure to the side of a spool of such fiber. Figure provided courtesy of Juefei Zhou of Washington State University.

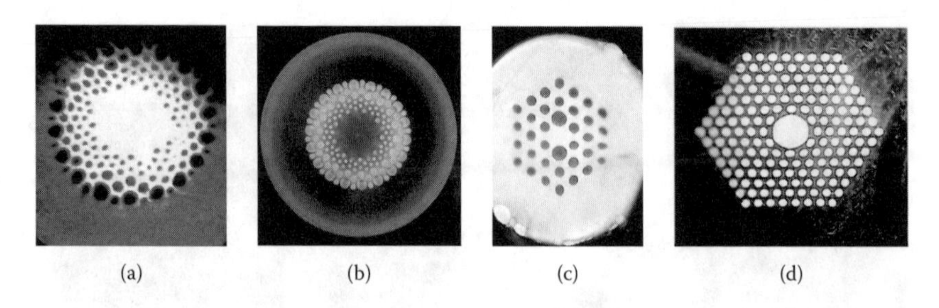

FIGURE 8.14 (a) Light guiding in a gradient photonic crystal polymer fiber waveguide and (b) the preform used for making such a fiber. (c) A birefringent photonic crystal polymer fiber and (d) a photonic crystal polymer fiber with a waveguiding air hole. All figures provided courtesy of and with permission of Jing Xie of Cactus Fiber Pty Ltd.

pressure lead to large changes in the transmitted intensities at the spectral edge. Stress and temperature sensors based on fiber Bragg gratings are discussed in Section 8.4.

It is interesting to note that all of the kinds of fiber that can be made using standard waveguides (i.e., core and cladding made of dye doped polymer) can also be made using microstructured fiber. In the fiber shown in Figure 8.12c, the region with the holes can be thought of as a lower refractive index area than the core. So, this case is analogous to a standard fiber with a core of elevated refractive index relative to the cladding. Graded-index fiber, on the other hand, is made with a parabolic refractive-index profile. Figure 8.14a shows light guiding in a fiber with changing hole diameter and spacing that mimics a graded-index fiber. Figure 8.14b shows the preform that was used to make this fiber.

Similarly, the array of holes can be designed to make a birefringent fiber. Figure 8.14c shows an example of such a fiber where the symmetry is broken by making two larger holes near the waveguiding core. Similarly, a birefringence holey fiber can be made using oblong holes or by breaking the hexagonal symmetry by deforming the structure along one axis.

Figure 8.14d shows a photonic crystal polymer fiber with a central air hole that defines the waveguiding region. Such fibers can be used to transmit high power laser light that is well above the damage threshold of the surrounding material; and, since most of the light intensity is in the hole, the optical loss due to the material surrounding the hole is smaller than it would be for propagation within that same material. Figure 8.15a shows a different design of a fiber that is made of an alternating refractive-index structure surrounding an air hole in a coaxial geometry.[127] By virtue of the large bandgap of such a structure, the inner wall acts as an omnidirectional dielectric mirror.

This structure was made by starting with a capillary tube and evaporating a thin layer of tellurium, followed by dip coating in a polystyrene/toluene solution to make a polystyrene layer, followed by another tellurium layer. The subsequent layers are made of alternating polyurethane and tellurium; and a total of nine

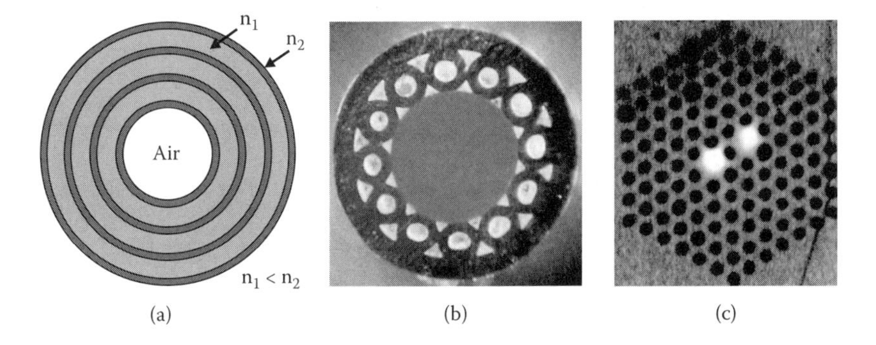

FIGURE 8.15 (a) A dielectric coaxial cable fiber waveguide and (b) a specialty holey fiber (provided courtesy of and with permission of Paradigm Optics). (c) A dual-core holey fiber (provided courtesy of and with permission of Jing Xie of Cactus Fiber Pty Ltd).

alternating layers are fabricated. The silica glass capillary tube is removed with concentrated hydrofluoric acid to make the final structure. The resulting fiber results in good-quality air guiding even around 90-degree bends.

A similar fiber with a coaxial omnidirectional dielectric mirror with nine alternating layers was fabricated for guiding high-power IR light.[128] The dielectric mirror was made with alternating layers of high-refractive index contrast by combining thin layers of arsenic triselenide (As_2Se_3) with a refractive index of 2.8 and thick layers (about 1 μm) of the high glass-transition temperature thermoplastic poly(ether sulphone) having a refractive index of 1.55. At a wavelength of 10.6 μm, the optical loss was found to be less than 1 dB/m, an order of magnitude lower than the loss of the intrinsic fiber material. Similarly, this fiber could transmit high-power CO_2 laser light over a couple of meters length, where the light exiting the fiber would burn holes in poly(ether sulphone), the same polymer used to make the fiber.

Figure 8.15b shows a fiber with a large core that guides red light. The core is held in place by 12 capillary tubes that fit on the inside of a larger hollow plastic tube. This is yet another example of the flexibility available for making novel structures.

As we discussed in Section 1.4.7, parallel waveguides can be made in a fiber; and light launched in one of the guides will hop back and forth between the two guides over a distance of propagation that is related to the distance between the waveguides. A dual-core fiber is the basis of one design of an all-optical switch, as described in Section 7.2.3. Similarly, a dual-core waveguide can be made in a holey fiber with two defects (i.e., two missing holes). Figure 8.15c shows an example of a dual-core holey fiber made by Cactus Fibers. There is clearly an overlap between the fields of the light propagating between the two guiding regions.

To conclude this section, given the flexibility of polymer fiber fabrication methods (drilling, stacking, redrawing, etc.), all sorts of microstructured fibers have been demonstrated and many others are possible. The kinds of fibers that can be made are limited only by the imagination.

8.3 PHOTOREFRACTION

The interference pattern between two plane waves of the same wavelength, λ, results in a sinusoidal intensity pattern. The simplest example is the case of two counterpropagating plane waves that form a standing wave with a periodicity of $\lambda/2$. In the presentence of a nonlinear material with an intensity-dependent refractive index of the form $\Delta n = n_2 I$ (see Section 7.2), where n_2 is proportional to the third-order susceptibility, this intensity modulation will result in a sinusoidal refractive index profile. As such, the two counterpropagating waves form a Bragg grating. So, all of the formalism that we derived in Section 8.1.1 will apply to the effect of the grating on an independent beam of light.

However, the interaction of the two beams is a more complicated matter. For example, an interesting question is if the two counterpropagating beams will be reflected by the grating. In the case where the refractive index change is proportional to the intensity, the geometry is symmetric, in that the amount of forward-traveling light reflected backward will be equal to the amount of backward-traveling light that is reflected forward. As a result, there is no observed change of intensities in the two counterpropagating beams. However, if there is a spatial phase shift between the refractive-index grating and the intensity profile, the symmetry is broken and energy will be exchanged between the forward- and backward-traveling beams.

We start this section by considering the theory of two-beam coupling and end with a description of photorefractive effects in a polymer optical fiber.

8.3.1 THEORY

8.3.1.1 Two-Beam Coupling

In this section, we describe the theory of two-beam coupling when the two beams enter the sample from the same side. There are various other geometries that are possible, but treating all of the cases is beyond the scope of this book. However, this case illustrates the method, which can easily be applied to the other cases.

Consider two beams (approximated by plane waves) that enter a sample from the same side, as shown in Figure 8.16. This will result in grating planes that are perpendicular to the x direction. Assuming that both beams are polarized perpendicular to the plane of the fiber, we can express the fields by

$$\vec{E}_j = \vec{A}_j \exp\left[i\left(\omega t - \vec{k}_j \cdot \vec{r}\right)\right], \text{ with } j = 1, 2. \tag{8.38}$$

The intensity is given by

$$I = \eta \vec{E} \cdot \vec{E}^* = \eta(\vec{E}_1 + \vec{E}_2) \cdot (\vec{E}_1^* + \vec{E}_2^*), \tag{8.39}$$

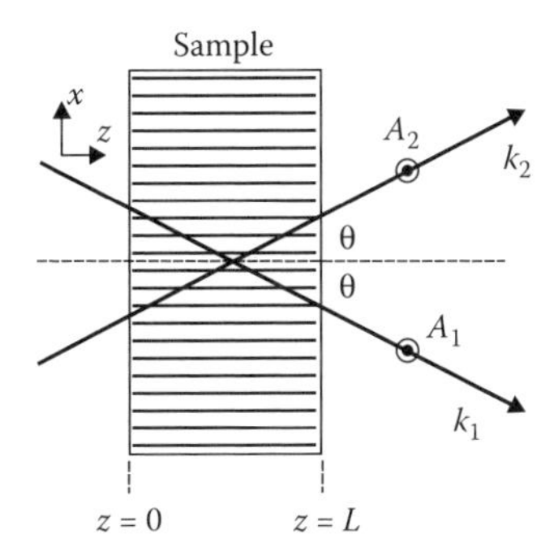

FIGURE 8.16 Geometry of two-beam coupling where both beams enter the sample from the same side.

where η depends on the choice of units. Substituting Equation (8.38) into Equation (8.39), we get

$$I = \eta(|A_1|^2 + |A_2|^2 + A_1^* A_2 e^{-i\vec{K}\cdot\vec{r}} + A_1 A_2^* e^{+i\vec{K}\cdot\vec{r}}), \tag{8.40}$$

where A_i is the complex magnitude of the vector $\vec{A}_i$ and $\vec{K}$ is the grating wavevector given by

$$\vec{K} = \vec{k}_1 - \vec{k}_2. \tag{8.41}$$

In analogy to our derivation in Section 8.1.1, we start with Maxwell's wave equation,

$$\nabla^2 E + \frac{\omega^2}{c^2} n^2 E = 0, \tag{8.42}$$

with

$$n = n_0 + n_2 I, \tag{8.43}$$

where I is given by Equation (8.40) and the fields given by Equation (8.38). If the two incident beams are of infinite extent (i.e., plane waves), the amplitudes A_1 and A_2 will only depend on z so in Equation (8.42) we make the substitution $\nabla^{\rightarrow}\partial/\partial z$. Furthermore, in the steady state, the amplitudes will be independent of time. Putting this all together, we get

$$\sum_{j=1}^{2} \left(\left[\frac{\omega^2}{c^2} \left(n_0^2 + 2n_0 n_2 I_0 \right) - \beta_j^2 \right] A_j - 2i\beta_j \frac{dA_j}{dz} \right) e^{-i\vec{k}_j\cdot\vec{r}}$$
$$+2\frac{\omega^2}{c^2} \sum_{j=1}^{2} \left(n_0 n_2 A_1^* A_2 e^{-i\vec{K}\cdot\vec{r}} + n_0^* n_2^* A_1 A_2^* e^{+i\vec{K}\cdot\vec{r}} \right) e^{-i\vec{k}_j\cdot\vec{r}} = 0, \tag{8.44}$$

where we define $\beta_j = (k_j)_z$ and $I_0 = \eta\left(|A_1|^2 + |A_2|^2\right)$. In arriving at Equation (8.44), we have used the slowly varying envelope approximation (so second derivatives of the amplitudes can be ignored) and we assume that the refractive-index change is small so we only keep terms to first order in $n_2 I$.

With the static field dispersion defined by

$$\beta_j^2 = \frac{\omega^2}{c^2}\left(n_0^2 + 2n_0 n_2 I_0\right), \tag{8.45}$$

the term in square brackets in Equation (8.44) vanishes. Now we are prepared to get two coupled equations by considering Fourier components of Equation (8.44) at wavevectors $\vec{k}_1$ and $\vec{k}_2$. Summing over j and projecting the wavevector at $\vec{k}_1$, we get

$$-2i\beta_1 \frac{dA_1}{dz} + 2\frac{\omega^2}{c^2} n_0^* n_2^* A_1 A_2^* A_2 = 0. \tag{8.46}$$

Doing the same for $\vec{k}_1$, we get

$$-2i\beta_2 \frac{dA_2}{dz} + 2\frac{\omega^2}{c^2} n_0 n_2 A_1 A_1^* A_2 = 0. \tag{8.47}$$

To uncouple Equation (8.46) and Equation (8.47), we express them in terms of the intensities of each beam. First note that according to Figure 8.16, under the assumption that the frequencies of the two beams are the same, the wavevectors both have the same projection on the z-axis, or $\beta_1 = \beta_2 = k\cos\theta$. Second, we assume that β is real. Finally, we express the refractive-index product $n_0 n_2$ as a real amplitude times a phase factor, or $n_0 n_2 \rightarrow n_0 n_2 \exp[-i\phi]$. With these simplifications, we multiply Equation (8.46) by A_1^* and add to it the complex conjugate of Equation (8.46) multiplied by A_1. This yields

$$\eta\beta \frac{dI_1}{dz} = 2\frac{\omega^2}{c^2} n_0 n_2 I_1 I_2 \sin\phi, \tag{8.48}$$

and similarly, Equation (8.47) leads to

$$\eta\beta \frac{dI_2}{dz} = -2\frac{\omega^2}{c^2} n_0 n_2 I_1 I_2 \sin\phi, \tag{8.49}$$

where we have defined $I_j = \eta \left|A_j\right|^2$, which is proportional to the power.

Adding Equation (8.48) and Equation (8.49), and integrating, we get the expected result that $I_1 + I_2 = I_0$. Using this relationship, it is trivial to solve each equation. For example, substituting $I_2 = I_0 - I_1$ into Equation (8.48), collecting all of the intensity terms to one side, and integrating, we get

$$I_1 = \frac{I_0}{1 + m\exp\left[-I_0 \alpha z\right]}, \tag{8.50}$$

where

$$\alpha = 2\frac{\omega^2}{c^2} n_0 n_2 \sin\phi / \beta\eta \tag{8.51}$$

and m is the ratio of the powers of the two beams before interacting in the material, or

$$m = I_2^{(0)} / I_1^{(0)}, \tag{8.52}$$

where $I_j^{(0)}$ is the incident power of beam j.

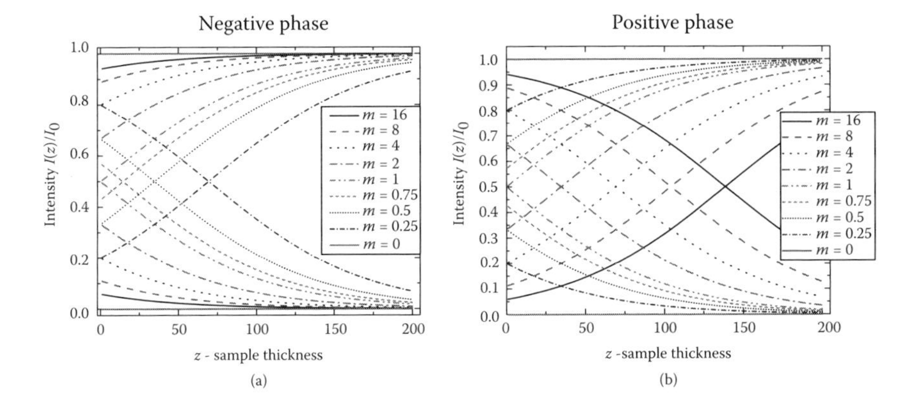

FIGURE 8.17 Intensity as a function of sample thickness for (a) negative phase angle ϕ and (b) for positive phase angle ϕ.

Figure 8.17 shows the intensity of each beam as a function of distance propagated through the sample for both a negative and positive phase ϕ. Note that from Equation (8.50), we get the important result that there is no exchange of energy between the beams when the intensity grating is in phase with the refractive-index grating ($\phi = 0$). Indeed, the maximum efficiency for two-beam coupling occurs when the intensity grating and refractive-index grating are 90° out of phase.

To understand the need for a phase mismatch, consider first the in-phase case. Since the geometry is symmetric with respect to interchanging the two beams, the conversion of energy from Beam #1 to Beam #2 must be equal to the conversion of energy from Beam #2 to Beam #1. Only when this symmetry is broken will there be a net exchange of energy. Note also that the positive phase and negative phase behavior is qualitatively different. For example, consider the solid curves. For a negative phase with an incident intensity ratio of $m = 16$, the power in Beam #1 starts high and continues to grow until all of the power propagates in Beam #1. For a positive phase at the same input power, the intensity of Beam #1 falls and all the power is converted to Beam #2.

In the rest of this section, we consider examples of interactions of two and three beams in polymer optical fiber. Since the calculations of these other cases are similar to the above treatment of two-beam coupling, we do not present calculations for each case since the reader can apply the methods explained in this section to any process.

8.3.2 Real-Time Holography in a Polymer Fiber

An azo dye, such as disperse red 1, can have two shapes — called conformations. The lowest energy state is the trans isomer and the cis isomer can be formed by exciting the trans form either thermally or optically. Since the cis isomer typically has a smaller polarizability than trans, the refractive index of a collection of trans isomers will be much larger than the refractive index of a collection of cis isomers.

An important property of one-dimensional azo dyes is that the refractive index is highly anisotropic. Thus, when a polarized beam of light excites an isotropic distribution of trans molecules in a polymer, the probability of conversion to the cis state is proportional to the square of the cosine of the angle between the molecular axis and the beam's electric field vector. So, the molecules along the electric field polarization will be depleted, and since the cis molecules are smaller that trans molecules, will thermalize into an isotropic distribution of orientations. Thus, they will relax to the trans state with isotropic orientation. However, those molecules that end up oriented along the light's plane of polarization will again be converted to the cis isomer. As such, the net effect is that the orientational distribution of the trans molecules becomes highly anisotropic, with a depletion of molecules oriented along the electric field. This effect is called angular hole burning. (See Section 9.2.2 for an extended discussion on this topic.) Sekkat and coworkers were the first to study and model the optical properties of a dye-doped polymer in the presence of an intense polarized laser source.[129–133]

It is found that photoisomerization and subsequent angular hole burning yields a large refractive-index change; but the time scale for the change is long. Therefore, it is an ideal system for studying nonlinear-optical properties of polymer fibers using inexpensive equipment such as Helium-Neon lasers and photodiodes.

Bian and coworkers demonstrated real-time holography in a PMMA fiber doped with DR1 azo dye.[134] Holographic storage in such photosensitive fibers offers many advantages over conventional films or bulk media, such as enhanced sensitivity, higher diffraction efficiency, and large-angle access sensitivity, which allows multiple holographic images to be stored in the same physical volume. A hologram is created by interfering light scattered from an object with a reference beam in the photorefractive material. So, the process is effectively a two-beam writing process, where a refractive-index grating in the material is formed by the reorientation of trans molecules and angular hole burning.

The multimode DR1-doped PMMA fiber used by Bian ranged in diameter from $400\mu m$ to $1,200\mu m$, lengths from $10\,mm$ to $50\,mm$ and DR1 concentrations from 0.05% by weight to 2%. We note that these parameters need to be optimized under the constraint that the light from the laser is not strongly depleted over the length of the fiber so that the whole volume of the fiber is used to form a grating.

Figure 8.18a shows the forward two-beam mixing geometry for making a transmission-type hologram as reported by Bian and coworkers.[134] The two beams in the material can also counterpropagate to form a reflection-type hologram. In the first set of experiments, the ring is photographed at the output plane on the screen when only beam A is turned on (Figure 8.18b) or only Beam B is turned on (Figure 8.18c). Subsequently, both beams are turned on for about 30 seconds to allow the refractive index grating to be written. Then when either Beam A or Beam B is blocked (Figures 8.18d and 8.18e, respectively), both circles are observed. This is due to the light scattering from the holographic refractive-index grating in a process similar to the one described in Section 8.3.1.

The above experiments are the simplest case of a grating written in a polymer fiber with two beams. However, the advantage of holography is the ability to store

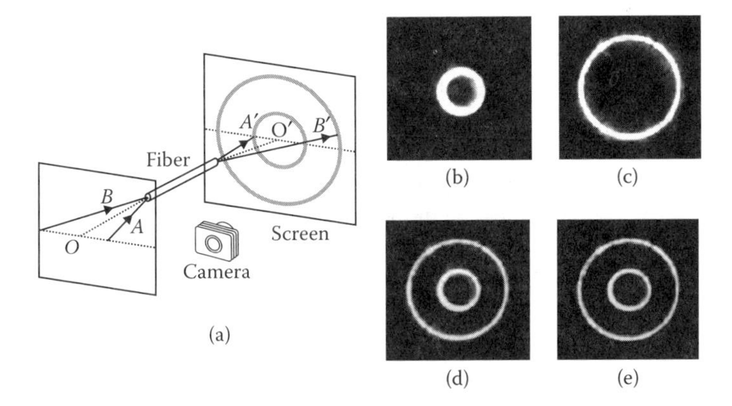

FIGURE 8.18 (a) The forward mixing geometry for a transmission-type hologram in a fiber leads to rings at the output plane. (b) Photograph of resulting ring when only Beam A is turned on and (c) Photograph of resulting ring when only Beam B is turned on. (d) Photograph at the output plane after the hologram is written in the presence of both beams, and Beam A is blocked, then (e) Beam B is blocked. Adapted from Ref. [134] with permission of the Optical Society of America.

images. Bian and coworkers used the transmission geometry to write an image in a polymer optical fiber using the experiment shown in Figure 8.19(a). The light exiting the mask that defines the object is focused down the axis of the fiber while the reference beam is launched into the fiber at an angle to the fiber axis after focusing. Figure 8.19(b) shows the image observed by the camera with no reference beam.

After the object beam and the reference beam illuminate the fiber for about 30 seconds, the hologram is written and the object beam is turned off. Figure 8.19(c)

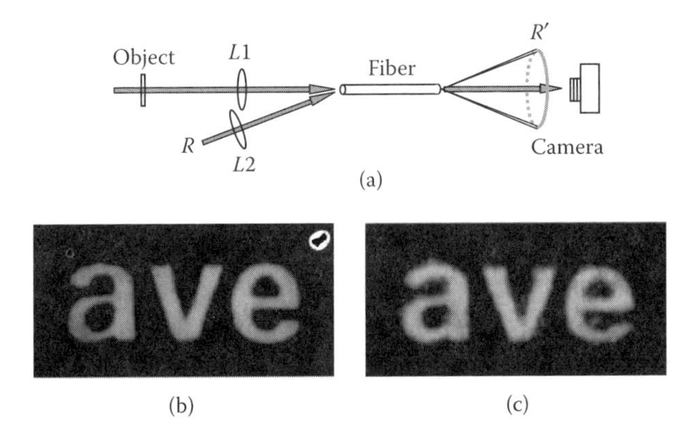

FIGURE 8.19 (a) An experiment for writing an image in a fiber using a transmission hologram. (b) A photograph of the object and (c) the reconstructed image using the reading beam. Adapted from Ref. [134] with permission of the Optical Society of America.

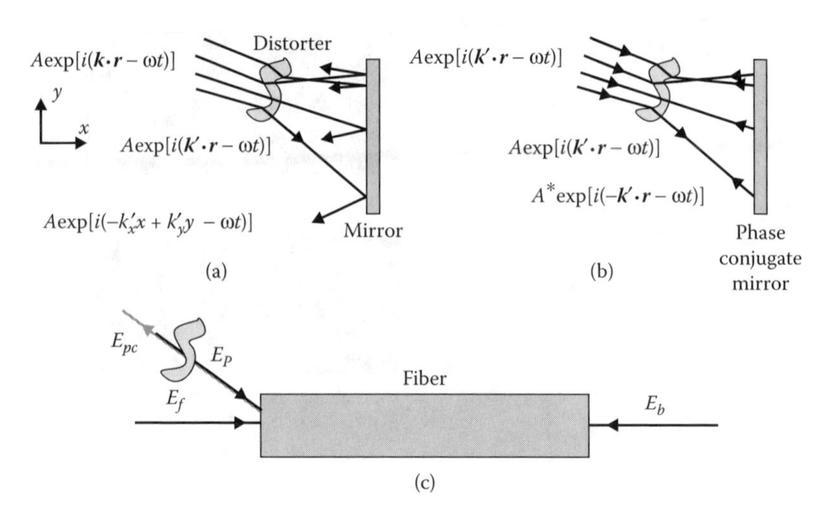

FIGURE 8.20 A plane wave incident on a transparent phase distorter followed by (a) a reflection from a mirror and (b) a reflection from a phase conjugate mirror. (c) A phase conjugate mirror in a polymer optical fiber.

shows the image produced in the presence of only the reference beam. Note that the ring due to the reference beam is well outside the camera's field of view. Multiple images can be written in the same fiber by changing the angles of the beams; and the images can be erased by exposing the fiber to an intense uniform beam. As such, photorefractive fiber can be used as an erasable and rewritable holographic storage device.

8.3.3 Phase Conjugation in a Polymer Fiber

To understand phase conjugation, we first consider reflection from an ordinary mirror. Figure 8.20a shows a plane wave incident on a blob of transparent material that distorts the wave. As the light passes through the material, it refracts by differing degrees based on the shape of the material at the points of entry and the points of exit. The rays can be traced by simply using Snell's Law. When a ray hits a mirror, the component of the wavevector perpendicular to the surface changes sign, which leads to the standard result that the angle of reflection is the same as the angle of incidence.

The light reflected from a phase conjugate (PC) mirror, as represented by Figure 8.20b, is given by the complex conjugate of the spatial part of the wave, or, $A\exp[i(\vec{k}\cdot\vec{r}-\omega t)] \rightarrow A^*\exp[i(-\vec{k}\cdot\vec{r}-\omega t)]$ so that the reflected wavevector is 180° relative to the incident wavevector. So, the reflected ray from a phase conjugate mirror retraces its path in reverse. Alternatively, we get the same result by replacing t with $-t$ and taking the complex conjugate, so the PC wave is sometimes called the time-reversed wave. Intuitively, this picture is consistent with the observation that the phase conjugate wave appears to travel backwards in time as one would observe when playing a movie in reverse. After the PC wave goes back through the distorter, the plane wave is reconstructed in the backward direction, again retracing the time-reversed path. Therefore, a photorefractive material can be used to correct abberations, as for example, one would get when light passes through a turbulent atmosphere.

Bian and coworkers demonstrated phase conjugation in DR1 dye-doped PMMA optical fiber in the geometry shown in Figure 8.20c.[135] To understand the process of the generation of the phase conjugate wave, we can imagine that the probe beam, with field amplitude E_p, interferes with the forward beam, with field amplitude E_f, to make a grating from which the backward wave of electric field amplitude E_b scatters to form the phase conjugate wave, which retraces the path of the probe beam.

Figure 8.21 shows the experiential geometry used by Bian to demonstrate phase conjugation in a DR1-doped PMMA polymer optical fiber. Beamsplitter BS 1 deflects a portion of the beam, which acts as a probe while BS 2 splits the remaining into two equal power beams that act as the forward and backward pump beams. Part of the probe beam is directed to a screen (S2) with a beam splitter so that the intensity profile of the probe beam can be recorded with a CCD camera. The photograph adjacent to screen S2 shows the intensity pattern of the incident probe beam.

All three beams are required to be present to write a grating that produces a reflected beam that is the phase conjugate of the probe beam. The reflected beam is measured at Screen S1. The photograph next to Screen S1 shows the intensity profile of the phase conjugate beam. If any one or more of the beams are absent at the start of the experiment, no reflected phase conjugate beam is observed.

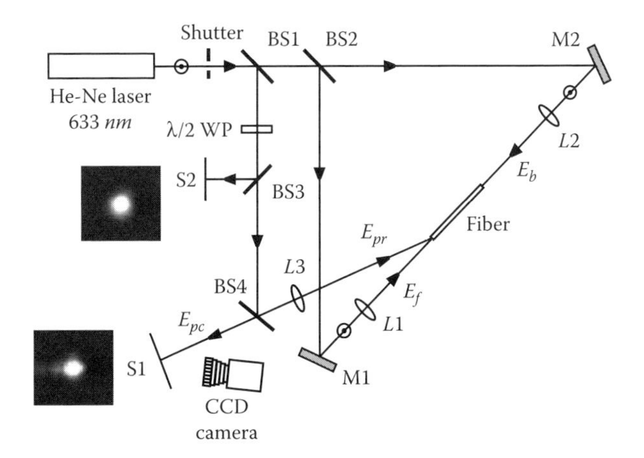

FIGURE 8.21 The experimental geometry for demonstrating phase conjugation in a polymer optical fiber. Adapted from Ref. [135] with permission of the American Institute of Physics.

The dynamics of the grating-writing process can be studied by measuring the intensity of the PC signal as a function of time after the shutter is opened. A bi-exponential function fits the data with a slow response time of about 50 seconds and a fast response time constant of about 5 seconds.[135] After the grating is written, the phase conjugate beam is observed to persist even after the probe beam is blocked. The decay of the PC signal as a function of time after the probe is blocked also shows a bi-exponential behavior with similar time constants. Such dynamics are commonly observed in dye-doped polymer.

Instead of using a beam as the probe, a probe image can also be projected (in a manner similar to real time holography, as discussed in Section 8.3.2). In such cases, the phase conjugate image is the same as the incident image even if the probe image first passes through a distorter. So, the images observed on screen S1 and S2 always appear identical.

8.4 STRESS AND TEMPERATURE SENSORS

8.4.1 Discrete Sensors

As demonstrated in Section 8.1.3, a Bragg grating is sensitive to both temperature and strain; so, to use such fibers in making temperature or stress sensors, these two contributions must be separately isolated. The best method for differentiation is to use two sensors in tandem that respond differently to temperature and stress.

Consider two fiber Bragg gratings made of different materials and placed in series. The change in spectrum of each to stress and temperature will be different and for a linear response can be represented in matrix form,[124]

$$\begin{pmatrix} \Delta\lambda_1 \\ \Delta\lambda_2 \end{pmatrix} = \begin{pmatrix} K_{1T} & K_{1\epsilon} \\ K_{2T} & K_{2\epsilon} \end{pmatrix} \begin{pmatrix} \Delta T \\ \Delta\epsilon \end{pmatrix}, \tag{8.53}$$

where $\Delta\lambda_i$ is the shift in the reflectance peak of grating i, ΔT the temperature change, and $\Delta\epsilon$ the applied strain. K_{iT} is a coefficient that relates the temperature change to the wavelength shift of the reflectance of grating i and $K_{i\epsilon}$ the constant of proportionality between the stress and wavelength shift. This is the most general formula that represents the change in reflectivity as a linear combination of temperature and strain.

We can solve for the change in temperature and strain of the sensor pair in terms of the change in reflectance wavelengths by inverting Equation (8.54),

$$\begin{pmatrix} \Delta T \\ \Delta\epsilon \end{pmatrix} = \frac{1}{K_{1T}K_{2\epsilon} - K_{2T}K_{1\epsilon}} \begin{pmatrix} K_{2\epsilon} & -K_{1\epsilon} \\ -K_{2T} & K_{1T} \end{pmatrix} \begin{pmatrix} \Delta\lambda_1 \\ \Delta\lambda_2 \end{pmatrix}. \tag{8.54}$$

Note that the "K" matrix depends on the two material constants. Clearly, if grating 1 and grating 2 were to be made of similar materials, the determinant would be small and the temperature and strain would not decouple. As such, polymer Bragg gratings, having very different material properties from silica, provide a simple way of making hybrid stress/temeprature sensors with silica polymer pairs.

8.4.2 Distributed Sensors

Distributed sensor systems provide the ability of measuring an environmental property at various locations. For example, one method for making a distributed temperature/strain detector is to place pairs of Bragg gratings along the length of an optical fiber. The two common methods for differentiating between the positions of the detectors is wavelength- or time-division multiplexing. In the latter, the delay between the launch of the interrogating pulse and reflected spectrum determines the position of the sensor being interrogated. In the former, each sensor is made with a different grating period so that the position can be determined by the reflectance wavelength provided that changes in the peak reflectance wavelength of any one sensor does not overlap the range of operational wavelengths of all the other sensors in the chain. (Using a series of Bragg gratings for making smart materials is described in detail in Chapter 9.)

Figure 8.22a shows an example of wavelength division multiplexing. The two fibers represent the two extreme strain states and the plots above show the reflectance spectrum for these two states at three different points in the fiber. The plot on the left shows the incident spectrum and the plot to the right the transmitted spectrum. Section 8.1.4.2 describes how the signal from such a system can be demultiplexed into a signal at different colors.

Spatial resolution of a time division multiplexed system is determined by the temporal resolution of the system. If the reflectance spectrum is resolved over a time $\Delta\tau$, the spatial resolution is

$$\Delta x = c\Delta\tau/\beta, \tag{8.55}$$

where β is the mode index and c the speed of light in vacuum. So, for a system that can take a spectrum over a time interval of $10\,ns$, the spatial resolution for a mode

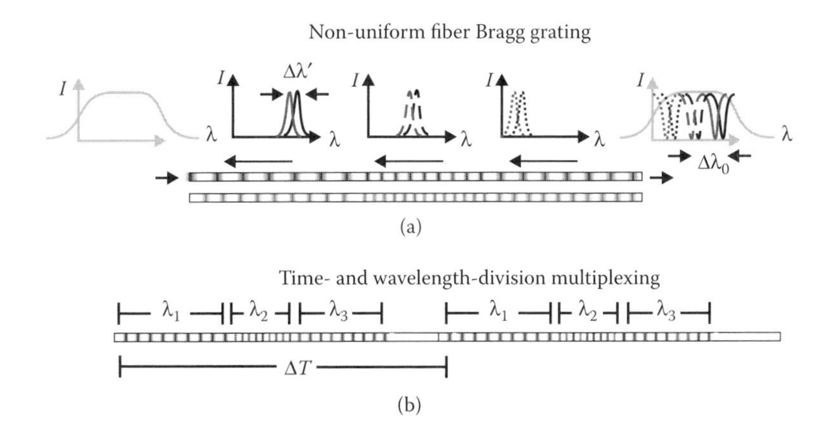

FIGURE 8.22 (a) Incident and reflected spectra at different points in a fiber designed for wavelength division multiplexing. The change in wavelength of the reflectance minimum of each grating due to temperature or strain is smaller than the spacing between the minima due to the other gratings. (b) Wavelength and time division multiplexing in the same fiber.

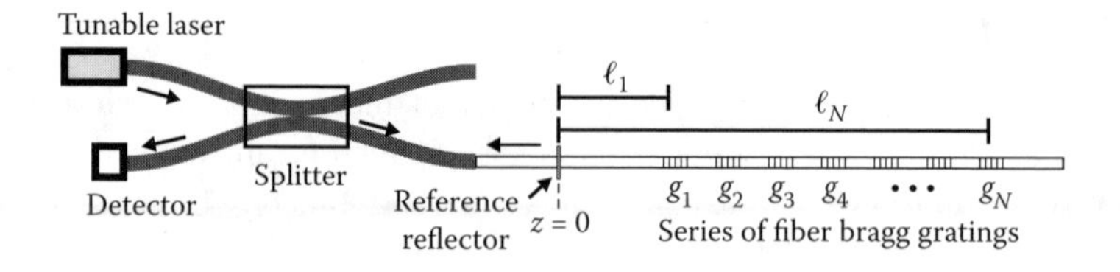

FIGURE 8.23 Optical frequency domain reflectometry apparatus as a distributed sensor.

index of 1.5 is $2\,m$. Wavelength division multiplexing gives the best resolution, being limited by the width of the laser source and the quality of the grating. From Equation 8.15, we had found that for a Bragg grating, all of the energy is converted to the backward wave over a grating length $\ell = \pi/(\sqrt{3}\epsilon\beta_0)$. For a wavelength of $500\,nm$ in a material of refractive index 1.5 and refractive index modulation depth 0.01, a grating of length $10\,\mu m$ is sufficient to observe complete reflection. So, position resolution is limited not by the grating size but the resolution of the spectrometer, the quality of the fiber, and the width of the reflectance peak.

Two methods of multiplexing can be combined to increase the number of sensing points. As an example, Figure 8.22b shows a fiber that combines wavelength- and time-division multiplexing. Large portions of the fiber are interrogated by time division multiplexing (labeled by ΔT) while subsections of these larger segments are interrogated with wavelength division multiplexing (labeled λ_1, λ_2, etc.). This dual multiplexing scheme allows for high-density multiplexing over long lengths of fiber.

There are many designs for distributed temperature sensors based on Bragg gratings. Perhaps one of the more interesting ones is Optical Frequency Domain Reflectometry (OFDR), reported by Froggatt.[136] Figure 8.23 shows the system. A tunable laser beam is launched through a $3dB$ fiber coupler (or circulator) and reference reflector into a fiber with a series of Bragg gratings. Each fiber grating, labeled g_1, g_2, $\ldots g_N$, acts as a reflector, so the light intensity measured by the detector results from interference between the light reflected from the reference reflector and light reflected from the Bragg gratings. Since the reflectivity of each Bragg grating is small, multiple reflections between the reference reflector and the grating are negligible.

Froggatt assumes that the grating can be expressed as a periodic function (i.e., the grating period) whose frequency is close to the wavelength of the light beam with a slowly varying complex enveloped function. Hence, Maxwell's wave equation, with a scalar field E, is of the form

$$\frac{\partial^2 E}{\partial z^2} + \beta_0^2 \left[1 + \frac{\Delta\epsilon_r(z)}{\epsilon_r}\right] E = 0, \tag{8.56}$$

with

$$\frac{\Delta\epsilon_r(z)}{\epsilon_r} = \kappa(z)\exp\left[i2\beta_0 z\right] + \kappa^*(z)\exp\left[-i2\beta_0 z\right], \tag{8.57}$$

where β_0 is the wave vector of peak reflectance of the grating. Note that Equation (8.57) represents the dielectric function over the whole set of Bragg gratings, so in the spaces between gratings, $\Delta\epsilon_r = 0$. The weakly varying envelope solution is then of the form,

$$E = A_f(z,\xi)\exp\left[i(\beta_0+\xi)z\right] + A_b(z,\xi)\exp[-i(\beta_0+\xi)z], \tag{8.58}$$

where ξ is a small parameter that describes the difference between the wavevector of the incident field, β, and the wavevector of peak reflectance, β_0, or $\xi = \beta - \beta_0$. Note that this derivation is similar to the one in Section 8.1.1 except that here, we do not take into account the fact that the wavevector can be a sinusoidal function of the position. As such, the calculation in this section applies only for small reflectivity: $|A_b| \ll |A_f|$.

In analogy to the description in Section 8.1.1, substituting Equation (8.57) and Equation (8.58) into Equation (8.56) leads to the coupled equations:

$$\frac{\partial A_f}{\partial z} = \frac{(\beta_0+\xi)}{2i}\kappa(z)A_b\exp\left[-2i\xi z\right] \tag{8.59}$$

and

$$\frac{\partial A_b}{\partial z} = -\frac{(\beta_0+\xi)}{2i}\kappa^*(z)A_f\exp\left[+2i\xi z\right]. \tag{8.60}$$

If depletion of the forward wave is negligible over the set of gratings in the fiber, we can make the approximation that $A_f(z,\xi) = 1$. Furthermore, if the gratings only fill a finite length of fiber, $A_b(\infty,\xi) = 0$. Using these conditions, the reflected wave at $z = 0$ (the position of the reference reflector) can be calculated by integrating Equation (8.60) with $A_f = 1$:

$$A_b(0,\xi) = -\int_\infty^0 \frac{(\beta_0+\xi)}{2i}\kappa^*(z)\exp\left[+2i\xi z\right]dz. \tag{8.61}$$

To calculate the intensity at the detector, we need to first determine the amplitude of the backward reflected wave from the grating and the field reflected from the reference reflector. If the reference reflector reflects a fraction a of the field, the total reflected field will be

$$A_b(0,\xi) = a + \left(1-a^2\right)^{1/2}\int_0^\infty \frac{(\beta_0+\xi)}{2i}\kappa^*(z)\exp\left[+2i\xi z\right]dz. \tag{8.62}$$

This leads to a power at the detector (normalized to the incident power) that is given by the square of the magnitude of Equation (8.62)

$$\begin{aligned}|A_b(0,\xi)|^2 &= a^2 + \left(1-a^2\right)\frac{(\beta_0+\xi)^2}{4}\left|\int_0^\infty \kappa^*(z)\exp\left[+2i\xi z\right]dz\right|^2 \\ &\quad + a\left(1-a^2\right)^{1/2}\frac{(\beta_0+\xi)}{2i}\int_0^\infty \kappa^*(z)\exp\left[+2i\xi z\right]dz \\ &\quad + c.c.,\end{aligned} \tag{8.63}$$

where $c.c$ refers to the complex conjugate of the previous term.

In order to evaluate Equation (8.63), we first note that the gratings are contained in the region $z > 0$. As such, $\kappa(z) = 0$ for $z < 0$, so we can change the bottom limit of integration to $-\infty$ without affecting $|A_b(0, \xi)|^2$. All of the integrals are clearly Fourier transforms, leading to

$$\begin{aligned} |A_b(0, \xi)|^2 &= a^2 + \left(1 - a^2\right) \frac{(\beta_0 + \xi)^2}{4} 2\pi \tilde{\kappa}^*(-2\xi)\tilde{\kappa}(2\xi) \\ &+ a\left(1 - a^2\right)^{1/2} \frac{(\beta_0 + \xi)}{2i} \sqrt{2\pi}\, \tilde{\kappa}^*(-2\xi) \\ &+ c.c., \end{aligned} \tag{8.64}$$

where $\tilde{\kappa}(\xi)$ is the Fourier transform of $\kappa(z)$. Next, we take the inverse Fourier transform of Equation (8.64)

$$\begin{aligned} P(x) &= \frac{1}{\sqrt{2\pi}} \int_{-\infty}^{\infty} d\xi \exp[+i\xi x]\, |A_b(0, \xi)|^2 = a^2\delta(x) \\ &+ \left(1 - a^2\right) \frac{\left(\beta_0 + \frac{1}{i}\frac{\partial}{\partial x}\right)^2}{4} 2\pi \cdot \frac{1}{\sqrt{2\pi}} \int_{-\infty}^{\infty} d\xi \exp[+i\xi x]\, \tilde{\kappa}^*(-2\xi)\tilde{\kappa}(2\xi) \\ &+ a\left(1 - a^2\right)^{1/2} \frac{\left(\beta_0 + \frac{1}{i}\frac{\partial}{\partial x}\right)}{2i} \sqrt{2\pi} \cdot \frac{1}{\sqrt{2\pi}} \int_{-\infty}^{\infty} d\xi \exp[+i\xi x]\, \tilde{\kappa}^*(-2\xi) \\ &+ c.c., \end{aligned} \tag{8.65}$$

where we have used the relationship $\xi \exp[i\xi x] = \frac{1}{i}\frac{\partial}{\partial x} \exp[i\xi x]$.

Let's first consider the second term on the right-hand side of Equation (8.65). To rewrite it, we start by using the deconvolution relationships derived in Chapter 5. Taking the reverse Fourier transform of Equation (5.98) to which we apply Equation (5.93), and making the substitutions $\bar{I}_s \rightarrow \tilde{\kappa}^*/\sqrt{2}$, $\bar{I}_p \rightarrow \tilde{\kappa}/\sqrt{2}$, $\omega \rightarrow \xi/2$, $t \rightarrow x'$ and $\tau \rightarrow x/2$, we get

$$\frac{1}{\sqrt{2\pi}} \int_{-\infty}^{\infty} d\xi \exp[+i\xi x]\, \tilde{\kappa}^*(-2\xi)\tilde{\kappa}(2\xi) = \frac{1}{2} \int_{-\infty}^{\infty} dx' \kappa\left(x'\right) \kappa^*\left(x' - \frac{x}{2}\right). \tag{8.66}$$

But, the right-hand side of Equation (8.66) is just the correlation between a grating and that same grating when shifted by $x/2$. As shown in Figure 8.23, $\kappa(z) \neq 0$ when $\ell_1 < z < \ell_N$; so the correlation function will not vanish when $\ell_1 < x' < \ell_N$ AND $\ell_1 < x' - \frac{x}{2} < \ell_N$. Figure 8.24 shows a schematic diagram of these two conditions. As is clear from Figure 8.24, the integral vanishes when $\ell_1 + x/2 > \ell_N$ or when $\ell_1 > \ell_N + x/2$. Defining $\Delta L = \ell_N - \ell_1$ as the length over which the Bragg grating is defined, these two conditions yield a vanishing self-correlation function when

$$|x| > 2\Delta L. \tag{8.67}$$

Therefore, assuming that the distributed sensor region is shorter than the distance to the reference reflector, Equation (8.67) holds so the integral in Equation (8.66) vanishes. Recalling that the envelope varies slowly compared to the

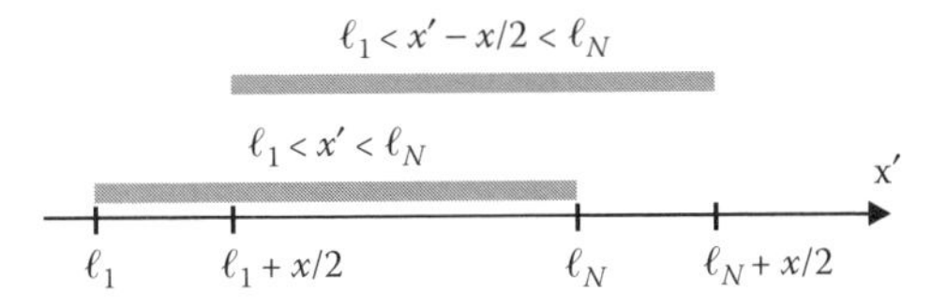

FIGURE 8.24 The integral in Equation (8.66) vanishes if the two gray bars do not overlap.

variation of the optical field, we can also make the approximation that $\frac{\partial}{\partial x}\kappa(x) \ll \beta_0\kappa(x)$. Putting this all together, Equation (8.65) becomes

$$P(x) = a\left(1 - a^2\right)^{1/2}\frac{\beta_0}{4i}\sqrt{2\pi}\cdot\left[\kappa\left(\frac{x}{2}\right) - \kappa^*\left(-\frac{x}{2}\right)\right]. \tag{8.68}$$

To use such a device for sensing, the wavelength of the tunable probe laser, ξ, is scanned and the power at the detector is measured to get $\tilde{P}(\xi)$. The result is then Fourier transformed to get $P(x)$. Using Equation (8.68), $\kappa(\frac{x}{2})$ can be determined. Recall that $\kappa(x)$ is the amplitude of the refractive-index profile of the grating, which is the parameter that changes under the influence of local stress or temperature. As such, since $\kappa(x)$ is related to the sensed parameter (i.e., the pressure or temperature can change the refractive-index profile), the parameter can be determined as a function of position.

Figure 8.25 shows an example of the data that one would expect from a distributed sensor. The gray regions show the locations of the fiber Bragg gratings and the dashed curve shows the parameter that is being sensed, T. If the refractive index amplitude of the Bragg grating is a linear function of the parameter that is being measured, the Fourier transform of the signal should be proportional to the position dependence, $T(x)$. However, since the Bragg gratings are written in discrete pieces, the areas between the gratings will show no net signal. The solid curve shows how the data would appear for an ideal sensor. In a typical sensor

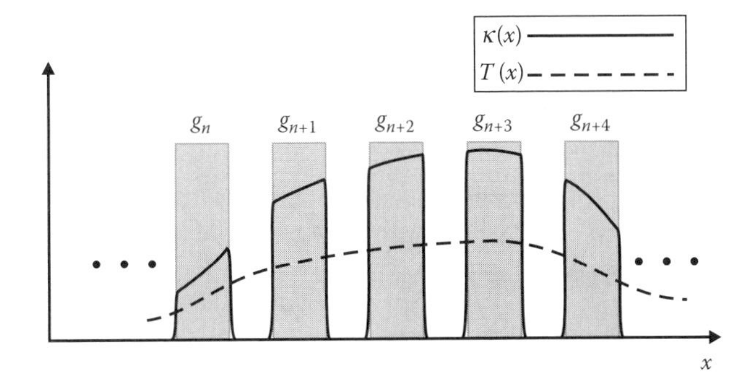

FIGURE 8.25 Simulated data from a distributed sensor. Gray areas show locations of fiber Bragg gratings. The dashed curve shows the parameter that is being sensed while the solid curve shows the Fourier transform of the signal.

system, the position resolution of the detector is given by the spacing between the discrete Bragg gratings.

While this type of system is at present being made by Luna Innovations using silica fibers, a hybrid polymer/silica system would enable the construction of sensors that could differentiate between temperature and stress sensing — as described in Section 8.4.1.

8.5 CHEMICAL SENSORS

8.5.1 Single-Agent Sensors

Fibers that combine silica glass waveguides and photosensitive polymer coatings are being used to make hybrid detectors. At present, a common method for making a chemical sensing fiber is to coat a glass fiber with a polymeric material whose fluorescence spectrum is changed by the presence of a particular chemical agent such as a nerve agent, pesticide, toxic waste molecule, or illegal substance. Figure 8.26 shows one design of a chemical sensor as reported by Jenkins and coworkers.[137–139]

The polymer coating is pre-sensitized to permit only one particular molecule to get near the fluorescing site. At the heart of the sensing coating are voids that are imprinted inside the polymer in the shape of the molecule to be sensed. A fluorescing molecule is placed near each of these voids. When a chemical agent molecule finds its way into the polymer, it fills the void, changing the observed fluorescence spectrum. Only one specific molecular shape will fit. Jenkins has shown that the selectivity is high so that a molecule that is similarly shaped to the target molecule will yield only a negligible signal.

Light is launched down the glass fiber to the polymer-coated section and excites fluorescence. Some of the fluorescence signal couples back into the fiber and

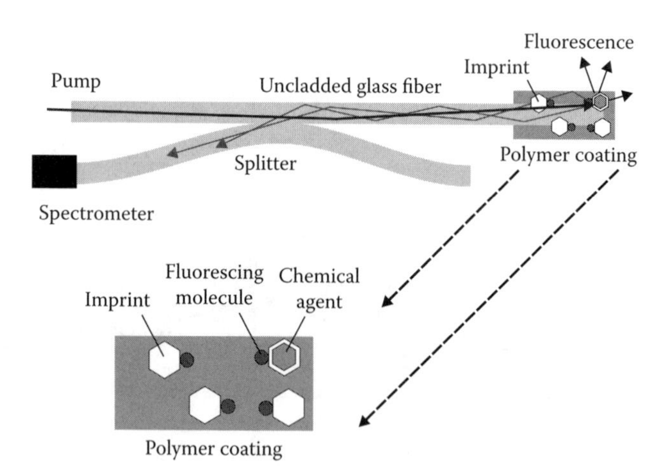

FIGURE 8.26 A chemical sensor made with an imprinted polymer coating at the end of a glass fiber.

is directed back toward the source, where a splitter redirects the signal to a spectrometer. Even for small amounts of a chemical agent, the spectrum is altered in a very specific way that signals the presence of that agent. Note that the cladding in the fiber needs to be removed where the polymer coating is applied so that the sensing coating defines the outer surface of the waveguide.

The imprinted polymer is made by first surrounding the target molecule with ligands. The encapsulated molecules are mixed with a monomer and partially polymerized (and cross-linked) into a monomer/polymer mixture. The glass fiber is coated with this partially polymerized liquid and the polymerization process is completed on the surface of the glass fiber. The coated fiber is soaked in a mixture of alcohol and water to cause swelling, which releases the target molecules and leaves behind empty voids. The sensor is then dried and ready for use.

8.5.2 Artificial Nose

While this device is good at sensing a single chemical agent, it is not easy to use this design to make a single fiber that will sense multiple agents. Holey polymer optical fibers with multiple cores may make multi-agent-specific sensors possible, as follows: Multimode holey fiber can guide a broad spectral range of colors, and the structure can be designed in a way such that the light propagates in the region of the holes. So, the whole length of a holey fiber can be used as the sensor. This in itself does not provide a big advantage over the glass-coated sensor, though the longer length and larger sensor volume increases the device's sensitivity. More interestingly, the holes that define the waveguides can be used to deliver air (or whatever material is to be sensed) to the region of the detector. The increased surface area that is exposed to the agent also increases the sensitivity and decreases the lag time in observing the signal.

The real advantage of the holey polymer optical fiber approach, however, is the potential for integration; that is, many devices can be defined in a single polymer optical fiber. For example, consider Figure 8.27, which is a device made of a photonic fiber as shown as a cross-section. Each waveguide (shaded area) has a

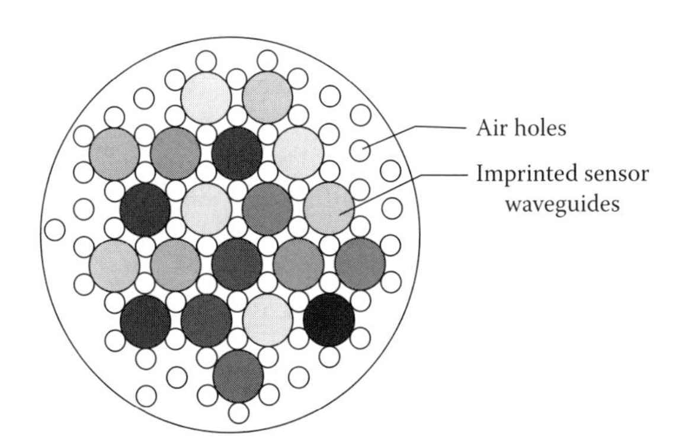

FIGURE 8.27 An array of chemical sensors made with imprinted waveguiding cores.

material with the imprint of a different target molecule. The fluorescence spectrum of each core could be measured by imaging the end of the fiber onto an array of fiber-optic spectrometers.

Such a device is an example of an artificial nose that can be designed to detect a set of agents; or alternatively, it could be used to determine the chemical composition of a particular environment. Applications would range from personal safety, security, and scientific instrumentation. Given how multi-functionality can be imparted to a fiber, complex integrated devices could be made that combine sensing, actuation, and logic in a highly distributed architecture.

8.6 APPENDIX — COUPLED WAVE EQUATION

In this appendix, we solve Maxwell's wave equation (Equation 8.6) for the field amplitudes using the sinusoidal refractive index profile given by Equation (8.3) and the electric field given by Equation (8.5). The second derivative of the electric field in Maxwell's equation is of the form $\frac{\partial^2}{\partial z^2}(A\exp[\pm i\beta z])$, where A represents either A_f or A_b. The first derivative is given by

$$\frac{\partial\left(A\exp\left[\pm i\beta z\right]\right)}{\partial z} = \left(\frac{\partial A}{\partial z} \pm iA(z\beta' + \beta)\right)\exp\left[\pm i\beta z\right], \tag{8.69}$$

where $\beta' \equiv \partial\beta/\partial z$. The second derivative yields

$$\begin{aligned}\frac{\partial^2\left(A\exp\left[\pm i\beta z\right]\right)}{\partial z^2} = &\left[\left(\underbrace{\frac{\partial^2 A}{\partial z^2}}_{small} \pm iA(z\beta'' + 2\beta') \pm i\underbrace{\frac{\partial A}{\partial z}}_{small}(z\beta' + \beta)\right)\right.\\ &\left.+\left(\frac{\partial A}{\partial z} \pm iA(z\beta' + \beta)\right)\left(\pm i(z\beta' + \beta)\right)\right]\exp\left[\pm i\beta z\right].\end{aligned} \tag{8.70}$$

We simplify Equation (8.70) using several approximations. First, we use the slowly varying envelope approximation, which allows us to ignore the second derivative of A. Second, we assume that the index modulation parameter ϵ is small so we ignore terms of order ϵ^2 and higher. Note that the slowly varying envelope approximation is in part valid because the degree of index modulation is small. Finally, we will also ignore terms that are products of ϵ and any derivative of the envelope function A. Using these approximations, the two terms labeled *small* in Equation (8.70) are ignored. In the second line of Equation (8.70), the product $\beta' \cdot \beta'$ can also be ignored since $\beta' \propto \epsilon$. Similarly, $\beta'\beta = \beta_0\beta'$ to the first order in ϵ. Using these conditions, Equation (8.70) becomes

$$\begin{aligned}\frac{\partial^2\left(A\exp\left[\pm i\beta z\right]\right)}{\partial z^2} = &\left[\left(\pm iA(z\beta'' + 2\beta') \pm i\beta_0\frac{\partial A}{\partial z}\right)\right.\\ &\left.+\left(\pm i\beta_0\frac{\partial A}{\partial z} \underline{\underline{- A(2z\beta'\beta_0}} + \underline{\underline{\beta^2}})\right)\right]\exp\left[\pm i\beta z\right].\end{aligned} \tag{8.71}$$

Note that the underlined term $-A\beta^2$ of Equation (8.71) — will cancel the second term in Equation (8.6) further along in this calculation. Anticipating this eventual cancellation, we ignore that term. Using Equation (8.4) for $\beta(z)$ and collecting terms, Equation (8.71) becomes

$$\begin{aligned}\frac{\partial^2 (A \exp[\pm i\beta z])}{\partial z^2} &= \left[\left(A(\pm iz\beta'' \pm 2i\beta' - 2z\beta'\beta_0) \pm 2i\beta_0 \frac{\partial A}{\partial z}\right)\right] \\ &\quad \times \exp[\pm i\beta_0 z] \\ &= \left[\left(A\epsilon\beta_0(\mp iK^2 z\cos(Kz) + 2K(z\beta_0 \mp i)\sin(Kz))\right.\right. \\ &\quad \left.\left.\pm 2i\beta_0 \frac{\partial A}{\partial z}\right)\right] \exp[\pm i\beta_0 z]. \end{aligned} \tag{8.72}$$

Substituting Equation (8.5) into Equation (8.6) with the help of Equation (8.72), we get

$$\begin{aligned}&\left[\left(A_b\epsilon\beta_0(-iK^2 z\cos(Kz) + 2K(z\beta_0 - i)\sin(Kz)) + 2i\beta_0 \frac{\partial A_b}{\partial z}\right)\right] \\ &\quad \times \exp[+i\beta_0 z] \\ &\quad + \left[\left(A_f\epsilon\beta_0(iK^2 z\cos(Kz) + 2K(z\beta_0 + i)\sin(Kz)) - 2i\beta_0 \frac{\partial A_f}{\partial z}\right)\right] \\ &\quad \times \exp[-i\beta_0 z] \\ &\quad = 0. \end{aligned} \tag{8.73}$$

We can decouple Equation (8.73) into two first-order differential equations as follows: First, we multiply it by $\exp[-i\beta_0 z]$ and integrate from $z = 0$ to $2\pi/\beta$ to get one equation. Then, we multiply Equation (8.73) by $\exp[+i\beta_0 z]$ and integrate over the same range to get the other equation. We assume that A and $\partial A/\partial z$ are constant over the range of integration.

To solve the resulting equations, we need to evaluate the following integrals:

$$\int_0^{2\pi/\beta_0} \sin(Kz)\exp[\pm 2i\beta_0 z]dz = \frac{1}{4\beta_0}\left(\frac{\pm 4\sin^2\left(\frac{\pi K}{\beta_0}\right) + i\frac{K}{\beta_0}\sin\left(\frac{2\pi K}{\beta_0}\right)}{1 - \left(\frac{K}{2\beta_0}\right)^2}\right), \tag{8.74}$$

and

$$\begin{aligned}\int_0^{2\pi/\beta_0} z\cos(Kz)\exp[\pm 2i\beta_0 z]dz &= \frac{\partial}{\partial K}\int_0^{2\pi/\beta_0} \sin(Kz)\exp[\pm 2i\beta_0 z]dz \\ = \frac{1}{4\beta_0}\frac{\partial}{\partial K}&\left(\frac{\pm 4\sin^2\left(\frac{\pi K}{\beta_0}\right) + i\frac{K}{\beta_0}\sin\left(\frac{2\pi K}{\beta_0}\right)}{1 - \left(\frac{K}{2\beta_0}\right)^2}\right), \end{aligned} \tag{8.75}$$

where the derivative is not compact enough to display here. Similarly,

$$\int_0^{2\pi/\beta_0} \cos(Kz)\exp[\pm 2i\beta_0 z]dz = \frac{1}{4\beta_0}\left(\frac{\pm 4i\sin^2\left(\frac{\pi K}{\beta_0}\right) - \frac{K}{\beta_0}\sin\left(\frac{2\pi K}{\beta_0}\right)}{1-\left(\frac{K}{2\beta_0}\right)^2}\right), \tag{8.76}$$

and

$$\int_0^{2\pi/\beta_0} z\sin(Kz)\exp[\pm 2i\beta_0 z]dz = -\frac{\partial}{\partial K}\int_0^{2\pi/\beta_0} \cos(Kz)\exp[\pm 2i\beta_0 z]dz. \tag{8.77}$$

Expressing the general result from Equation (8.73) is straightforward, but algebraically messy. We therefore leave this to the reader for those specific cases that are of interest. As an example, Section 8.1.2 treats the Bragg condition, where $K = 2\beta$.

In closing, it is important to stress that many approximations are made in arriving at Equation (8.73), which need to be more rigorously examined if a qualitatively accurate result is required. Furthermore, several subtleties have been brushed aside. However, in the limiting cases, the qualitative behavior of the theory of energy exchange between forward and scattered beams matches the experiments. More importantly, the theory provides a means for attaining physical intuition about how Bragg gratings work.

9 Smart Fibers and Materials

9.1 SMART MATERIALS

9.1.1 INTRODUCTION

A smart material has the ability to change shape or other properties in response to stimuli such as heat, electric field, light, etc. Such behavior is useful for making devices without individual moving components. For example, a motor usually has wire windings, magnets, and gears. Materials that have these functions built in convert the stimulus directly to mechanical energy, so they can be made much smaller and more reliable. Indeed, the materials that Bell used to make musical tones from light pulses is an example of a smart material (see Chapter 1). In his experiments, the acoustical waves are generated in the material when energy is absorbed from a light beam.

9.1.2 PHOTOSTRICTION

An example of a modern-day smart material is lead lanthanum zirconate-titanate (PLZT). Because this material is an electret, the microscopic dipole moments are aligned. When light is absorbed, the electrons in the material move preferentially in the direction of the dipole moment, yielding typical potential differences in centimeter-sized materials of 10,000 volts. This potential difference results in a large force on the material causing it to expand. This effect is called photostriction. The mechanism of photostriction is much more efficient than the mechanisms responsible for the production of sound in Bell's Experiment.

Uchino [140–142] has used photostriction to make an optical speaker using experiments similar to Bell's (see Chapter 1). While photostriction is very efficient, it takes a substantial time for the electric fields to build up inside the material. Photostrictive speakers are audible only below $80Hz$. [140–142]

A more intriguing application of photostriction is the optical crawler, which has two PLZT legs that support a plastic top (see Figure 9.1). The legs are each a double layer structure of oppositely oriented dipole moment so that light exposure causes differential expansion. By alternately pulsing the legs, the device crawls like an inch worm. Typical crawlers move at $1mm$ per minute. In the same vein, Camacho-Lopez and coworkers demonstrated that small elastomer chips will float on water and will "swim" away from regions exposed to light. [143] In the dark, the photomechanical swimmers no longer change shape in response to light, therefore they remain stationary.

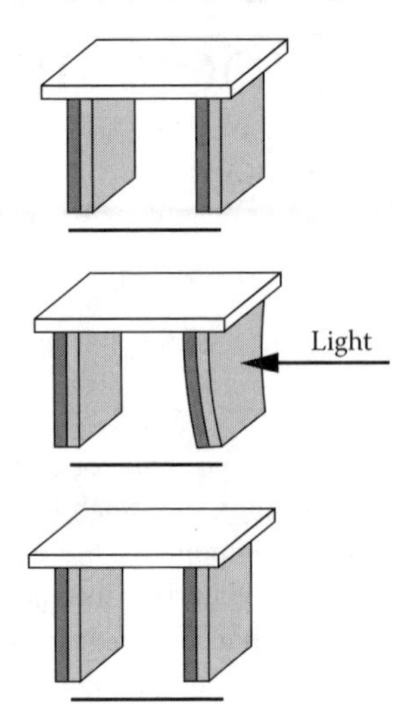

FIGURE 9.1 A crawler based on photostriction.

9.1.3 Smart Structures

A smart structure is made of highly interconnected smart materials so that they interact with each other and with external stimuli. A smart material, on the other hand, is equivalent to a smart structure where the components are microscopic or even single molecules that interact with each other. Figure 9.2 shows a schematic diagram of such a structure or material.

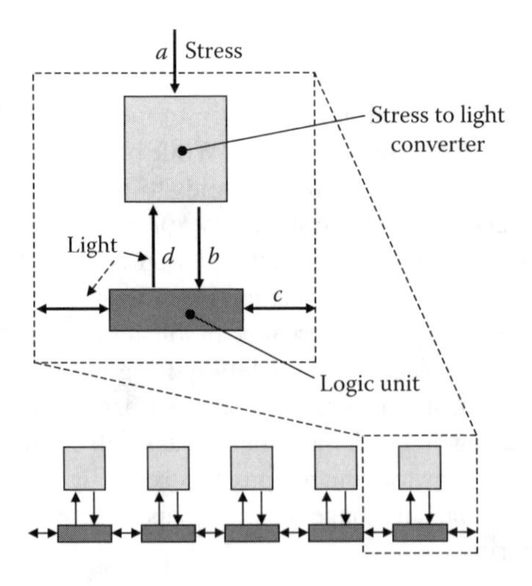

FIGURE 9.2 A smart structure or material that includes highly interconnected logic units, sensors, and actuators.

A smart material or structure needs to be highly interconnected if it is to perform complex operations. Light is the ideal carrier of information due to its high bandwidth and its ability for parallel processing. Details of how such a smart material could be made using polymer fiber is left to a later section of this chapter. To make a smart structure that is interconnected with light, several issues (actuation, logic, and sensing) all need to be based on light-sensitive materials.

Figure 9.2 shows a structure comprised of components that are interconnected with each other and the surroundings. Note that each component need not be identical. Furthermore, because the information carried by the light can be encoded into different colors, each unit in the structure can send information to a select subset of the others; and which units would be sensitive to a given color could be changed as the state of the overall system changes.

Consider one individual unit. (a) The stress sensor converts an applied external stress to a light signal. (b) The light is directed to the logic unit, which operates on the light to modulate it, and changes its color or its phase. (c) In addition, the logic unit is fed light from all the other units. As such, the state of the unit is determined from a combination of the light input from its associated sensor as well as all the other units. (d) The logic unit then sends out a signal to the other units and to the sensor, which now acts as an actuator. Due to this highly interconnected scheme, very complex logic can be performed. In contrast, an electronic logic unit acts on data serially, where the logic unit performs only simple computations. Even today's parallel computer systems are not nearly as interconnected as the possibilities offered in a light interconnected system. The smart system can adapt and perform functions that are much more complex by virtue of its associative architecture.

Many applications of such materials can be envisioned, some of them bordering on science fiction. Imagine, for example, a smart skin on an aircraft wing. As the wing on an aircraft experiences turbulence, a nonuniform stress profile develops. A smart material made of a two-dimensional array of sensors/actuators could detect the stress pattern and compensate with the ideal wing profile to prevent a bumpy ride. The same concept could be used to make smart wallpaper that prevents sound from penetrating by setting up an out-of-phase acoustical disturbance. Simpler smart structures could be used as vibration isolators for sensitive optics or on satellite platforms to prevent the build-up of mechanical vibrations, which are not easy to dissipate in space.

In the following chapter, we discuss progress that has been made in making optical sensors, actuators, and logic using polymer fibers. Note that the sensor discussed here need not be limited to stress, but can include sensitivity to selected chemicals, light, heat, or any other stimuli.

9.2 PHOTOMECHANICAL EFFECTS

A photomechanical effect is simply a light-induced change in the mechanical properties of a material. For example, light can soften a material, change its density, or cause it to bend. The focus of this chapter is on the length change, or more

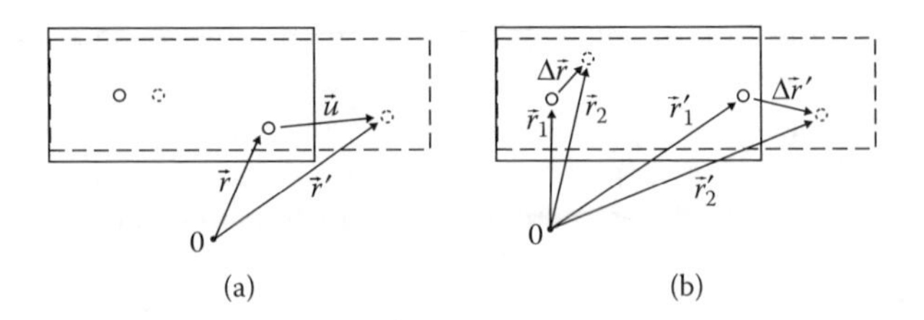

FIGURE 9.3 (a) When a material is stretched (dashed outline) from its equilibrium shape (solid lines), different points in the material are displaced by a different amount. (b) The displacement between any two points in the material changes upon stretching.

generally, the change in the physical dimensions of a fiber in the presence of light. The length change can be a nonlinear function of the intensity. Furthermore, the effect at one point in the material can depend on the intensity of light in another point and at a previous time. So, a full theory of the photomechanical response requires a nonlinear and nonlocal treatment. In this section, after the most general theory is presented, several mechanisms are described as special cases of the theory. Subsequently, experiments that characterize the photomechanical effect and applications are discussed.

9.2.1 Theory

9.2.1.1 Theory of Strain

When viewed on the microscopic level, a shape change of an object results in two parts of the material moving relative to each other. As such, it is useful to mentally draw a dot inside the material, and follow it as the material is deformed. Consider Figure 9.3a, which shows a material before it is stretched (solid outline) and after it is stretched (dashed outline). A point in the material that starts at coordinate $\vec{r}$ ends up at $\vec{r}'$, so its displacement upon stretching is

$$\vec{u} = \vec{r}' - \vec{r}. \tag{9.1}$$

Now we consider two points, 1 and 2, that are close together and separated by

$$\Delta\vec{r} = \vec{r}_2 - \vec{r}_1, \tag{9.2}$$

as illustrated in Figure 9.3b. When the material is stretched, the two points become separated by

$$\Delta\vec{r}' = \vec{r}'_2 - \vec{r}'_1. \tag{9.3}$$

As shown in Figure 9.3, both the position and the separation between any two points depend on the initial (unstretched) position. Thus, we seek to determine how the displacement between two points changes upon stretching as a function of the initial position, $\vec{r}$. Differentiating Equation (9.1),

$$\Delta\vec{r}' = \Delta\vec{r} + \Delta\vec{u} \tag{9.4}$$

and calculating the norm of Equation (9.4), we get

$$
\begin{aligned}
\Delta r'^2 = \Delta\vec{r}' \cdot \Delta\vec{r}' &= \Delta r^2 + \Delta u^2 + 2\Delta\vec{u} \cdot \Delta\vec{r} \\
&= \Delta r^2 + \Delta u_i \Delta u_i + 2\Delta u_k \Delta r_k \\
&= \Delta r^2 + \left(\frac{\partial u_i}{\partial r_j}\Delta r_j\right)\left(\frac{\partial u_i}{\partial r_k}\Delta r_k\right) + 2\frac{\partial u_k}{\partial r_j}\Delta r_j \Delta r_k \\
&= \Delta r^2 + \left(\frac{\partial u_i}{\partial r_j}\frac{\partial u_i}{\partial r_k} + 2\frac{\partial u_k}{\partial r_j}\right)\Delta r_j \Delta r_k \qquad (9.5)
\end{aligned}
$$

where we have written the equations in Cartesian components using the fact that when u is a function of $\vec{r}$, then $\Delta u(\vec{r}) = (\partial u(\vec{r})/\partial r_i)\Delta r_i$ (Δr_j is the jth Cartesian component of $\Delta\vec{r}$). Also note that we use Einstein summation convention (indices that appear twice in a product are summed).

Because j and k can be interchanged without affecting Equation (9.5), we can write Equation (9.5) in the symmetric form,

$$
\Delta r'^2 = \Delta r^2 + 2u_{jk}\Delta r_j \Delta r_k, \qquad (9.6)
$$

where u_{jk} is the strain tensor. Comparing Equation (9.6) with Equation (9.5), the strain can be expressed as

$$
u_{jk} = \frac{1}{2}\left(\frac{\partial u_i}{\partial r_j}\frac{\partial u_i}{\partial r_k} + \frac{\partial u_k}{\partial r_j} + \frac{\partial u_j}{\partial r_k}\right). \qquad (9.7)
$$

To understand the meaning of Equation (9.6), we consider the results in one dimension,

$$
\Delta x'^2 = \Delta x^2 + 2u_{11}\Delta x \Delta x = (1 + 2u_{11})\Delta x^2. \qquad (9.8)
$$

Recall that $\Delta x'$ is the separation between two points after the material is stretched and Δx is the separation before the material is stretched. If we consider these two points to define the two ends of the material, then Δx and $\Delta x'$ are just the lengths of the material before and after stretching. We call these L and L', respectively. Furthermore, if the strain is small (i.e., change in length is small compared to the original length), Equation (9.8) can be written as

$$
L' = \sqrt{1 + 2u_{11}}L \approx (1 + u_{11})L. \qquad (9.9)
$$

Rewriting Equation (9.9), we get

$$
u_{11} = \frac{L' - L}{L}. \qquad (9.10)
$$

For small strain in one dimension, the strain is what one would intuitively expect: the change in length per unit length.

The one-dimensional approximation holds for fibers that are uniformly illuminated by a waveguiding mode. Furthermore, most photomechanical effects yield strains that are on the order of 10^{-4} or less, so Equation (9.10) will apply to most processes of interest.

9.2.1.2 Nonlocal Response

The purpose of this subsection is to show how one would develop a model for light-induced stress for the various mechanisms of photomechanical effects. For example, if light induces a stress on the surface of a fiber, the strain in the material can be determined from classical elasticity theory. If, on the other hand, light-induced forces act throughout the whole volume of the material (in elasticity theory, called body forces), the theoretical development extends beyond standard elasticity theory. Often, numerical codes must be used to solve even the simplest problems. Finally, body forces applied in one part of the material might affect the strain in another part of the material. These are handled using nonlocal response theory.

In this subsection, we develop the nonlocal photomechanical response to illustrate how one goes about tackling this type of problem. Limiting cases of this model are applied to the various mechanisms in the sections that follow.

The starting point is to determine how strain in a material is influenced by light. The form of this interaction depends on the mechanism. In our general treatment here, we define the nth photomechanical susceptibility tensor $\xi^{(n)}_{ij,kl\ldots}$, which relates the electric field in the material, $\vec{E}$, to the strain, and is of the form

$$\begin{aligned} u_{ij}(\vec{r},t) = & \int_{-\infty}^{t} dt' \int_{V} dV' \xi^{(1)}_{ij,k}(\vec{r}-\vec{r}\,',t-t')E_k(\vec{r}\,',t'), \\ & + \int_{-\infty}^{t}\int_{-\infty}^{t} dt'dt'' \int_{V}\int_{V} dV'dV'' \xi^{(2)}_{ij,kl}(\vec{r}-\vec{r}\,',t-t';\vec{r}-\vec{r}\,'',t-t'') \\ & \times E_k(\vec{r}\,',t')E_l(\vec{r}\,'',t'') + \cdots . \end{aligned} \tag{9.11}$$

The integrals are over all times previous to the measurement time and the volume of the material. As such, every part of the material and the electric field at all previous times can contribute to the strain at a given point. The *comma* in the subscript of the response function in Equation (9.11) separates the indices corresponding to the strain tensor components and the components of the applied field. $E_k(\vec{r}\,',t')$ represents the kth Cartesian component of the electric field at time t' and position $\vec{r}\,'$.

Equation (9.11) is the most general result. If the applied electric field is due to light, the frequency is so high that the material cannot keep up. Thus, the first term is small. The second term, however, is proportional to the square of the electric field and has a nonzero average value that is proportional to the intensity of the light. This term and all the even-order terms do not vanish. If, in addition, the response is local and the electric field is polarized along one axis, Equation (9.11) reduces to

$$u = \xi^{(2)}\frac{I}{a} + \xi^{(4)}\left(\frac{I}{a}\right)^2 \cdots \tag{9.12}$$

where I is the intensity and a is a system-of-units-dependent constant that relates the intensity to the electric field ($I = a < E^2 >$).

Many mechanisms can be treated by the response function given by Equation (9.11). First, the photomechanically susceptibility needs to be determined for that particular mechanism. Subsequently the integrals need to be evaluated or approximated.

9.2.2 Mechanisms

In this section, we treat some of the more common photomechanical mechanisms. Some of them have been experimentally observed, while others are only theoretically predicted.

9.2.2.1 Photothermal Heating

The photothermal heating mechanism is in general a nonlocal one. Consider a Gaussian beam in a slightly absorbing medium. The amount of energy absorbed and converted to heat in the center of the beam is higher than around the edges. Thus, when the beam is first turned on, the temperature profile will match the intensity profile. Through photothermal expansion, the strain will also follow the intensity profile. As the heat diffuses from the central hot region, the temperature profile flattens out so that the strain profile is broader than the intensity profile. When the sample expands, the density of material drops so that the absorption coefficient near the center drops. As a result, the absorbed power at the center of the beam drops relative to the $t = 0$ value. It is clear that if the intensity is high enough, these processes can lead to a complex photomechanical time dependence, so photothermal heating is both nonlocal in space and time and the general theory has to be used. Figure 9.4 shows the intensity profile and strain at $t = 0$ and $t > 0$ for a thermal process.

9.2.2.2 Photo-Isomerization

Photoisomerization is the process by which the shape of a molecule changes when it is exposed to light. When such molecules are embedded in a polymer matrix, the shape change causes a strain that deforms the polymer.

There are several classes of molecules that undergo isomerization. The most common ones are the azo dyes. These dyes have two nitrogen atoms that are connected together with a double bond that acts as a bridge between the two parts

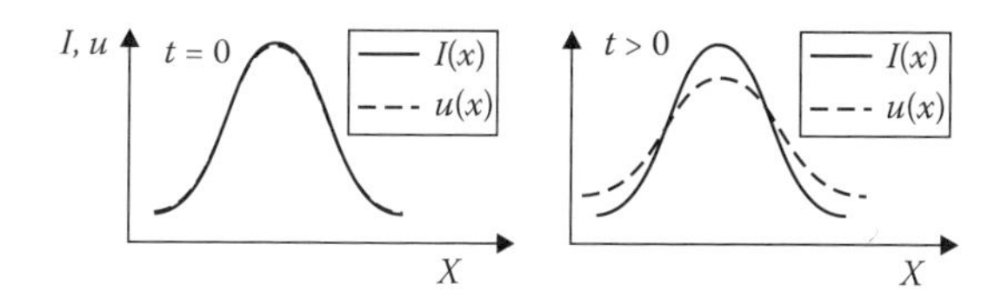

FIGURE 9.4 The intensity profile and stress profile at $t = 0$ and $t > 0$ during photothermal heating.

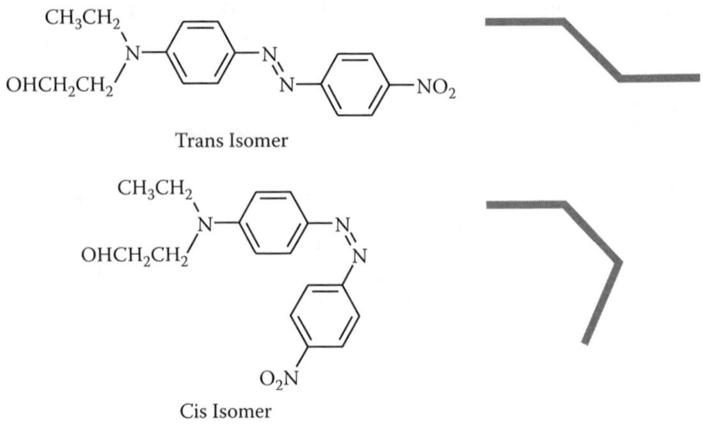

CH_3CH_2 N N N NO_2 $OHCH_2CH_2$

Trans Isomer

CH_3CH_2 N N N $OHCH_2CH_2$ O_2N

Cis Isomer

FIGURE 9.5 The trans and cis isomer of Disperse Red #1 azo dye. The right part of the figure shows a stick figure representation.

of the molecule. The azo bridge allows one part of the molecule to rotate around the other part with ease. Figure 9.5 shows the azo dye whose trade name is disperse red azo dye number one, or DR1. The trans isomer is the lowest energy state, so the cis isomer decays to the trans state. At non-zero temperature, the relative population of the two isomers depends on the temperature. The right part of the figure shows a stick picture of each isomer.

Upon exposure to light whose energy per photon matches the transition energy to the first excited state, the molecule gets excited. From the excited state, it can either de-excite back to the ground state or to the cis isomer ground state. Since the cis lifetime can be relatively long when the molecule is embedded in a polymer (on the order of minutes), a substantial cis population can form. Since the polarizability of the cis isomer is different from the trans one, the process can be monitored by measuring the refractive index or the absorption spectrum of the material.

The microscopic mechanism that leads to material deformation can be pictured using Figure 9.6. For purposes of illustration, we assume that initially, all of the molecules are in their lowest energy state (trans) and isotropically oriented (top left pane in Figure 9.6). The orientational distribution function, shown to the right, represents the probability of finding the molecule oriented along θ, where the angle is defined in the right middle pane. When the polarized light interacts with the material, those molecules that are aligned with the polarization of the light are excited, and some of them de-excite into the cis state. Because the cis molecules are shorter than the trans molecules, they can orientationally diffuse into random alignment (middle left pane in Figure 9.6). Those cis isomers that are now randomly oriented decay back into a randomly oriented trans population. Since the light selectively converts molecules that are aligned with the light's polarization to randomly aligned trans isomers, the net effect is that the orientational distribution function decreases along the light's polarization axis and increases perpendicular to it, as shown in the bottom right pane of Figure 9.6.

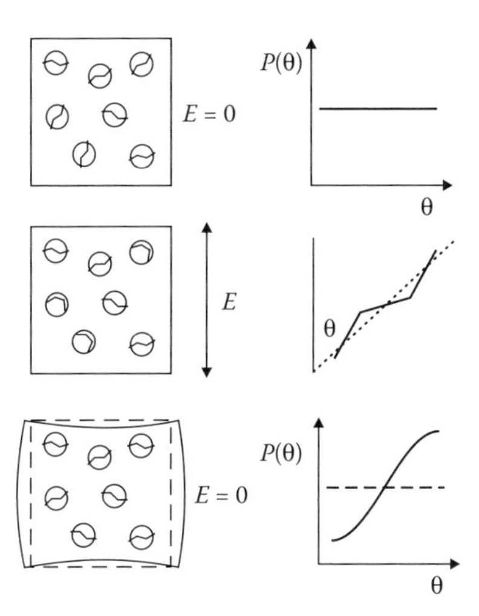

FIGURE 9.6 The three left panes show the angular distribution of the dyes before, during, and after a light beam illuminates the material. The upper and lower right panes show the angular distribution function while the middle pane shows the definition of the angle of the trans isomer.

Polymers contain free volume, which can be pictured as a distribution of voids. In our cartoon version of a polymer, the voids are spherical. While this is not an accurate portrayal of the process, it is a good metaphor, as follows: When a dye-doped polymer is formed, the free volume shrinks around the molecules, rendering them re-orientationally immobile. When the smaller cis isomers reorient and decay to the trans state, they apply an internal force to the polymer along the long axis of the molecules. This is what results in a strain. From this picture, the length of the material should decrease along the polarization of light and increase perpendicular to it. The dashed line represents the original shape and the solid drawing, the final shape.

For a light beam propagating in a dye-doped fiber, the length of the fiber should thus decrease. In contrast, the photothermal heating mechanism results in an increase of the length. Since both mechanisms are of a comparable time scale (milliseconds), dynamical data should be able to determine the relative contribution of each. The one complication that needs to be considered when using this approach is coupling between the two mechanisms. Clearly, if the density of the material changes due to thermal expansion, the absorbed energy density will decrease, affecting the isomerization mechanism.

9.2.2.3 Electrostriction

Electrostriction literally translates as electric-induced squeezing. When an electric field is applied near an interface between two materials, a surface charge layer is induced. An electric field at the surface thus results in a force. Inside the material, the light will induce a polarization, so an electric field gradient, when coupled to the induced dipoles found in the volume, also results in a net force with a stress perpendicular to that surface. Each mode in an optical fiber waveguide has a distinct intensity profile, where the intensity at the interface between the core and cladding usually falls off approximately exponentially. Thus, electric field gradients are present. Figure 9.7a shows both the intensity profile and gradient of the intensity of a single mode waveguide.

Figure 9.7b shows a fiber in equilibrium under static illumination. The net stress inward around the surface of the cylindrical fiber causes the diameter to decrease and the length to increase. Figure 9.7c shows a fiber guiding a short pulse of light. If the fiber response were instantaneous, the diameter would be pinched in the illuminated region. In cases where the response of the fiber is slow compared with the temporal pulse width of the light, the fiber's physical dimensions behind the light pulse would be a complex function of time and position. In the calculation below, we consider the static case with the purpose of developing an intuitive view of the process. The more general nonlocal calculation, which we will not consider, requires the elastic and dielectric response of the medium to be incorporated in a theory that couples Maxwell's equations with acoustics. At this time, such a calculation has not been done. However, Buckland and Boyd have performed a similar calculation for the change in refractive index due to electrostriction. [144]

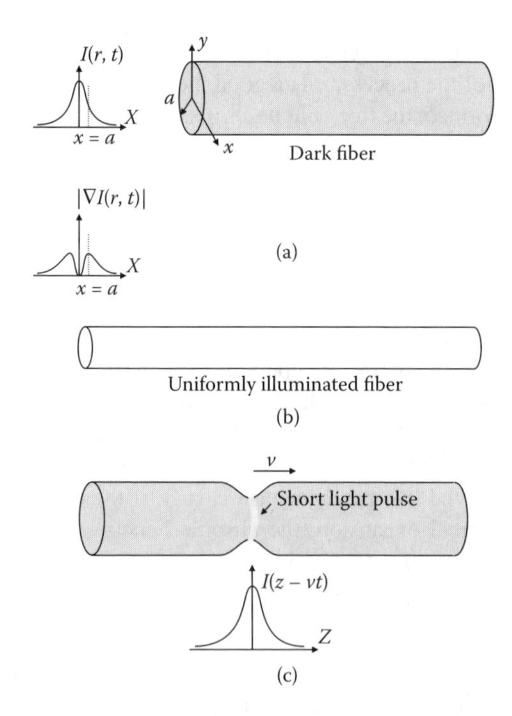

FIGURE 9.7 (a) A dark fiber, (b) uniformly illuminated fiber, (c) and fiber with pulsed light source.

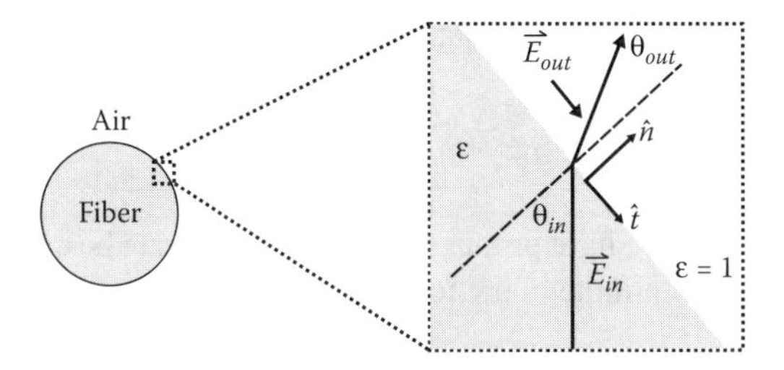

FIGURE 9.8 A close-up of a cylindrical dielectric surface.

The first case that we treat is the force at an interface due to the induced charge on the surface. (As before, we use Gaussian units but also display the final results in SI units.) We consider a planar interface between air and a cylindrical dielectric as shown in Figure 9.8. A small portion of the cylindrical surface is considered so that we can approximate it as a plane. The electric field at the surface will induce a bound charge density, which according to Gauss's law, is related to the discontinuity of the normal component of the electric field,

$$\sigma = \frac{1}{4\pi}\left(\vec{E}_{out} - \vec{E}_{in}\right) \cdot \hat{n}. \tag{9.13}$$

The normal component of the force per unit area on the induced charge is then given by

$$f_N = \sigma \frac{\left(\vec{E}_{out} - \vec{E}_{in}\right) \cdot \hat{n}}{2} = \frac{1}{8\pi}\left(E_{out,n}^2 - E_{in,n}^2\right), \tag{9.14}$$

where $E_{out,n}$ and $E_{in,n}$ represent the normal components of each electric field.

To calculate the force, we need to determine the electric field outside the fiber in terms of the field just inside. To do so, we use the boundary conditions that the normal component of the electric displacement is continuous and that the tangential component of the electric field is continuous. This yields

$$\epsilon E_{in} \cos\theta_{in} = E_{out} \cos\theta_{out} \tag{9.15}$$

and

$$E_{in} \sin\theta_{in} = E_{out} \sin\theta_{out}, \tag{9.16}$$

respectively. We can use Equations (9.15) and (9.16) to solve for the angle of the output field relative to the interface normal in terms of the incident angle,

$$\tan\theta_{out} = \frac{1}{\epsilon}\tan\theta_{in}, \tag{9.17}$$

and the output field magnitude in terms of the input field magnitude

$$E_{out} = E_{in}\sqrt{\epsilon^2 + \left(1 - \epsilon^2\right)\sin^2\theta_{in}}. \tag{9.18}$$

Finally, using Equation (9.18) in Equation (9.14)

$$f_N = \frac{1}{8\pi} E_{in}^2 \left(\epsilon^2 - 1\right) \cos^2 \theta_{in}. \tag{9.19}$$

For a given mode, the light propagating down a fiber can be expressed in terms of the electric field amplitude in the form

$$E_{in} = E(\rho, \phi) \cos(Kz - \omega t). \tag{9.20}$$

Because the mechanical response is slow compared with an optical cycle, the force acting on the fiber surface is proportional to the square of the time-averaged field,

$$\left\langle E_{in}^2 \right\rangle = \frac{1}{\frac{2\pi}{\omega}} \int_0^{\frac{2\pi}{\omega}} E^2(\rho, \phi) \cos^2(Kz - \omega t) dt = \frac{1}{2} E^2(\rho, \phi). \tag{9.21}$$

The Poynting vector relates the electric field amplitude to the intensity, and is of the form

$$I = n \frac{c}{8\pi} E^2(\rho, \phi), \tag{9.22}$$

so using Equation (9.22), (9.21), (9.20), and (9.19); the time-averaged stress in the fiber is given by

$$\boxed{\sigma_n \equiv \langle f_n \rangle = \left(\frac{n^4 - 1}{n}\right) \frac{I(\rho, \phi)}{2c} \cos^2 \theta_{in}} \quad \begin{array}{c}\text{Gaussian Units} \\ \text{and SI Units}\end{array}, \tag{9.23}$$

where we have used the relationship between the refractive index and the dielectric function, $\epsilon = n^2$.

At this point, it is important to note that, while the above calculation started with equations in Gaussian units, the results are the same for SI (MKS) units. In the former, the intensity is expressed in units of $erg/cm^2 \cdot s$ and the speed of light in units of cm/s; leading to a surface stress in units of $dyne/cm^2$. In SI units, the intensity is given in units of W/m^2 and the speed of light is given in m/s leading to a surface stress in units of N/m^2.

It is instructive to make hand-waiving approximations to Equation (9.23). First, let's assume that the intensity at the core-cladding interface (at $\rho = a$) is the average intensity in the waveguide, or $I(a, \phi) = \bar{I}$. Second, we will assume that the internal angle is the critical angle, given by $n \sin \theta = 1$. Using these assumptions, Equation (9.23) becomes

$$\sigma_n = \left(\frac{\left(n^4 - 1\right)\left(n^2 - 1\right)}{2n^3 c}\right) \bar{I}. \tag{9.24}$$

As an example, let's evaluate Equation (9.24) for a typical multimode fiber that is guiding a laser pulse near the threshold for optical damage. The typical optical damage threshold for PMMA polymer is on the order of a few GW/cm^2 (note that laser intensity is most often quoted in units of gigawatts per square centimeter). So, for an intensity of $I = 2GW/cm^2 = 2 \times 10^{13} W/m^2$ and a refractive index $n = 1.48$, the surface stress is given by

$$\sigma_n = 4.6 \times 10^4 N/m^2. \tag{9.25}$$

Depending on the mode propagating in the fiber core, the force can be radial (i.e., inward or outward everywhere around the fiber), or quadrupolar (where the force is maximum on one radial axis and vanishes on the radial axis that is perpendicular to this axis.) In the former, the fiber's radius changes evenly for all azimuthal angles leaving the fiber circular in cross-section. In the quadrupolar case, the fiber radius changes along just one axis — yielding an elliptical fiber. Many other interesting cases are possible. For example, if the waveguide is excited by a superposition of modes that is asymmetric about the center axis of the fiber (i.e., about $\rho = 0$), the differential force can lead to fiber bending. As such, electrostriction in a fiber can be used for several kinds of optical actuation.

It is instructive to estimate the strains induced in a typical polymer fiber material such as PMMA. The bulk modulus for thermoplastics is on the order of $1GN/m^2 = 1 \times 10^9 N/m^2$. (The bulk and shear modulus, for example, are usually equal to within a factor of three.) There are many possible types of deformations, and each associated strain will be different. Aside from geometric factors that are within an order of magnitude of unity, the strain is the ratio of stress to bulk modulus. For a laser pulse near the damage threshold, and for a modulus of $1GN/m^2$, the strain is about $u = 4.6 \times 10^{-5}$. Thus for a fiber of $0.05m$ length, the induced length change for a longitudinal deformation is about $2.3\mu m$. Since this is larger than the wavelength of light in the visible and near IR, such changes can be easily measured with laser interferometry.

The problem, however, is somewhat complicated by the fact that the elastic modulus depends on frequency. Furthermore, a full treatment requires that we consider nonlocal forces, which are beyond the scope of this book.

Next we consider the force inside the material in the presence of an electric field gradient. (Note that we will again do the calculation in **Gaussian units** but we will also present the results in SI units.) Figure 9.9 shows a small volume element in the presence of a field gradient. Note that the tensor nature of the problem makes it difficult to visualize, because the applied field, dipole moment, field gradient, and force can in general be in four different directions. As such, the figure is intended to be general enough to provide an accurate visual aid, but by no means does it show the most general possible case. (For example, we show the force to be perpendicular to the dipole moment because a rectangular volume element is the simplest case to treat.) If an electric field, $\vec{E}$, is applied, the induced dipole moment, $\vec{p}$, in a volume V of material is

$$p_i = V\chi_{ij}^{(1)} E_j, \tag{9.26}$$

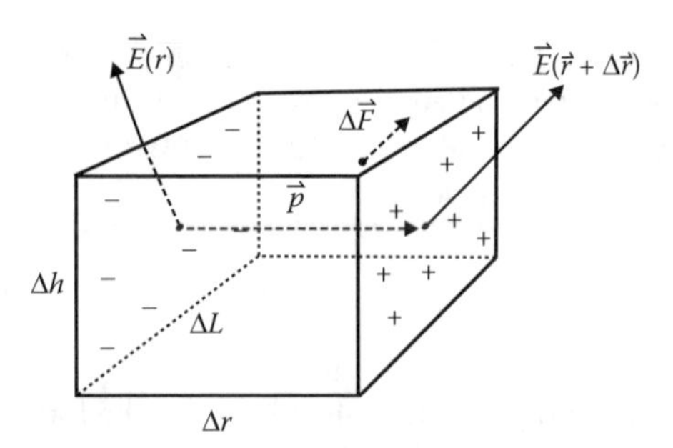

FIGURE 9.9 An electric field applied to a small volume yields a dipole moment. In a field gradient, the force on each polarity of charge is different, yielding a net force.

and the force on that volume of material, $\Delta\vec{F}$, is given by

$$\Delta F_k = p_i \frac{\partial E_i}{\partial x_k} = V \chi_{ij}^{(1)} E_j \frac{\partial E_i}{\partial x_k} = \frac{1}{2} \frac{\partial}{\partial x_k} \left[\chi_{ij}^{(1)} E_i E_j \right] V, \tag{9.27}$$

where all of the above calculations are expressed in component form and where we use the summation convention.

As seen in Figure 9.9, the volume element is given by $V = (\Delta h \Delta r)\Delta L = \Delta A \Delta L$, and the component of the force along the ΔL direction is shown. Using Equation (9.27), the Lth-component of the force can be expressed as

$$\Delta F_L = \frac{1}{2} \frac{\partial}{\partial L} \left[\chi_{ij}^{(1)} E_i E_j \right] \Delta A \Delta L. \tag{9.28}$$

On any given surface element, the stress, f, is defined as the force per unit area. Equation (9.28), on the other hand, gives the net force on the volume — which is related to the difference in stress between the back and front surface along L, or Δf. The stress difference along L is

$$\Delta f_L = \frac{\Delta F_L}{\Delta A} = \frac{1}{2} \frac{\partial}{\partial L} \left[\chi_{ij}^{(1)} E_i E_j \right] \Delta L. \tag{9.29}$$

Integrating Equation (9.28), we get the internal stress:

$$\boxed{f_L = \frac{1}{2} \chi_{ij}^{(1)} E_i E_j} \quad \text{Gaussian Units.} \tag{9.30}$$

In SI units, the same result is given by

$$\boxed{f_L = \epsilon_0 \frac{1}{2} \chi_{ij}^{(1)} E_i E_j} \quad \text{SI Units.} \tag{9.31}$$

More appropriately, the stress we calculate above should be called f_{LL}, which has the meaning that the stress on the surface with a normal vector along L results in a force along L. It is possible that the resulting force on a surface is not along the unit normal. In the case when the force is parallel to the surface, it is called shear and quantified by terms such as f_{xy}. In contrast, f_{xx} results in a compression.

When light is coupled into the fiber so that it is brightest at the center and monotonically decreases away from the center in the radial direction, the largest intensity gradient is at the surface of the fiber, $\rho = a$. Since the stress on the fiber's surface is related to the field gradient, and the field gradient is along the radial direction, this case results in a decrease in the fiber diameter and an increase in the fiber length. In the following discussion, we calculate the degree of length change due to a uniformly illuminated fiber, or when the light intensity changes over time scales that are long compared with the material response time.

Consider a slice along a fiber of length dz, which is placed in an electric field $E(\rho)$. If the optical pulse intensity varies slowly along the length of the fiber, there is no electric-field-induced stress in the z-direction. The electrostrictive stress, then, is along the radial direction only. For a cylindrical shell at radius ρ, the compressive force per unit area along the radial direction, $f_{\rho\rho}$ (we use Gaussian units for this example), is

$$f_{\rho\rho} = \frac{1}{2}\chi^{(1)} E^2(\rho). \tag{9.32}$$

The change in radius induced by the radial stress results in a strain along the fiber axis, u_{zz}, and is given by

$$u_{zz} = -\frac{2\sigma}{Y}\sigma_{\rho\rho} = -\chi^{(1)}\frac{\sigma}{Y}E^2(\rho), \tag{9.33}$$

where we have used Equation (9.32). σ is Poisson's ratio and Y Young's modulus. Note that Equation (9.33) is the quasi-static approximation and applies when the duration of the pulse is long compared with the response time of the material. (The negative sign takes into account the fact that a decrease in fiber diameter leads to an increase in length.) Because the electric field depends on ρ, the length change is calculated by integrating the strain both over ρ and z. As in the previous case, because the material response is negligible over one optical cycle, the net length change of the fiber will be given by the time average $\langle E^2 \rangle$. Similar to the previous calculations leading to Equation (9.23), for a sinusoidal wave of amplitude E, this yields $\langle E^2 \rangle = E^2/2$ and the stress and strains in Equations (9.32) and (9.33) can be expressed in terms of the average intensity in the fiber,

$$\boxed{\langle f_{\rho\rho} \rangle = \sigma_{\rho\rho} = \frac{(n^2 - 1)}{2nc}\bar{I}} \quad \begin{array}{c}\text{Gaussian Units}\\ \text{and SI Units}\end{array}, \tag{9.34}$$

and

$$\boxed{\langle u_{zz} \rangle = -\frac{(n^2 - 1)}{n}\frac{\sigma}{Y}\frac{\bar{I}}{c}} \quad \begin{array}{c}\text{Gaussian Units}\\ \text{and SI Units}\end{array}, \tag{9.35}$$

respectively.

We can compare the stress induced by field gradients to the stress induced by surface charge by taking the ratio of Equation (9.34) and Equation (9.24), which yields

$$\frac{\sigma_{pp}}{\sigma_n} = \frac{n^2}{n^4 - 1}. \tag{9.36}$$

For a plastic fiber of refractive index $n = 1.48$, $\sigma_{pp} = 0.58\sigma_n$, so both types of stress are comparable.

9.2.2.4 Molecular Reorientation

When a polarized light beam interacts with a material, dipole moments are induced in the molecules that make up the material. The field can then interact with the induced dipoles to orient them, with the resulting internal forces leading to a distortion of the bulk material. In this section, we calculate the effect of molecular reorientation for the very simplest material: a non-polarizable polymer host doped with one-dimensional polarizable molecules. The molecules within any such material will have a distribution of orientations. We will focus on one molecule. The effect that reorientation has on the material can be calculated by summing over the effects of the individual molecules.

Figure 9.10a shows a molecule before the electric field is applied in its initial orientation given by the angle θ_0. In the classical picture, the molecule is held in that orientation by the polymer. Because the polymer is a viscoelastic medium, if an external force tries to reorient the molecule, the polymer exerts an elastic spring-like force in the opposite direction. When an electric field is applied, the electron cloud in the molecule is distorted, inducing a dipole moment. The electric field then couples to this dipole moment and exerts a torque. The molecule comes to rest when the elastic forces balance the applied electric torque. The internal strain distorts the polymer, leading to a change in the physical dimensions of the bulk. Figure 9.10b shows the reoriented distorted molecule.

To calculate this process, we assume that the molecules are non-interacting and that the applied field is identical to the local field. (It is straightforward, albeit more messy, to generalize these assumptions.) For a one-dimensional molecule of

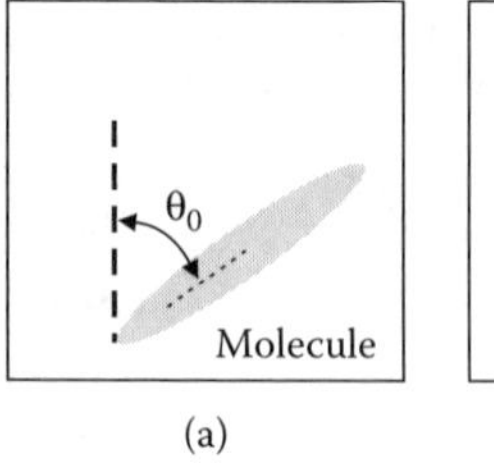

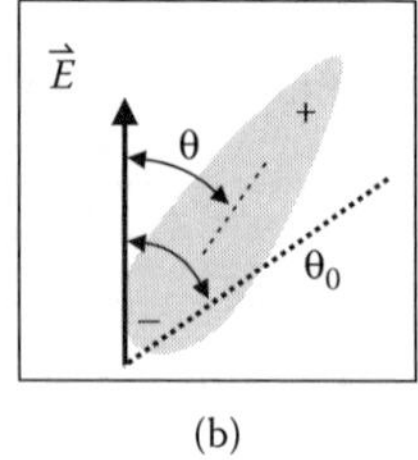

FIGURE 9.10 a) A one-dimensional molecule without a field applied. b) When an electric field is applied, a dipole moment is induced, causing the molecule to reorient along the field inasmuch as the polymer allows it to do so.

polarizability α_{33} along its major axis (with all other components equaling zero), the induced dipole moment in the i axis is

$$p_i = \alpha_{i3} E_3. \tag{9.37}$$

The electric potential energy is then given by

$$U_E = -\vec{p} \cdot \vec{E} = \alpha_{i3} E_3 E_i = \alpha_{33} E_3^2, \tag{9.38}$$

where the last equality follows because α_{33} is the only non-zero tensor component for a 1-dimensional molecule.

If the electric field is due to a beam of light that is polarized at an angle θ relative to the long axis of the molecule, the potential energy is

$$U_E = -\alpha_{33} E_3^2(t) \cos^2 \theta. \tag{9.39}$$

For an optical beam, the electric field oscillates so quickly that the molecule cannot keep up. We thus average the potential energy function over one optical cycle and express the electric field in terms of the intensity, using the same procedure described in Section 9.2.2. This results in an electric potential energy function of the form

$$\langle U_E \rangle = -\frac{4\pi \alpha_{33}}{nc} I \cos^2 \theta. \tag{9.40}$$

If the molecules' original orientation is given by θ_0, the total potential energy of the molecule in the field is given by

$$\langle U(\theta) \rangle = -\frac{4\pi \alpha_{33}}{nc} I \cos^2 \theta + \frac{1}{2} k \left(\theta - \theta_0\right)^2, \tag{9.41}$$

where k is the micro-elasticity of the polymer matrix. The equilibrium angle of the molecule occurs when the net force vanishes. Differentiating Equation (9.41) with respect to angle, the equilibrium condition yields

$$\frac{4\pi \alpha_{33}}{nc} I \sin 2\theta_{eq} = k \left(\theta_{eq} - \theta_0\right). \tag{9.42}$$

Note that Equation (9.42) is a transcendental function, so it needs to be solved numerically. Figure 9.11 shows a graphical representation of the solution. The sinusoidal curve is the function on the left-hand side of Equation (9.42) and the dashed lines are the right-hand side for two different values of the microscopic elasticity. Note that the θ-intercept of the lines is $\theta = \theta_0$ while the slope is proportional to the elasticity. The equilibrium angle is given by the intersection between the line and the curve. Note that as the polymer elasticity gets larger (i.e., the polymer gets more rigid), the slope of the line increases and the equilibrium angle approaches the zero-field value, $\theta_{eq} \to \theta_0$. Therefore, as the polymer becomes more rigid, the molecules reorient less in the presence of a fixed-intensity light source.

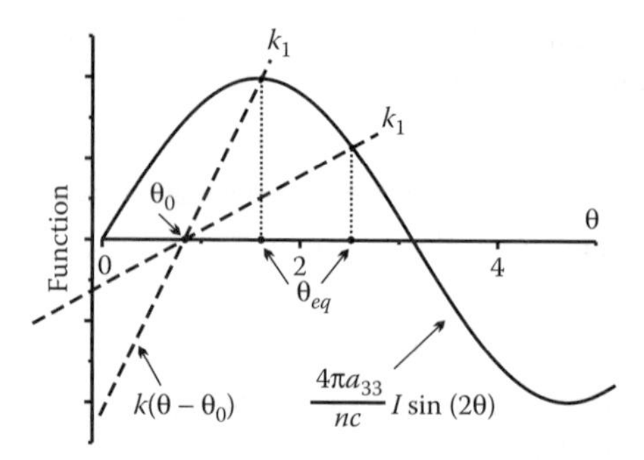

FIGURE 9.11 A plot of the left-hand and right-hand sides of Equation (9.42). The dashed lines show $k(\theta - \theta_0)$ for two different values of k. The intersection between the line and the curve yields the equilibrium angle θ_{eq}.

To understand how the reorientational process leads to a length change, we consider an initially isotropic distribution of orientations as shown in the left-hand part of Figure 9.12. The torque exerted by the electric field is proportional to $\sin 2\theta$. As such, molecules that are oriented between $\theta = 0$ and $\theta = \pi/2$ are oriented generally along the electric field while those between $\theta = \pi/2$ and $\theta = \pi$ are aligned generally opposite to the electric field. The net effect is that the long axes of the molecules get aligned along the field. This type of order is called axial order because the long axes are oriented along the field direction, but the dipole moments can be randomly oriented (i.e., no polar order). The same type of order was seen with the photoisomerization mechanism as described in Section 9.2.2.2. In analogy to photoisomerization, the orientational order is quantified by the second-order Legendre polynomial.

Let's consider the molecules that are near θ_0. Using the approximation that $\sin 2\theta \approx 2\theta$, Equation (9.42) gives an equilibrium angle,

$$\theta = \frac{\theta_0}{1 - \frac{8\pi\alpha_{33}}{knc} I}, \tag{9.43}$$

which for small intensity yields

$$\frac{\Delta\theta}{\theta} = \frac{8\pi\alpha_{33}}{knc} I, \tag{9.44}$$

where we have defined $\Delta\theta = \theta - \theta_0$. The change in length of the material is proportional to the shear strain $(\Delta\theta/\theta)$, so within our approximations the length change is a linear function of the intensity. (The constant of proportionality is, however, complicated to determine so we will not present such a calculation here.) Furthermore, the length change is proportional to the polarizability and inversely proportional to the polymer elasticity — as expected.

Note that the above treatment of the problem is not meant to be an accurate theory of the process; rather, it serves to illustrate its key features. For example, we

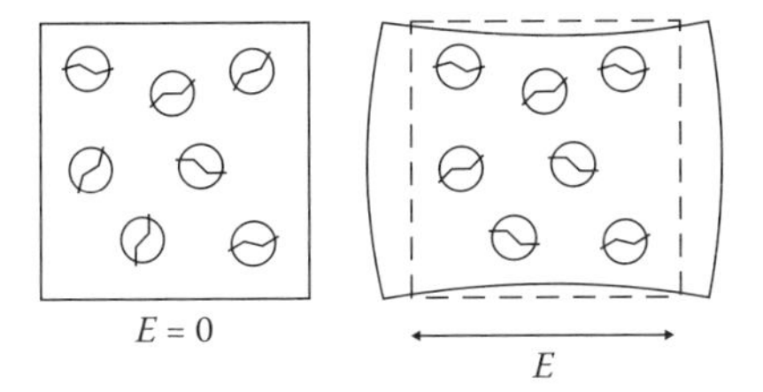

FIGURE 9.12 Molecules that are initially isotropically oriented (left) become axially oriented in the direction of light polarization (right), which yields a length change in the material.

have ignored the effects of temperature, the orientational distribution of the chromophores, the distribution of sites (the elasticity is different for each molecule), and local fields. Many of these issues are described in more detail in the literature. [48,72,145,146]

9.2.2.5 Electron Cloud Deformation

Electron cloud deformation is one of the fastest processes. Unfortunately, it leads to the smallest length change. There are certain combinations of parameters in nature that tend to be relatively constant over large ranges in the parameters themselves; and the ratio of the magnitude and time response of the photomechanical effect is probably one of these. This situation is similar to that of the third-order susceptibility, which itself can vary over many orders of magnitude; but the ratio of the susceptibility to the time scale of the response varies by just a couple orders.

Put simply, since the electronic structure of the atoms and molecules that make up a material determine bulk structure, light-induced electron cloud distortion leads to a length change. A calculation of this length change depends on the details of the electronic structure of the particular atoms or molecules. Since such calculations are both complicated and tedious, and since no general principles can be determined, we will not treat this problem here.

9.2.3 Experimental Examples

There are clearly a wealth of mechanisms that can lead to a length or shape change in a fiber, and each of these could potentially be used to build all sorts of novel devices that integrate sensors, logic, and actuators that operate solely under the power of light. As such, this is potentially a very fruitful area of research that has for the most part remained untapped. The experimental examples that are described below should therefore be considered as just a small sample of the science and technology that could result if this area of research were to be pursued with the same vigor as more fashionable present-day research.

9.2.3.1 The First Demonstration of an All-Optical Photomechanical System — Nano-Stabilization

It is fun to speculate what kinds of new technologies would be made possible if photomechanical effects were put to use. Ironically, the first demonstration of a photomechanical system — an active motion stabilization device — included the integration of all of the major device classes: sensors, logic, actuation, and information transmission. [147] Furthermore, all of these were all-optical components, so no electrical components were used.

Figure 9.13 shows a schematic diagram of the nano-stabilization system. A $30cm$ long photomechanical fiber made of PMMA fiber is suspended vertically with a mirror attached to the bottom end. This mirror, of mass $9g$, defines one arm of a Michelson interferometer. The second mirror is rigidly fixed in place. A $250mW$ continuous krypton laser beam at $647.1nm$ is launched into the interferometer. The light intensity leaving the interferometer depends on the length of the hanging fiber. A small fraction of this light is reflected to a detector while the remaining light is directed into the hanging fiber. About $5mW$ reaches the input face to the fiber. Through the photomechanical effect, the change in the length of the fiber from its equilibrium length is proportional to the intensity of the light in the fiber. Note that once the light passes along the length of the fiber, it is absorbed by the back side of the hanging mirror; so, light from the fiber does not enter the interferometer.

The principle of how this device operates is as follows: The interferometer effectively acts as a position sensor of the hanging mirror. When the fiber length changes, the mirror moves and the light exiting the interferometer changes. This light, when directed into the hanging fiber changes its length, which once again changes the light output from the interferometer, and so on, and so on .... The

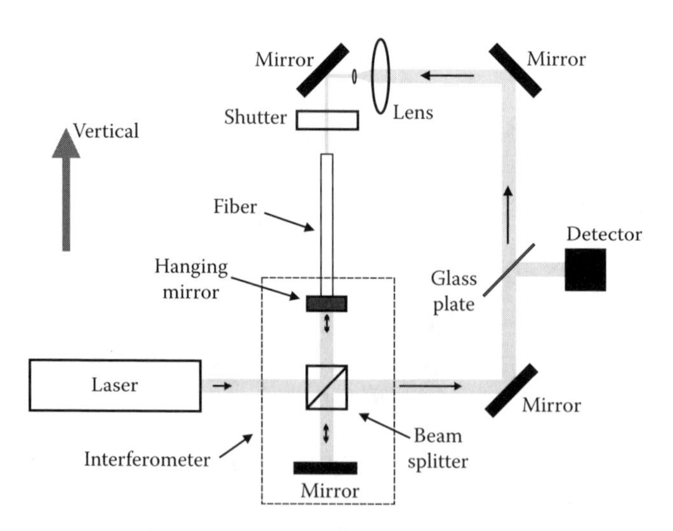

FIGURE 9.13 An all-optical nano-stabilization system. Note that the apparatus is mounted vertically. Adapted from Ref. [147] with permission of the American Institute of Physics.

system is inherently nonlinear due to this optical feedback loop. Note, however, that the light entering the fiber can be blocked by a shutter, thus turning off the optical feedback.

Figure 9.14 shows the output intensity as a function of time as measured with the detector. When the shutter is closed, the interferometer drifts by one wavelength (647*nm*) over about 200 seconds. When the shutter is opened, the interferometer output changes abruptly and then becomes constant. Judging from the output intensity, the interferometer no longer drifts. Just after about 300 seconds, the intensity jumps to another constant level. The implication is that the fiber length increases, and that the mirror's position remains stable over time. Indeed, the authors reported that the intensity leaving the interferometer did not change over long periods of time, signifying that the fiber length is being actively kept constant by optical feedback to the photomechanical fiber.

The two smaller graphs at the right-hand side of Figure 9.14 show a magnified view of the two apparent discontinuities in the data. The data is accumulated at a rate of 10 points per second. The turn-on time of the photomechanical effect (from the upper-right plot) is thus determined to be about 200*ms*. The second discontinuity has a similar response time.

Most nonlinear systems with feedback are known to show a rich set of phenomena. One of these is called optical multistability, where a system can be in any one of many distinct states for the same parameters of the system. Since the laser intensity did not change over the duration of the experiment presented in Figure 9.14, the implication is that the system allows for at least two different length states of the fiber for a fixed laser intensity given by the two flat regions

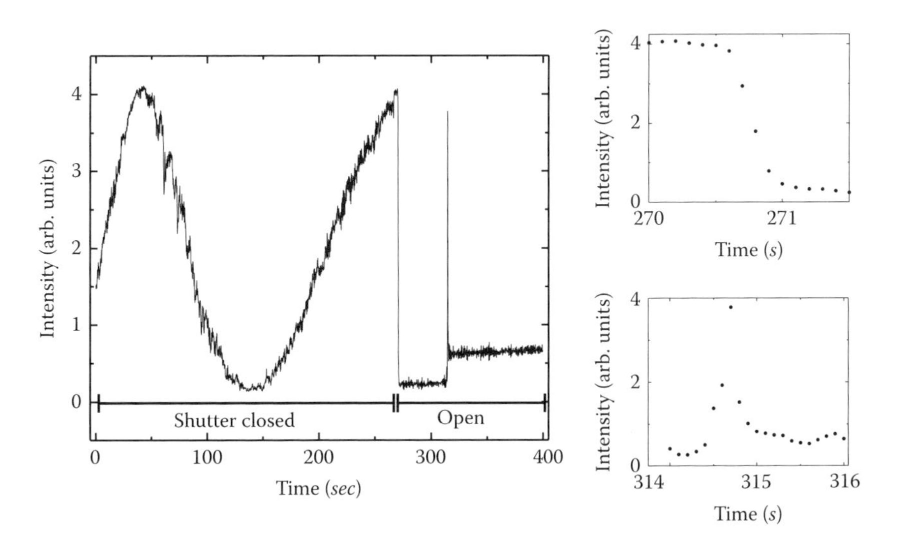

FIGURE 9.14 Interferometer output intensity as a function of time with shutter closed, then opened. The two figures on the right show an expanded view of two different time regions. Adapted from Ref. [147] with permission of the American Institute of Physics.

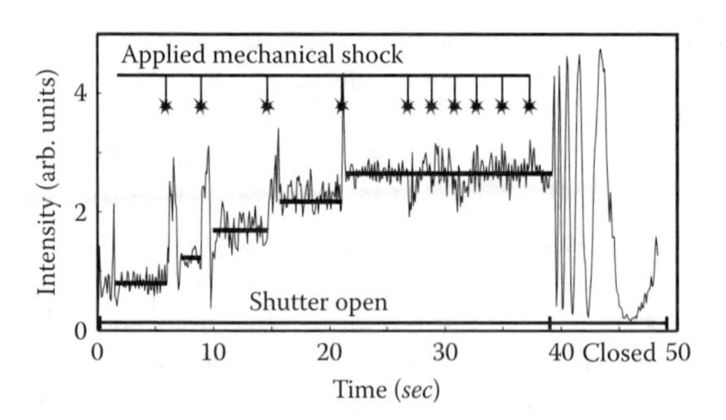

FIGURE 9.15 Interferometer output intensity as a function of time while the system is shocked with an external impulse. Adapted from Ref. [148] with permission of Taylor and Francis.

after the shutter is open. (See Section 7.2.4 for more details on the theory of multistability.)

To test the hypothesis of multistability, the output of the interferometer was measured as a function of time with the shutter open. Figure 9.15 shows a continuation of the data of Figure 9.14 for longer times. During this run, the mount holding the hanging fiber was hit with a mechanical impulse several times. The times at which these shocks are applied are labeled above the data in Figure 9.15. At each mechanical impulse, the system makes a transition to the next equilibrium state until the sixth state is reached, at which point no mechanical impulse is strong enough to cause the system to jump to another state. At about $t = 40s$, the shutter is closed to turn off the optical feedback. The length of the fiber relaxes to the dark state, as can be seen by the interferometer fringes in the last 10 seconds of the experiment.

Several features in the data give us a clue about the physical process that leads to the observed behavior. First, Welker showed that the impulse required for a transition between states gets larger as the system climbs the steps. Indeed, the very first transition was observed to occur spontaneously. Most likely, a small perturbation to the system was a cause. In contrast, the 6th state is so stable that no amount of force could cause the state to change. A second feature is the spikes between the steps. Note that these peaks get narrower as the steps get higher.

The explanation proposed by Welker and Kuzyk [147] to describe this observation was based on the feedback between the interferometer and the photomechanical length change. Figure 9.16 shows a graphical solution to the problem. The straight line shows the fiber length's dependence on the intensity while the sinusoidal curve shows the interferometer output as a function of fiber length change. The solution to the coupled equations is given by the intersection points between the two curves. For a positive photomechanical effect (length increases as a function of intensity), the solid points show the stable solutions. The other intersections are unstable solutions, which can be understood as follows.

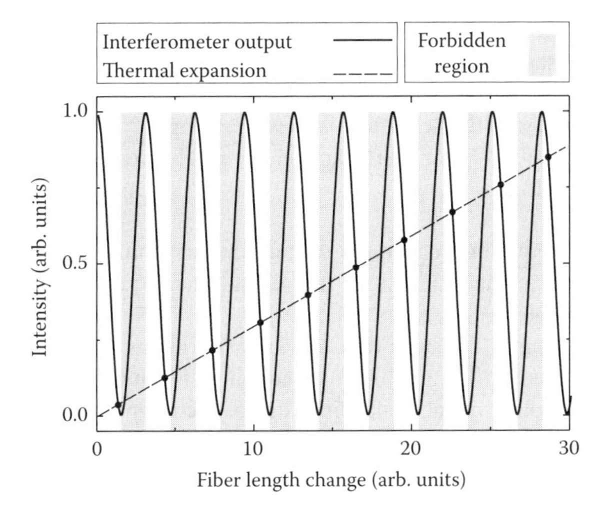

FIGURE 9.16 Interferometer intensity as a function of time (sinusoidal curve) and intensity as a function of length (line). Adapted from Ref. [147] with permission of the American Institute of Physics.

Consider any one of the stable points. If the fiber length decreases due to a small perturbation, the interferometer output intensity increases, thus causing the fiber length to increase. As a result, the feedback process will keep the fiber length from increasing. The shaded regions correspond to the unstable regions. If the system is found there, a small increase in length is reinforced, so a runaway process causes the fiber length to increase until the interferometer output peaks. Once the system passes just over a peak, the fiber length increases until the system reaches the next stable equilibrium point. Note that the equilibrium point near the origin is very close to the minimum; thus small perturbations can bring the system into the forbidden region, forcing the system to jump to the next stable point. As the system walks up the steps, the distance from the minimum gets greater, so a larger shock needs to be applied in order to increase the fiber length enough to get the system into the gray region and then to the next step (i.e., equilibrium length) through a photomechanically driven length increase. Also note that there is a tendency of the system to walk up the steps and not down because the distance to the next minimum is less than the preceding maximum until the mid-point equilibrium is reached. Based on this argument alone, it is reasonable to assume that the system will tend to be the most stable at the center point, that is, at an interferometer intensity that is half the maximum value; because for this case, the system must be given the largest possible kick to get it to either the previous or the next stable equilibrium.

There is, however, one more important consideration, and that is the strength of stabilization. The effective force supplied by the photomechanical effect depends on the slope of the sinusoidal curve at the equilibrium point. Clearly, if the slope is large, small changes in the length of the fiber yield large changes in intensity, causing a strong restoring force. This explains why the system settles at the mid-point and why the stabilization effect here is so strong.

The observations in Figure 9.15 can therefore all be explained with the help of Figure 9.16 as follows:

1. The first equilibrium point is so near the first minimum that the system spontaneously jumps to the second stable point as observed in Figure 9.14.
2. The spacing in the steps (i.e., the intensity jumps) are approximately uniform, as predicted by the intersection points in Figure 9.16.
3. Peaks are observed at the leading edge of each step, which corresponds to the interferometer output passing through its maximum value before reaching the next equilibrium.
4. When the feedback to the fiber is turned off by closing the shutter, the interferometer passes through the same number of fringes as there are steps.
5. The peaks in the interferometer get narrower for the higher steps because the stabilization forces get larger. (Note that the spontaneous jump in Figure 9.14 seems to be the most narrow, but that is because the time scale is compressed.)
6. The system is the most stable when the interferometer output is at half its peak value.

Based on the understanding of the process, it is also useful to point out an important feature of this all-optical stabilizer. From the RMS signal fluctuations in the intensity, we can estimate the position tolerances of the stabilization process. We find that the RMS deviation in intensity corresponds to a position accuracy of about $3nm$. For the $30cm$ length of fiber used in these experiments, this corresponds to position stabilization to one part in 10^8.

Polymer fibers thus offer the potential for unique devices that would be competitive with electronic analogues. The disadvantage of this particular system is its large size. The $30cm$ fiber was suspended from a heavy mount so that the whole system is too bulky for many applications. In the next section we discuss how this device can be miniaturized.

9.2.3.2 Mesoscale Photomechanical Devices and Multistability

In Section 9.2.3, we saw how feedback in an interferometer leads to active stabilization. The most straightforward method for miniaturization is to build the interferometer directly into the fiber. A Fabry-Perot interferometer provides the most simple geometry.

The top part of Figure 9.17 shows a generic fiber interferometer with reflecting ends. Methods of making end reflectors are also diagramed. The principle of operation is as follows: The light waves bouncing inside the fiber, to (E_+) and

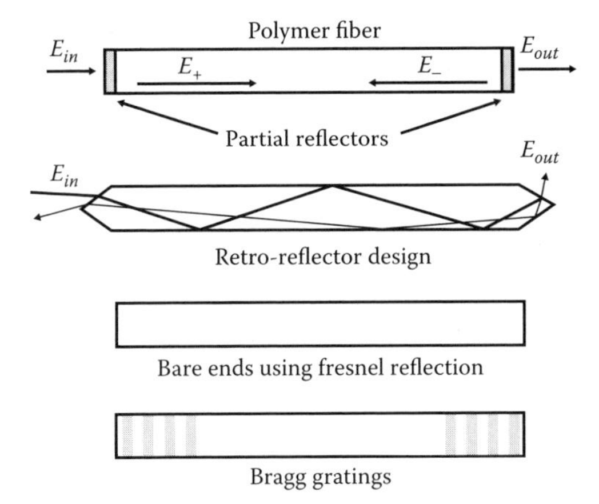

FIGURE 9.17 Partial reflectors are placed at the ends of a fiber to make a Fabry-Perot interferometer (top). Types of reflectors include jagged ends that form a retro-reflector, bare ends that result in Fresnel reflections, and Bragg gratings — which are recorded using light beams.

fro (E_-), make a standing wave. If an integral number of wavelengths fit between the reflectors, constructive interference results, leading to the highest possible intensity inside the fiber. Under this condition, most of the light is transmitted through the fiber. If the length of the fiber changes, the standing wave is affected, and the transmittance changes.

When the fiber is made with a photomechanical material, the length depends on the intensity, but, since the intensity depends on the length, a feedback loop is formed, leading to stabilization of the length of the fiber. Since polymer optical fibers can be made less than $1\mu m$ in diameter, this technology could be used to make nanoscale smart actuator devices. Figure 9.18 shows a $900nm$ diameter fiber made by Paradigm Optics. The highly robust polymer used is Zeonex. The Zeonex fiber is made by preparing a preform with the Zeonex material in the core of a PMMA cladding, pulling the preform into a fiber, and subsequently dissolving away the PMMA. The larger fiber behind the nanofiber in Figure 9.18 is the PMMA "substrate" fiber. This fabrication process can be used for making novel all-optical actuator systems. For example, arrays of nano-actuators could be made from multicore preforms.

Systems that have multiple states for a given set of parameters are called multistable. For the bulk position stabilizer, there were several equilibrium length states of the fiber for a fixed laser intensity. Each one of these length states corresponds to a different intensity inside the hanging fiber. For the Fabry-Perot fiber, the same will be true: For one incident intensity, there are many possible length states corresponding to different intensities inside the fiber.

In the case of the fiber interferometer, it is far easier to measure the transmitted intensity as a function of incident intensity than to directly measure the

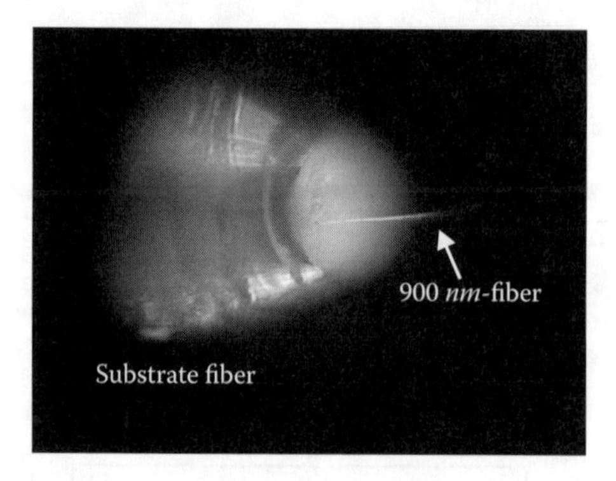

FIGURE 9.18 A nanofiber made by Paradigm Optics. See http://www.paradigmoptics.com/products/POF/nanoscalefiber.htm for more information. (Reproduced with permission of Paradigm Optics.)

length change. The response was measured by Welker and coworkers on a PMMA fiber doped with DR1 dye, of about $1cm$ length and a few hundred micrometers in diameter. [149] Because these fiber devices straddle the macro- and microscopic scales, the authors referred to these fiber interferometers as mesoscale photomechanical units, or MPUs. Figure 9.19 shows the transmittance function. The maximum incident intensity is about $18mW$ and the laser diode source intensity is adjusted by computer control of the supply voltage. At the higher intensities, a distinct open loop is observed as predicted from theory. A chaotic region appears at the medium energy range, which Welker later attributed to feedback in the laser diode.

Based on literature values of the thermal expansion coefficient and the temperature dependence of the refractive index, the authors concluded that intensity change observed in the output due to a length change in the fiber was larger than the

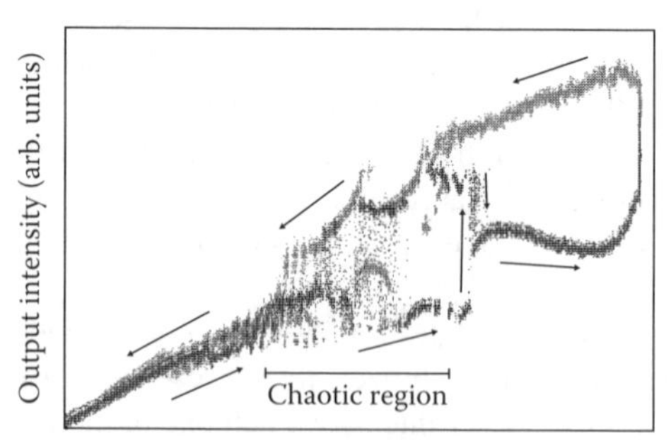

FIGURE 9.19 The transmitted intensity as a function of incident intensity for a mesoscale photomechanical unit. The arrows indicate the history of the trace. Courtesy of David Welker and M. G. Kuzyk.

change in transmittance due to a refractive index change in the material. As such, they concluded that the observed hysteresis was mostly due to the photomechanical effect. While the authors also speculated that photo-isomerization may have played a role, they did not estimate the length change due to this mechanism.

Aside from its small size, an additional advantage of the miniaturized fiber interferometer is that, if the photothermal mechanism is responsible, the smaller volume of the device makes its response much faster. In terms of experimental verification, the all-fiber device experiments are more difficult to interpret than the bulk device since the external interferometer is not sensitive to a change in the refractive index of the hanging fiber. In contrast, the output intensity of the all-fiber device is a convolution of both a length change and a refractive index change. As such, the problem is too complicated to treat fully. However, the multistable response of a fiber due to the length change mechanism and the refractive index change mechanism are both treated in Section 7.2.4.

The shape of the hysteresis loop, which we call the hysteresis function, can be calculated from the transmittance of a Fabry-Perot interferometer. There are two contributions: absorption and interference. The amount of light lost depends on the absorbance α_0 and the length of the fiber, L. The absorbance can depend on the intensity through the imaginary part of the intensity-dependent refractive index, n_{I2}, and the the length depends on the intensity through the photomechanical effect ($\Delta L/L_0 = u$ according to Equation 9.12). The interference condition inside the interferometer depends on the length of the fiber and the refractive index. Both of these also depend on the intensity. As such, the intensity of the light inside the interferometer depends on the optical path length between the reflectors — which depends on the distance between the ends and the refractive index of the fiber.

While the problem of calculating the hysteresis curve is complex, we can do so graphically — which leads to a better understanding of how multistability arises. Consider the left portion of Figure 9.20. The curve shows the interferometer output intensity as a function of the distance between the fiber ends for a fixed incident intensity. With a photomechanical effect, the length is a function of intensity. For the first term in Equation 9.12, the length change of the fiber due to the photomechanical effect is a linear function of the intensity inside the fiber. Therefore, if we invert this equation, we get a straight line that represents $\Delta L \propto I$. The inverse of this function would then represent the intensity inside the fiber as a function of ΔL.

For a fixed fiber length, the intensity transmitted by the interferometer, I_T, is directly proportional to the intensity inside the fiber. The interferometer output, however — which also depends on the Fabry-Perot cavity length — is a sinusoidal function of the length. Both of these functions are represented on the left-hand portion of Figure 9.20. As described in Section 7.2.4, we can use a graphical method to solve for the bistable and multistable solutions, as follows.

The slope of the photomechanical line decreases as the incident intensity is increased. Therefore at low incident intensity, the solution for the length and transmitted intensity is given by point a. Clearly, there is only one solution. As the intensity is increased from this point, the length continues to increase as shown on the right-hand portion of Figure 9.20 (Note that because Figure 9.20 is drawn

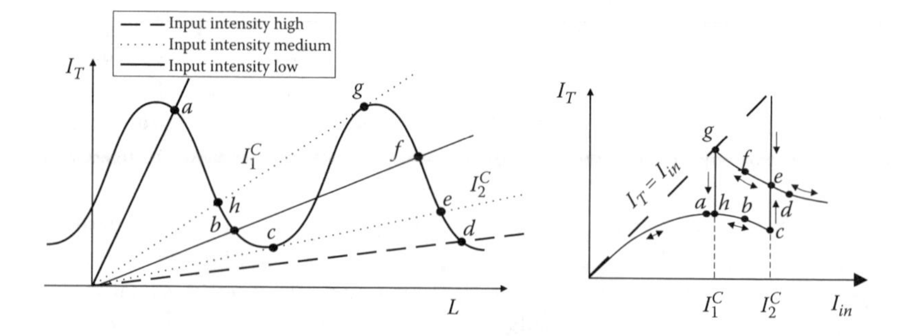

FIGURE 9.20 (Left) The curve shows the interferometer output intensity as a function of fiber length and the lines represent the inverse of the length of the fiber as a function intensity inside the fiber. (Right) The resulting hysteresis. The points labeled with the same letter on each graph represent the same state of the system.

graphically by hand, it is not an accurate representation of the shapes of the curves, but accurately reflects the critical points and the multistable behavior.) Point h is a critical point where a second solution appears at point q. At this incident intensity, the system becomes bistable. Because nature does not like discontinuities, as the input intensity is increased from point h, the system moves smoothly to point b rather than making the jump to point q.

As an aside, one might notice that the line defined by points b and f has a third solution that is not labeled. The only way this state can be reached is with a discontinuous jump or by external intervention (for example, a force could be applied to the fiber to coax it into this state). Note that this point lies inside the loop on the right-hand side of Figure 9.20. As such, we ignore these unstable points in our analysis.

As the journey of increased intensity continues, the system travels to point c. Beyond the critical point c there is again only a single solution d. At this critical point, the system jumps to a longer length state and a higher transmitted intensity given by point e. If the intensity is decreased from this point, the states trace out points e, f, g, h, and then a. As with the upward journey, when the intensity at the input is decreased, the system makes a discontinuous jump only when there is no other choice.

We have considered only one oscillation of the output intensity. As the input intensity gets high (and the slope gets lower) the photomechanical line will intersect the sinusoidal curve in many places, leading to multistability. All of the salient features of multistability can be understood with the graphical method.

We pause here to appreciate the implications of such a response. The photomechanical fiber (PMF) is multistable in both intensity and length, so it has both for optical and mechanical memory. The output of the fiber depends on its length, so it is a stress sensor. The length depends on the light intensity, so it can act as an actuator. And, all of these properties can interact with each other. As such, there are many complex features that can be built into this one unit. Compare this to an electrical device that has revolutionized our world, the transistor.

Basically, the only properties being used are electrical bistability and nonlinearity. Imagine the rich behavior of a PMF, coupled with the fact that light can carry much more information than electrons; and that optical architectures can be made to be more sophisticated than electronics. Clearly, combining such complex smart units together into bigger structures and complex architectures could lead to new devices that go well beyond present-day concepts. Examples of some of unique applications follow.

9.2.3.3 Optical Tunable Filter and All-Optical Switching

When the light intensity in a fiber Fabry-Perot cavity is changed, the induced length and refractive index change result in a change in the transmittance. Optical switching devices, in which one beam turns another one on and off usually operate on the refractive index change mechanism. In a mesoscale photomechanical unit, the length change mechanism is dominant. In such a device, a strong pump beam can be used to change the separation between the end mirrors, which will affect the transmitted intensity of a weak probe beam.

Welker used a $1cm$ long MPU of $110\mu m$ diameter made of a poly(methyl methacrylate)(PMMA) polymer doped with about 0.5 weight percent of Disperse Red 1 Azo dye to make an optically tunable filter, [150] which in principle can be used for optical switching if the finesse of the cavity were appropriately tuned. The end reflectors in this device were formed with conical retroreflectors, which provided a reflectivity of about 5%–10%.

Figure 9.21 shows the experimental setup. Two laser diode beams of different wavelengthes are coupled into the MPU. The control laser (labeled pump and

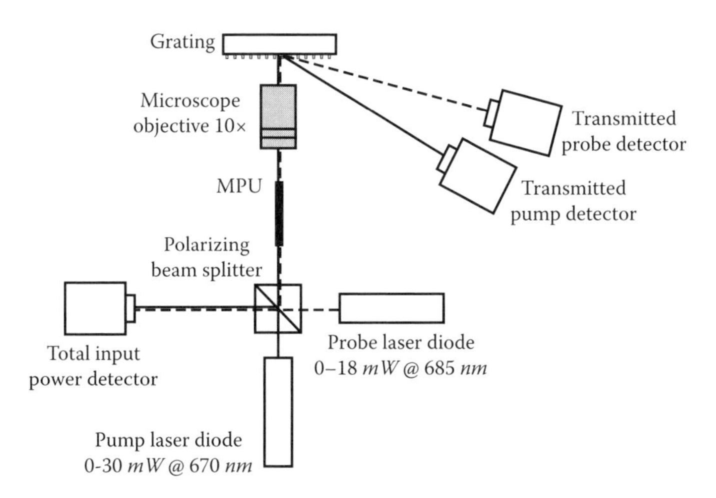

FIGURE 9.21 A pump and probe laser are directed into an MPU. A grating is used to separate the beams so that the intensity of each can be measured with a detector. Adapted from Ref. [150] with permission of the American Institute of Physics.

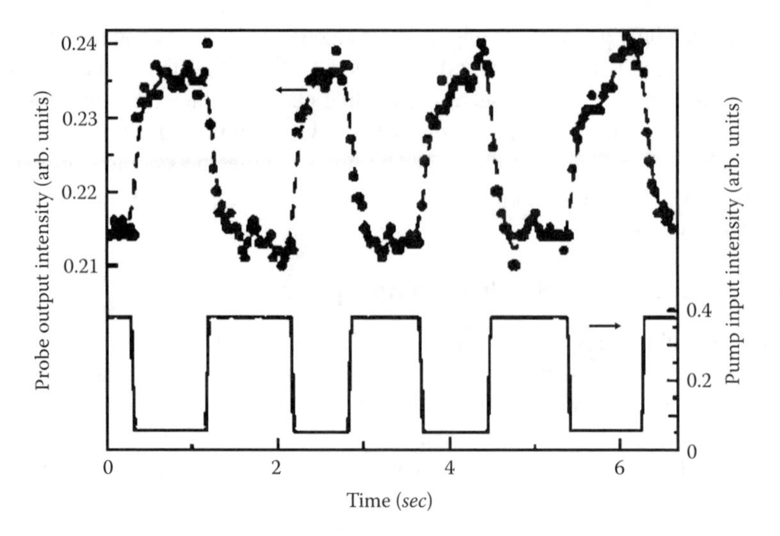

FIGURE 9.22 The pump laser (bottom graph) and probe beam transmitted by the MPU (top). Adapted from Ref. [150] with permission of the American Institute of Physics.

shown with a solid line) intensity is varied while the signal laser (labeled probe and shown as a dashed line) is kept at a constant intensity. The light leaving the MPU is collimated with a microscope objective and directed to a reflection grating that separates the two beams and directs them to two different detectors. The experiment is computer interfaced so that all voltages read by the detectors are digitized; and the intensity of the laser diode is controlled by sending a voltage to the diode controller.

Figure 9.22 shows the intensity of the incident control laser measured by the detector (bottom solid curve) and the transmitted intensity of the probe laser (points and dashed curve as a guide to the eye) as a function of time. The incident $28mW$ pump intensity is turned on and off. The $3mW$ probe intensity is on continuously. The modulation of the probe beam is only about 10% — consistent with the low reflectivities of the fiber end used in this experiment. The authors found that the turn-on time and decay of the probe signal is as expected for a photothermal mechanism. Based on the magnitude of modulation and the known polymer properties (such as specific heat and absorbance), the authors estimated that the photothermal length change was responsible for about 2/3 of the signal.

This was one of the first papers that suggested that the photo-cis/trans isomerization mechanisms could yield a length change and that it might contribute to the observed signal. Their conclusion was based on the consistency of the photoconversion rate of the trans isomer to the cis isomer with their results. In particular, Dumont and coworkers found that the isomerization process, followed by reorientational decay, occurs over a couple of minutes for an on-resonance illumination power of $30mW/cm^2$ in DR1-doped PMMA. [151–153] Based on an analysis of the response times, the laser pump power, and the fact that the pump laser was

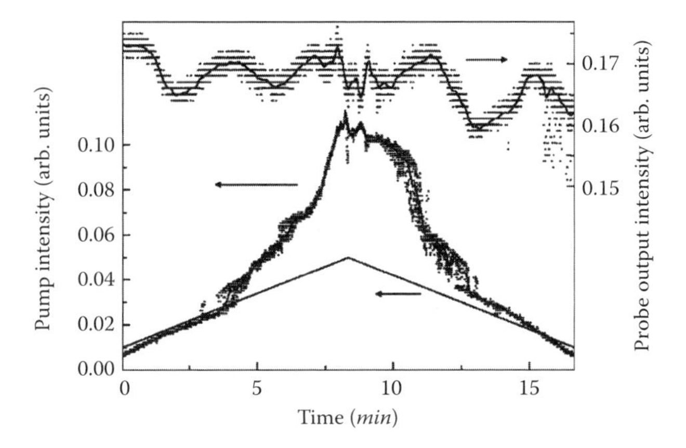

FIGURE 9.23 The incident pump laser, transmitted pump laser, and transmitted probe beam as a function of time. Adapted from Ref. [150] with permission of the American Institute of Physics.

off-resonance, Welker and coworkers concluded that the isomerization mechanism was not likely to be a dominant contributor to the 1 second pulsed laser measurement.

Figure 9.23 shows the pump power launched into the MPU (triangular curve), the pump power exiting the MPU, and the probe power exiting the MPU as a function of time. The $3mW$ probe beam is on continuously and the pump is ramped to $28mW$ at a rate of $0.058mW/s$. From the asymmetry in the pump output, it is clear that hysteresis is present. If the output is plotted as a function of input, a curve similar to the one shown in Figure 9.19 is found. The transmitted probe oscillates asymmetrically as a function of the pump intensity, showing that the probe is sensing the phase shift induced by the pump as well as the bistability of the system. (If the phase shift induced on the probe by the MPU was linear in the pump intensity, the transmitted probe would oscillate sinusoidally as the pump intensity is ramped.) The MPU response is clearly a nonlinear function of the pump intensity. In contrast to the pulsed measurement shown in Figure 9.22, the ramped experiment in Figure 9.23 is over a longer time scale, so the authors propose that the photoisomerization mechanism contributed substantially to the observation.

9.2.3.4 Stabilizing a Sheet with an MPU

While MPUs are small, they can be used to suppress vibrations in a macroscopic object such as a mylar sheet. [154] This demonstration is similar to the one described in Section 9.2.3, except instead of using a large bulky external interferometer, the whole system is built into a $2cm$ long and $110\mu m$ diameter MPU made of Disperse Red 1 (DR1) dye/doped poly(methyl methacrylate). As such, the unit provides sensing, logic, and actuation all in one small package.

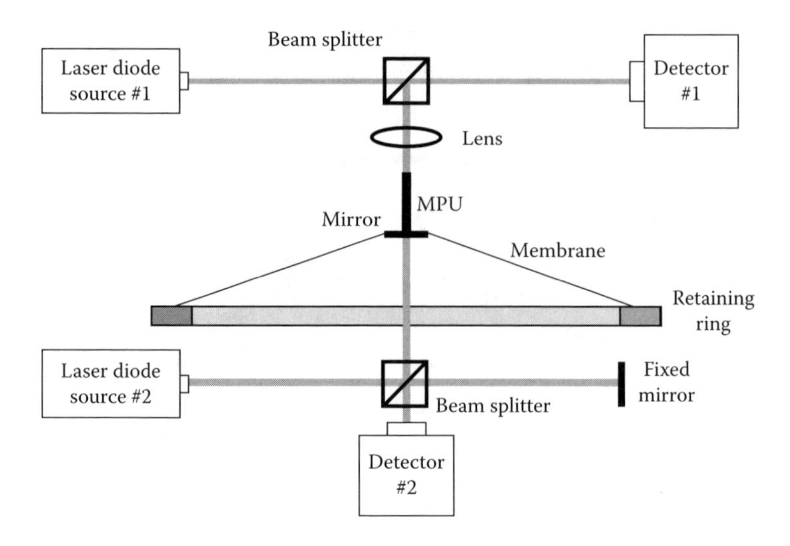

FIGURE 9.24 The experiment used to test the characteristics of vibration suppression of a mylar sheet with an MPU (top). Adapted from Ref. [154] with permission of the Optical Society of America.

Figure 9.24 shows the experiment that was used to test the MPU's ability to suppress vibrations in a large mylar sheet. The mylar sheet is stretched across a metal ring of about a $10cm$ diameter. A mirror is attached to the underside of the membrane while the MPU is attached above (the amount of stretch of the membrane shown in the figure is greatly exaggerated). A laser diode (Source #1) is launched into the MPU through a beam splitter and lens. The light that is reflected back from the MPU is directed through the upper beam splitter to Detector #1. This detector monitors the state of the MPU. The position of the center of the membrane is measured by reflecting the light of the hanging mirror and using this as one arm of a Michelson interferometer. Detector #2 is used to measure the interferometer output power. Note that the light in the MPU cannot pass through the mirror on the underside, and the light from Laser Diode #2 cannot pass through the mirror into the MPU; so, the interferometer is totally isolated from the MPU and therefore does not influence it.

It is interesting to note that in order to stabilize vibrations in the sheet, the MPU must act against both the tension in the membrane and the weight of the mirror. On the other hand, the mass of the mirror makes it more difficult to excite vibrations in the sheet. Since the sheet is essentially a drumhead, the noise in the room is sufficient to excite its various modes.

Figure 9.25 shows a typical Fourier spectrum of the noise in the sheet as measured by the interferometer with the MPU pumped with the diode laser and with a dark MPU. The amount of stabilization provided by the MPU is between $3dB$ and $5dB$. A similar amount of stabilization is found out to $5kHz$. [154] $5dB$ is probably a lower limit of the amount of stabilization that is possible since the reflectivity of the MPU ends were low, resulting in low Finesses.

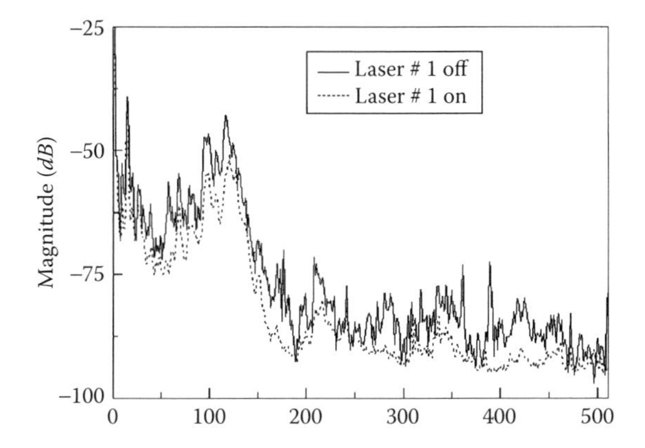

FIGURE 9.25 The Fourier spectrum of the light power measured by Detector #2 with Laser Diode #1 on and off. Adapted from Ref. [154] with permission of the Optical Society of America.

Figure 9.26 shows an array of cubic MPUs that could be used to make a smart platform for holding objects that cannot tolerate vibration. An array could be formed using microphotography and the light source below could be made from a vertical cavity surface emitting laser (VCSEL) array. For example, a vibration-isolation table could be made by placing a breadboard on top of such a structure. Alternatively, the array could hold a lens or serve as a platform for mechanical targeting. The profile of the array could be controlled by adjusting the light intensity pattern from the laser array. With electrical systems, such as those that contain piezoelectric transducers, wires need to be connected, so mechanical vibrations can be transmitted through the wires. In an optical system, the light source is not in physical contact with the system, making mechanical isolation better.

Since the MPUs are mechanically multistable for a fixed input intensity to all the units, the length state could be made to change in response to either a mechanical impulse or burst of light. Therefore, if the MPU is supporting, for example, an adaptive optics system, it can be kicked occasionally to make minor

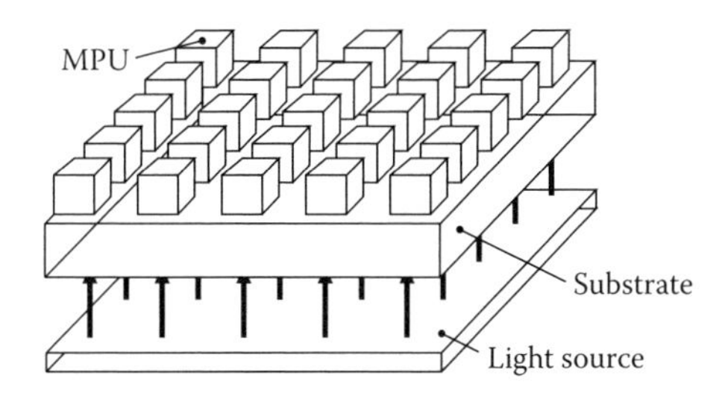

FIGURE 9.26 An array of MPU devices could be used for vibration isolation.

adjustments to the height profiles of the array. This could be useful for tracking slowly moving objects with high precision.

9.2.3.5 Transverse Photomechanical Actuation

Up to this point, only changes in length of a photomechanical fiber are considered. If light is launched into a fiber off-axis, a differential photomechanical effect will result in fiber bending. This gives an additional two degrees of freedom to an optical actuator. The most efficient way to accurately control the bending process is by defining one or more off-axis waveguides in a fiber. Section 9.3.5 describes this idea in more detail. In this section, we show results of experiments on a multimode fiber under the off-axis launching condition.

Bian and coworkers first demonstrated photomechanical bending in a PMMA polymer optical fiber. [155] Figure 9.27 shows a photograph of the end of a fiber under a microscope. The overlayed grids are spatially fixed. The plus sign marks the corner of a non-illuminated fiber. When light is launched into the fiber off-axis, it both bends and changes length. The diagram above the photographs shows a schematic representation of the experiment.

To quantitatively determine the angle of deflection, Bian attached a lightweight mirror (made from a small chip off the corner of a metalized microscope cover slip) to the end of the fiber and measured the deflection angle of a second laser that reflects from this mirror. In the geometry of the experiment, the deflection angle was only sensitive to bending. Note that the photograph in Figure 9.27 was taken before the mirror was attached.

Figure 9.28 shows the measured deflection angle of the fiber end as a function of time. The light is turned on at the point marked with the arrow. The time dependence is observed to be a biexponential function. The faster mechanisms are consistent with photothermal heating both in the rise time as well as the magnitude of the

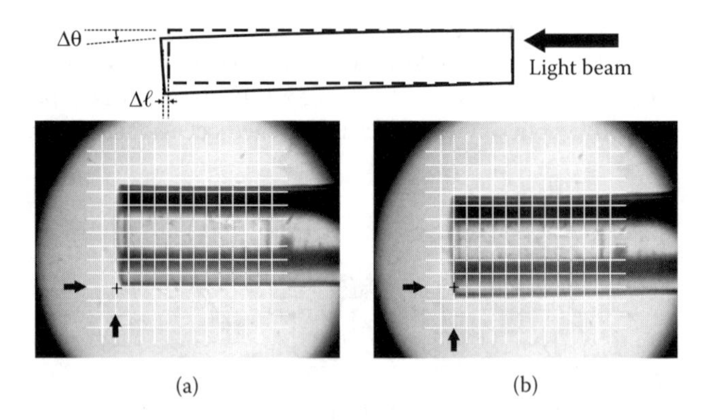

FIGURE 9.27 Light launched into a fiber off-axis causes it to bend and change length (top). (a) Photograph of the end of a PMMA fiber with light off and (b) with light on. The "+" in each photograph is at the same physical point. Adapted from Ref. [155] with permission of the Optical Society of America.

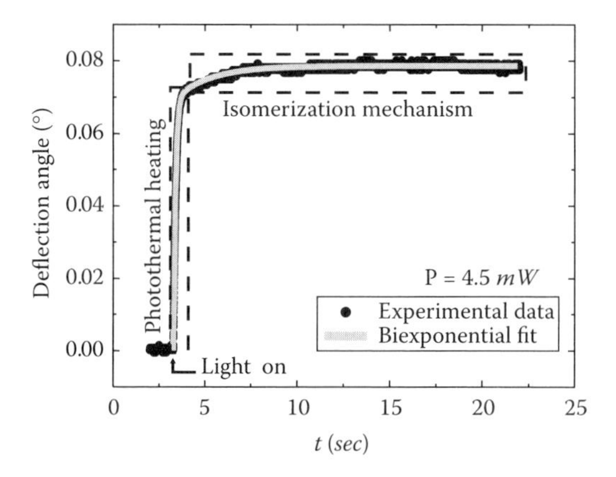

FIGURE 9.28 The angle of a fiber as a function of time after the light beam is turned on. Figure courtesy of Dr. S. Bian and the author.

deflection angle. The slower mechanism is consistent with the photoisomerization mechanism. The MPU used in these studies had a $d = 600\mu m$ diameter and $L = 2.6mm$ length. The excitation source was a HeNe laser at $\lambda = 633nm$, the spot size is $\omega = 20\mu m$, and the material used to make the fiber is DR1/PPMA with an absorption coefficient is $\alpha = 3.0mm^{-1}$.

The fact that the slower response is due to heating is determined by balancing the cooling rate of the fiber with the energy fed into the fiber by photothermal heating. Given the heat capacity of the fiber, the power of the laser source, and the thermal conduction of the polymer and surrounding air; the time to reach equilibrium can be calculated.

A simpler back-of-the-envelope calculation can be done as follows: Let's assume that the amount of time it takes the fiber to reach equilibrium is experimentally determined to be τ. For a power P that is coupled into the end of the fiber, the amount of energy absorbed over that time is $E = P\tau$, where we assume that the fiber length is chosen so that almost all of the light is absorbed. Assuming, for example, that about half of the fiber is illuminated, the increase in temperature of that half can be estimated from its heat capacity, C, $\Delta T = CE$. Finally, using the thermal expansion coefficient, α, the length change of that half of the fiber, ΔT, is given by $\Delta \ell = \ell \alpha \Delta T$. Finally, the bend angle will be given by $\Delta\theta = \Delta\ell / r$, where r is the radius of the fiber.

Putting all of this together, Bian finds that for a photothermal heating process, the ratio of the deflection angle to the response time — when half the fiber is illuminated — is given by

$$\frac{\Delta\theta}{\tau} = \frac{\ell \alpha P}{r}. \tag{9.45}$$

Using such a calculation, Bian showed that the measured deflection angle was consistent with the response time.

9.3 THE FUTURE OF SMART PHOTONIC MATERIALS

9.3.1 INTRODUCTION

Smart materials or structures have the ability to respond to stimuli with a predetermined functionality. Both the stimulus and response can be chemical, acoustical, optical, thermal, mechanical, etc. There are many types of components that can be used to make a smart material or structure: piezoelectric materials — which change shape in response to an electrical field; magnetostrictive materials — that respond to a magnetic field; fiber sensors — in which the light intensity transmitted or reflected by the fiber depends on stress, temperature, or chemical environment; liquid crystals — whose optical reflectivity can be altered by the application of a voltage or light; shape memory alloys — which change elasticity and shape with temperature or light exposure; and many others. An attractive feature of a smart material is that sensing, decisions, and response can take place at one physical point. The response of the material can either be local or collective: Each point can act independently or as an element in a larger interacting association. Availability of a pallet of smart material types would no doubt lead to the design of smart structures or materials with many new and exciting applications.

The basic elements of a conventional smart mechanical device are sensors, logic units, and actuators. The sensor converts information (such as stress or strain) into an electrical impulse that is analyzed with a logic unit (such as an integrated circuit), which responds with an electrical output that drives a motor or transducer. Light-encoded information can be processed much more quickly and with greater flexibility than electronics. Imparting the power of optical computing to standard smart devices, however, would add two more components: a converter that changes electricity to light and another that changes light to electricity. All-optical devices would be unique because mechanical information can be directly encoded into light and the light can be directly converted to mechanical motion. An additional benefit of MPU-type devices is that sensing, logic, and actuation are combined into a single mesoscopic monolithic unit. When such units are combined into a smart material, they can form a highly interconnected network in which interactions between large numbers of units form the basis of complex computing functions. The intelligence built into such a material would potentially be orders of magnitude faster and more sophisticated than electronics technology. Because photomechanical devices have been demonstrated to work, they bridge the technological gap needed to develop an all-optical technology base.

9.3.2 MAKING A SERIES OF MPUS IN A SINGLE FIBER

The technology exists today to make a highly interconnected structure of MPUs that behave in a complex interrelated way as shown schematically in Figure 9.2. Figure 9.29 shows an example of a smart fiber that is made with a chain of MPUs. Such a chain can be made by forming a series of Bragg grating reflectors (see

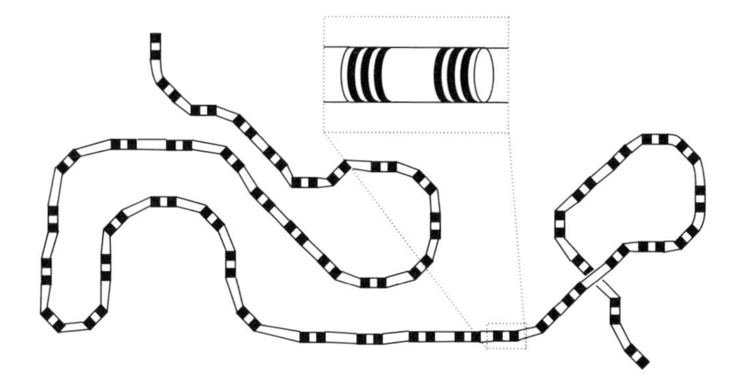

FIGURE 9.29 A chain of MPUs make a smart fiber.

Section 8.1) in a single fiber, making the process compatible with fiber drawing and spooling technology.

Consider, for example, a long length of fiber on two spools as shown in the top part of Figure 9.30. A grating can be written at one point, the fiber moved to the next position, where another grating is formed. The fiber spooling system could be programmed to write gratings at the appropriate positions to make a series of MPUs. The bottom portion of Figure 9.30 shows how two MPUs are written in a section of fiber. Second, the grating-writing beam-steering prisms and beam splitters can be mounted on programmable tilt mounts so that the angle between beams can be adjusted to make gratings with different spacings. Finally, the photomechanically active dopant can be bleached in regions where no photomechanical effect is desired. As such, the fabrication process could be used to make a series of MPUs of varying sizes that are each sensitive to distinct wavelengths. Furthermore,

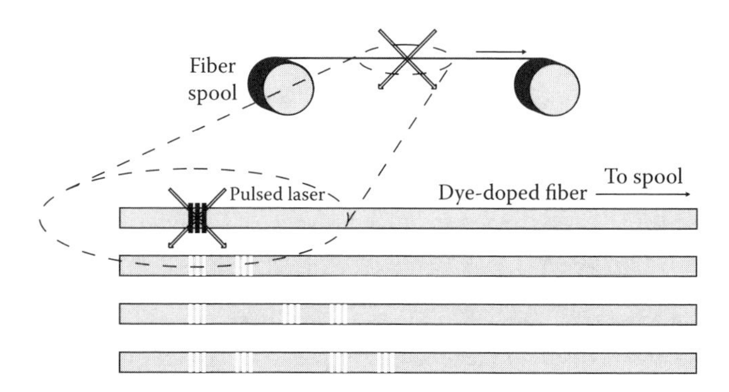

FIGURE 9.30 Making a series of MPU devices in a polymer optical fiber by writing a sequence of Bragg gratings. The pictures of the fibers are at different times to show how the gratings can be written as the fiber is advanced by the spooling system.

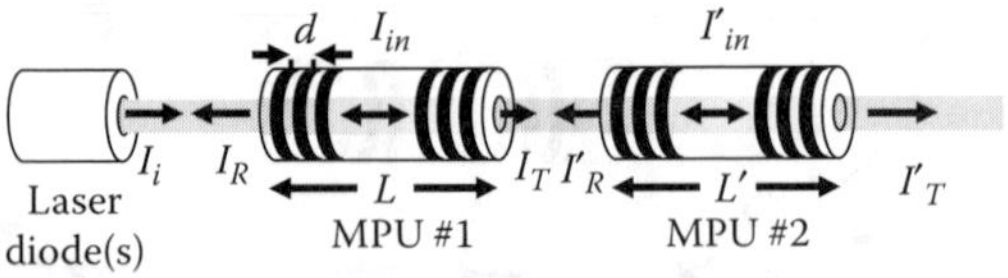

FIGURE 9.31 When two MPUs are placed in series, the transmitted intensity through the first one is incident on the second one, and the reflected intensity from the second one is incident on the first one.

regions that are not photomechanical can be made to act as sensors or simply as a transmission fiber. The process is therefore amenable to making all sorts of devices in a piece of fiber without the need for splices.

9.3.3 The Two-MPU System

Before thinking about more complex systems, it is useful to first consider the case of the two-MPU case, as shown in Figure 9.31, to build an intuition of the types of devices that can be made. The operational characteristics of each MPU depends on the wavelength of the laser source. There are two major geometrical factors that determine the color dependence: the periodicity of the refractive index grating and the separation between the reflectors. Equation 7.77 gives the transmitted intensity of the Fabry-Perot interferometer. If $\phi' = 0$, we get that the transmitted intensity is maximum when the optical path length, $n\ell$ is a half integer times a wavelength, or $n\ell = N\lambda/2$. The Bragg grating, on the other hand, with a periodicity of optical path length nd will *reflect* light when $nd = N\lambda/2$. So, to make an MPU to operate at a given color, both of these conditions need to be met. Given these design considerations, we begin by considering various possible devices based on pairs of MPUs in series.

9.3.3.1 CASE I: Photomechanical MPU #1; Passive MPU #2; Light Source with Wavelengths λ_1 and λ_2

Let's consider a two-MPU device in which the first one is photomechanical and the second one is passive. Two colors of light are launched into the fiber. At low light levels, MPU #1 is designed to block (or reflect) λ_1 and MPU #2 is designed to block λ_2. If the intensity of the beam of wavelength λ_1 is always low enough not to induce a photomechanical effect (the MPU could be designed to be less sensitive to the color λ_1 by proper choice of dopant), the second more powerful beam at λ_2 could be used to make MPU #1 pass the weaker beam at wavelength λ_1. The second MPU is designed to pass λ_1 but block λ_2; so this device would act as an all-optical modulator of the weak beam by the strong one. Since MPU #2 is passive, it acts as a filter to remove λ_2, so the bright beam never passes.

If information is being carried on the bright beam, this device would convert the signal from wavelength λ_2 to λ_1.

9.3.3.2 CASE II: Passive MPU #1; Photomechanical MPU #2; Light Source with Wavelength λ

Let's assume that the first MPU does not have a photomechanical effect, but acts as a stress sensor that is set to not pass light when no stress is applied. Furthermore, the bias is set so that a small increase (or decrease) in the MPU sensor length results in a decrease (or increase) in the photomechanical MPU so as to put the system into a push-pull configuration. So, when a stress is applied to the first MPU to increase its length, it causes the second one to decrease its length causing the total fiber length to stay the same. This device pair acts as a position stabilizer. While a single MPU will behave in the same manner, this design allows the point of stress to have a different location from the point of the length change.

9.3.3.3 CASE III: Photomechanical MPU #1 and MPU #2; Light Source with Wavelength λ

This is by far the most complicated situation. First, there is the issue of coherence. If each MPU is much shorter than the coherence length of the laser and the two MPUs are separated by a distance that is greater than the coherence length of the light source, then the MPUs will each show multistability but there will be no interference between the reflected beam from MPU #2 and the the light inside MPU #1. This is by far the easiest case to handle. If the coherence length of the laser is much longer than all length scales of the device (i.e., separation between MPUs, grating spacing within MPUs, etc.), then the system will behave as three coupled MPUs (the fiber between each adjacent grating pair acts as an MPU). While this second case leads to an incredibly complex response (consider that each MPU acts as a sensor, multistable device, and actuator) that could provide highly intelligent functionality, it is too difficult to conceptualize and will not be treated here.

There are many ways that such a device could be used. First, one can imagine a pure actuator mode. MPU #1 is designed with Bragg reflectors that only reflect wavelength λ_1 and MPU #2 is designed to only operate at wavelength λ_2. As such, each device can be independently actuated by shining light of the appropriate color. Similarly, the state of each MPU can be individually interrogated, making it possible to use the two sections of fiber as individually addressable stress or temperature sensors. Bragg grating temperature sensors using wavelength multiplexing are common in glass fibers. What makes the polymer fiber device interesting is that each sensor also acts as an actuator/stabilizer.

The most interesting situation is where the Bragg gratings in both MPUs are designed to reflect light of the same color. Then, the intensity inside MPU #2 comes from the output of MPU #1; but the intensity in MPU #1 enters both from the laser source and the reflected light from MPU #2. In order to treat this problem, we make the following assumptions:

1. The length of each Fabry-Perot cavity in the fiber is short compared with the coherence length of the source. The requirement ensures that the light bouncing to and fro inside the cavity will interfere with itself.
2. The separation between interferometers is greater than the coherence length. Under this assumption, the light that exits MPU #2 to the left (I'_R) and enters MPU #1 from the right will not interfere with the light entering MPU #1 from the left (I_i). However, the light that inters MPU #1 from the right will still interfere with itself inside the MPU. The average intensity inside the MPU will thus be the sum of the average intensity due to I_i and I'_R. (If the coherence length were longer than the separation between MPUs, the electric fields from each beam would interfere inside MPU #1. While this case can be treated with our methods described, the resulting interference leads to many complications that are well beyond the scope of this book.)
3. The separation between MPUs is large enough for MPU #1 to reach equilibrium before the light from MPU #2 is reflected back into MPU #1. (This will allow us to solve all the fields in an series of time steps.)

To gain an understanding of coupled MPUs, we will solve the problem graphically. Consider the diagrams in Figure 9.32. Diagrams *a* and *b* show MPU #1 and MPU #2 before the reflected light from MPU #2 reaches MPU #1. Diagram *c* shows MPU #1 after the light that is reflected from MPU #2 reaches MPU #1. Diagram *c* does not show the intensities I_i, I_R, and I_T. The plots in Figure 9.33 show the intensities in the MPUs corresponding to the diagrams; so diagram *a* is associated with plot *a*, diagram *b* with plot *b*1 and *b*2, etc. The dashed line represents the response when there is no MPU, which states the obvious — with no MPU, the output intensity equals the input intensity.

The strategy for solving the problem is to use the response functions of the MPUs to graphically calculate the intensities by using the output of one response function as the input to the next one in the same time sequence that occurs in the

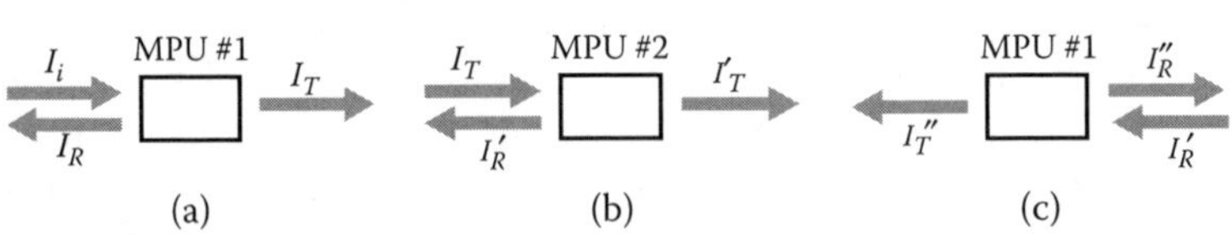

FIGURE 9.32 The time sequence of the graphical solution to two coupled MPUs. Diagram *a* shows the MPU #1 just after the field is turned on, prior to the time the reflected light makes it back from MPU #2. Diagram *b* shows the light in MPU #2 prior to the reflected light from MPU #2 reaching MPU #1. Diagram *c* shows MPU #1 after the light from MPU #2 reaches it. In Diagram *c*, the intensity from the incoming light source, as well as the reflected and transmitted beams from the light source, are not shown.

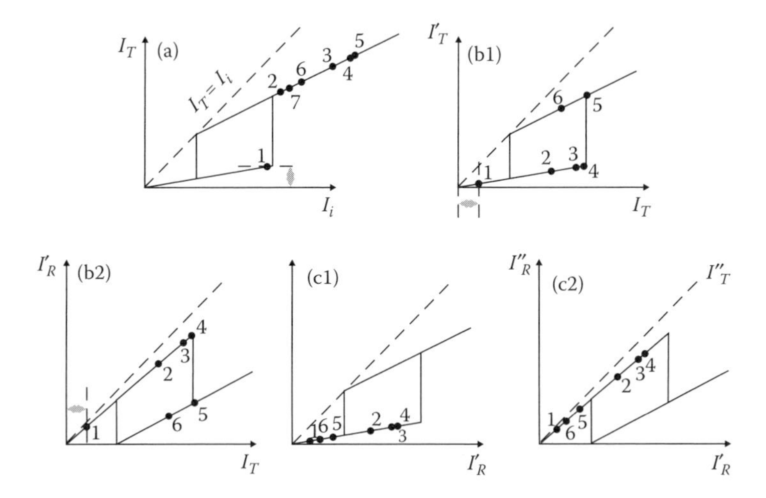

FIGURE 9.33 The plots show the various intensities in the MPU for these three time steps as diagramed in Figure 9.32.

physical process. For simplicity, we assume that the response function of each MPU is the same. This does not affect the generality of the result presented.

The sequence to the solution is as follows:

1. The light intensity is turned on to a level such that the output intensity, I_T, is represented by point 1 on plot a of Figure 9.33. The length of the vertical gray arrow represents the output intensity of MPU #1.
2. The output from MPU #1 is now the input of MPU #2. The gray arrow in plot $b1$ shows the incident intensity as determined from the output of MPU #1. As such, the transmitted intensity of the second MPU, I'_T, is given by point 1 labeled on plot $b1$.
3. The reflected intensity from MPU #2, I'_R, will be the difference between the incident and transmitted intensities (i.e., the difference between the dashed line in Plot $b1$ and the hysteresis curve). Plot $b2$ shows the reflected intensity where the gray arrow again shows the incident intensity coming from the output of MPU #1.
4. The intensity reflected from MPU #2, I'_R, is now incident on MPU #1. To understand what happens, let's — for the moment — assume that the beam I'_R does not induce a nonlinearity in MPU #1. If this were the case, the transmittance of MPU #1 would be the same as at point 1 on plot a. (This is so because the MPU is symmetric making the transmittance and reflectance independent of which side the light enters.) But, since the beam I'_R does induce a nonlinearity in the material, and we assume that the induced length change is proportional to the sum of the average intensities from the two incident beams on MPU #1 (i.e., $I_i + I'_R$), the

transmitted intensity due to beam I'_R is

$$I''_T = I'_R \frac{I_T(I_i + I'_R)}{I_i + I'_R}. \quad (9.46)$$

Point 1 on plot $c1$ shows an estimate of the transmitted intensity due to I'_R according to Equation (9.46).

5. Point 1 of Plot $c2$ shows the part of I'_R that is reflected by MPU #1 back at MPU #2. The total intensity moving to the right from MPU #1, $I'_R + I_T$, is now the sum of the intensity that is reflected from MPU #2 and the transmitted intensity from the laser source. (Because I'_R and I_T are not coherent, the intensities add, rather than the electric fields.) Point 2 on plot a shows the new transmitted intensity. Note that the feedback from MPU #2 has caused MPU #1 to jump to the upper branch of the hysteresis curve. This procedure is reiterated to produce the other points by considering the sequence a, $b1$, $b2$, $c1$, $c2$, for Point 2 and then repeating it for Point 3, etc.

The above graphical approach is not quantitatively accurate but is useful for showing the kinds of complex behavior that can be observed. This procedure can be implemented numerically to more accurately trace out the evolution of the system. The important result is that the output from this two-MPU system can behave chaotically.

The above response has only considered the effect of light on the system. If an external stress is applied to the fiber, more complex behavior results. For example, assume that the two-MPU system starts in equilibrium. If a small stress is applied to one of the MPUs, the resulting feedback to the other one can cause it change its mechanical state. So, we can imagine observing mechanical action at a distance since a stress on one part of the fiber results in an almost instantaneous response on the other other end, limited in time only by the speed of light.

9.3.4 Smart Threads and Fabrics

A series of MPUs defined in a single fiber would have a complex smart response based on the states of the whole collection of MPUs. (The response is so complex that understanding such a system would require intensive modeling and the development of new numerical analysis techniques.) As such, we call this a smart thread. The smart threads, in turn, can be combined into more complicated systems. For example, the threads can be woven together to form a smart fabric, which can then be embedded in another material to form a composite. Figure 9.34a shows parallel smart threads and Figure 9.34b an example of a smart fabric that is woven out of smart threads. The fabric can be embedded in a composite, as shown, or used as a free-standing material. Furthermore, the threads can be individually powered with light or chained together, depending on the functionality required. Figure 9.34c shows a sandwich structure made from two composite woven fabrics that cover the surface of an object.

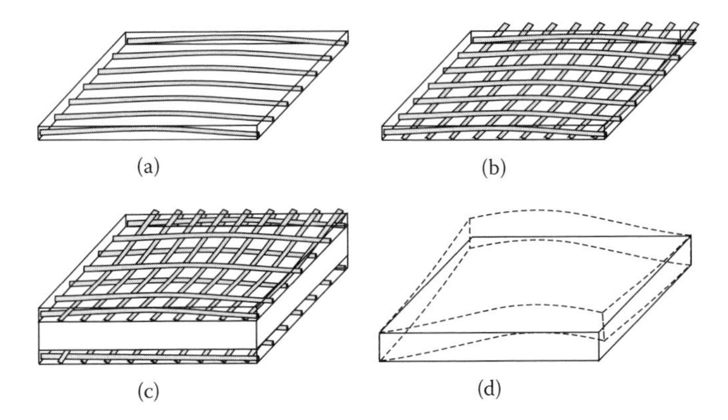

FIGURE 9.34 a) Parallel smart threads; b) Smart fabric composed of smart threads; c) Smart fabric built into a composite smart structure; and d) A shape change induced in the smart structure.

The sandwich structure can be bent by shining light into the fabric. Similarly, the shape of the structure can be actively stabilized against vibrations. While each unit in the smart structure reacts to local stress, the information is passed to nearby MPUs, which can react before the stress or pressure reaches the outlying units. Consider, for example, a smart skin that covers an aircraft wing. When the leading edge of the wing experiences a pressure change or turbulence, the smart fabric can be designed to readjust the trailing edge of the wing before the pressure wave reaches it.

9.3.5 Transverse Actuation

There are many other possible geometries that may be more suitable for a given application or may offer more mechanical leverage. Transverse motion of a fiber, for example, may be more appropriate in some applications than a length change. Such a device could be formed using an asymmetric fiber waveguide, in which the axis of a high refractive index guiding region (that is defined with a doped dye region) is offset from the fiber axis. In such a fiber, light is confined to a waveguide that is composed of a dye-doped region, which serves the dual purpose of elevating the refractive index to allow waveguiding and to impart the photomechanical effect to that region. A light beam will cause uniaxial deformation of the guide. With no deformation in the surrounding region, the differential stress will cause the fiber to bend in the transverse direction (similar to the action of a bimetallic strip to temperature changes).

This concept can be generalized to make a smart positioner with two degrees of freedom. Figure 9.35 shows such a device. When light is launched into one of the guides, it will bend in a direction away from that guide. This gives an up-down motion. If both guides are simultaneously illuminated, the fiber will elongate yielding the left-right degree of freedom. It is clear that full three-dimensional actuating would be possible with a three-core fiber in which the cores form an

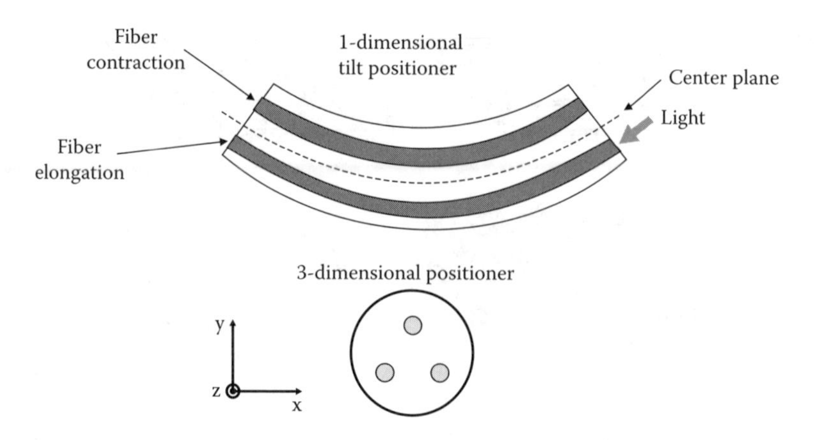

FIGURE 9.35 a) A two-dimensional transverse positioner, shown with light entering the bottom core causing it to expand. Illumination of both cores causes the fiber to change length. b) A three-core structure that produces two-dimensional control of the fiber in the *xy*-plane due to differential expansion and control in the *z*-direction by light-induced fiber elongation.

equilateral triangle, as shown in the bottom part of Figure 9.35. Illumination of all three cores would result in a change in the fiber length while any arbitrary direction in the *xy*-plane can be accessed with uneven illumination of the cores. Additionally, such a fiber could be incorporated as part of a smart thread inside a smart structure.

Each MPU is capable of nanometer level accuracy in positioning. The smart sheet, in which the surface shape could be controlled to within this same level of accuracy, would form the basis for many devices. They could be used to support fragile objects, to change the shape of optics for reconfigurable operation in targeting, to make sound suppressing skins, to act as a smart platform, and to be used as a stable positioner in high-precision manufacturing.

Such devices could operate in two modes. In the low-intensity mode, the shape of the object would be directly related to the stress and light intensity inside the smart sheet. In the high-intensity mode, the system could be mechanically or optically deformed, and — by virtue of the system's multistability — retain its new shape. This would form the basis of optical and shape memory materials. The ability to reconfigure surface shape with a light pulse would enable smart materials to be easily reprogrammed from an external source. Furthermore, it would be intriguing if a non-dissipative strain wave could be excited on a surface by a light pulse. This would allow for programming a dynamical response into a system that is powered with a continuous light source.

9.3.6 Crack Formation Sensing and Prevention

Small lengths of smart threads could be designed for highly specialized applications. For example, a smart thread could be built into a sensor unit that detects

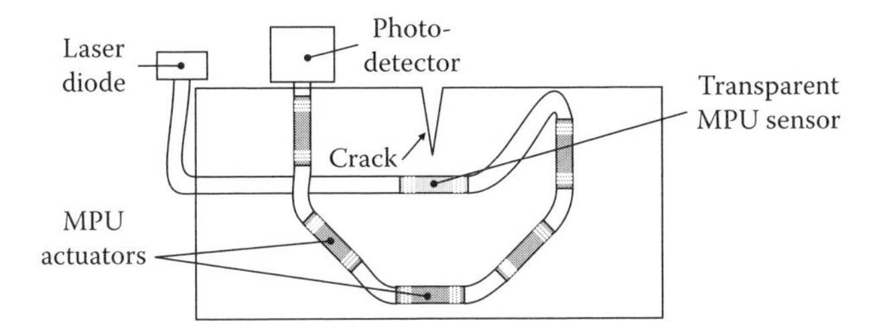

FIGURE 9.36 An MPU sensor near a crack site with distributed MPU actuators.

local stresses that may result in material fracture along with an array of distributed actuators that respond to the buildup of internal stresses. The actuators can be placed away from the potential cracking point in a geometry that maximizes the actuating stress that opposes crack formation.

Figure 9.36 shows an example of such a device. A transparent MPU is placed near a potential site for crack formation and acts only as a sensor. The distributed set of actuators is made from a photomechanical material. Aside from responding to local and distributed stress from the other actuators, the system can be designed to respond more strongly to small changes in strain due to the sensor unit's proximity to the crack. The output of the smart thread can be read by a photodetector that is externally metered for the purpose of assessing material fatigue. The photodetector could also be removed and the output of the thread sent to another set of devices that are distributed further from the crack site.

9.3.7 Noise Reduction

Vibration suppression leads to reduced acoustical noise radiation. Such noise cancellation systems are commercially available, which includes one or more microphones to detect noise and a control system that activates a set of speakers to create a noise field of opposite phase to the original noise. The MPU can play the role of microphone (by sensing the pressure), the logical control system, and speaker (by creating sound of the correct phase to cancel the noise), simultaneously. Therefore, it would be possible to develop structures that create a low noise environment, automatically. This possibility has several potential dual-use applications.

One can imagine "smart wallpaper" made of a distribution of MPUs that would reduce noise indoors. There are countless industrial situations that would benefit from such wallpaper. Furthermore, such wallpaper could also be used to reduce the amount of sound leaving a mechanical area that creates noise pollution. A smart fabric could be programmed for highly specialized applications. As an example, if the spectral distribution of noise in a particular area is fixed, or if the position of its source is known, the smart wallpaper could be programmed to reduce noise from only the noise sources but allow speech to pass. That would enable individuals to verbally communicate in areas with high levels of background noise. Smart skins and wallpaper would clearly have many other applications such as echo suppression (that could be used to design better concert halls or for stealth applications) and intruder surveillance.

9.3.8 Conclusion

In summary, smart materials composed of a large number of MPUs in which all MPUs interact with each other and with local stress will have a high level of intelligence. With a large number of units, the computational power of such an associative network would surpass existing technology by orders of magnitude. Perhaps, such smart materials would be more accurately called genius materials. Often, however, there is a fine line between genius and insanity. For a physical system of devices, this division is characterized by the onset of chaos. It is therefore important to study the conditions for which a given architecture becomes chaotic. Architectures need to be designed to either avoid chaos or to use it to our advantage. Whatever the outcome, smart materials that are optically interconnected offer the potential of very exciting science-fiction-like applications as well as broad opportunities for lots of new science.

10 Conclusion

It may very well be that optics provide superior performance to electronics, and more importantly, allow for unique architectures that cannot be implemented with electronics. The steady growth of lightwave technologies in a large variety of applications attests to its advantages.

The first applications to use waveguiding were the simplest, and included fibers for transmitting signals over long distances; endoscopes for imaging; and light pipes to gather photons or provide illumination. These are examples of passive applications and are analogous to bulk optics, such as mirrors and lenses, which control the propagation of light. Polymer fibers have practical advantages over glass fibers because of their low cost, light weight, and ease of splicing. For these reasons, plastic fibers have made inroads into local area networks and are expanding their influence into longer-distance applications.

Historically, the next step in bulk optics was to make small structures that could be used to better manipulate light. These include dielectric coatings for antireflection applications and gratings for color separation. Analogously, fibers have evolved to incorporate multiple cores and Bragg gratings, which similarly can be used to select certain wavelengths or modes. Because polymers are more flexible and easy to stretch, these properties make them better than silica glass fibers for sensors and tunable filters.

Photons in fibers with periodic structures are called photonic crystals and can be made to mimic the behavior of electrons in a crystal lattice. Taking the argument further, nonlinear-optical interactions of light beams allow the control of one beam with another, in analogy to a transistor. These are examples of active devices that have a much broader range of applications and give the "smarts" to optical logic. Some of the largest nonlinear optical susceptibilities are found in dye-doped polymers, so polymer fibers may play a dominant role as active device materials. And, given the higher throughput of waveguides and speed of the electronic nonlinear-optical response, active polymer fiber devices should be superior to transistors.

There is no doubt that the transistor has revolutionized our world. Active polymer fiber devices could be the basis for the next revolution. However, an even more far-reaching and exciting revolution will be made possible once photomechanical effects are made more efficient and integrated with sensing and logic. Such components in a distributed fiber network would enable new technologies the likes of which are now unimaginable. Smart materials are just one example.

Given the large intellectual base and technological infrastructure encompassing electronics, there is no reason to believe that electronic devices will ever be abandoned. More likely, new hybrid devices that combine optics with electronics will evolve into things what will make the early 21st century appear primitive in comparison. The basic physics behind future technological developments will most likely include many of the concepts covered in this book.

Given the kinds of phenomena that have been observed in polymer fibers, one's mind can imagine applications that, while difficult to implement, are well within the realm of the possible. Consider the smart threads discussed in Section 9.3.4. Imagine that these threads fill three-dimensional space in a massive tangle, yet are thin and dense enough to make the bulk material appear smooth. Consider further that the various parts of the threads are highly interconnected and include logic elements and actuators at all points inside the material. When powered by light, this bulk material could be made to morph into different shapes at will.

Just as a series of pictures on a piece of film can be projected onto a screen to show motion on a two-dimensional plane, the highly interconnected smart material could be made to go through a series of shapes leading to three-dimensional animations. For example, automobile designers could use such a morphing smart material to continually change the shape of a model to test its aerodynamics or aesthetics; a chair could be made to automatically change shape to accommodate any body type; and an exact replica of an individual could be made from information sent from a distance. Having your uncle Joe's three-dimensional solid-body facsimile sitting next to you in casual conversation would be much cooler than picture phones, which transmit only two-dimensional images.

Now imagine that the smart threads are not only interconnected with light and include various embedded optical devices, but also include electrodes to supply electrical current, and organic semiconductors, to make electronic circuits. Such a smart material could be programmed to become an iPod, which by voice command could turn into a DVD player and then a camera. Such a material would be the ultimate putty that could be transformed into many different objects.

It is an enjoyable exercise to imagine what can be. Such musings are the first step that points the way to the next revolution. With a firm grounding in the physics of light-matter interactions and a creative imagination, the future possibilities are endless.

Bibliography

1. A. G. Bell, "On the Production and Reproduction of Sound by Light," Proceedings of the American Association for the Advancement of Science **29**, 115–136 (1881).
2. L. Rayleigh, "On the Passage of Electric Waves Through Tubes or the Vibrations of Cylinders," Phil. Mag. **43**, 125 (1897).
3. D. Hondros and P. Debye, "Elektromagnetishe Wallen an Dielektrishen Dr*ä*hten," Ann. Physik **32**, 465 (1910).
4. O. Schriever, "Elektromagnetishe Wallen an Dielektrishen Dr*ä*hten," Ann. Physik **63**, 645 (1920).
5. H. H. Hopkins and N. S. Kapany, "A Flexible Fiberscope Using Static Scanning," Nature **173**, 39–41 (1954).
6. J. L. Baird, Brit. Pat. Spec. No. 20, 969/27 (1927).
7. E. Snitzer, "Cylindrical Dielectric Waveguide Modes," J. Opt. Soc. Am **51**, 491–498 (1961).
8. E. Snitzer and H. Osterberg, "Observation of Dielectric Waveguide Modes in the Visible Spectrum," J. Opt. Soc. Am **51**, 499–505 (1961).
9. C. Emslie, "Review Polymer Optical Fibres," J. Materials Science **23**, 2281–2293 (1988).
10. A. A. Griffith, "The Phenomena of Rupture and Flow in Solids," Phil. Trans. R. Soc. **221a**, 163 (1921).
11. T. C. Hager, R. G. Brown, and B. N. Derick, "Automotive Fibers" Trans. Soc. Automotive Eng. **76**, 581 (1976).
12. T. Kaino, M. Fujiki, S. Oikawa, and S. Nara, "Low-Loss Plastics Optical Fibers," Appl. Optics **20**, 2886–2888 (1981).
13. S. Oikawa, M. Fujiki, and Y. Katayama, "Polymer Optical Fiber with Improved Transmittance," Electron. Lett. **15**, 830–831 (1979).
14. Y. Ohtsuka, T. Senga, and H. Yasuda, "Light-Focusing Plastic Rod with Low Chromatic Aberration," Appl. Phys Lett. **25**, 659 (1974).
15. Y. Ohtsuka, T. Sugano, and Y. Terao, "Studies on the Light-Focusing Rod 8: Copolymers of Diethylene Glycol *bis*(Allkyl Carbonate) with Methacrylic Ester of Flourine Containing Alcohol," Appl. Opt. **20**, 2319 (1981).
16. Y. Ohtsuka and Y. Koike, "Determination of the Refractive-Index Profile of Light-Focusing Rods: Accuracy of Method Using Interphako Interference Microscopy," Appl. Opt. **19**, 2866–2867 (1980).
17. E. Nihei, T. Ishigure, and Y. Koike, "High-Bandwidth, Graded-Index Polymer Optical Fiber for Near-Infrared Use," Appl. Opt. **35**, 7085–7090 (1996).
18. M. G. Kuzyk, U. C. Paek, and C. W. Dirk, "Guest-Host Fibers for Nonlinear Optics," Appl. Phys. Lett. **59**, 902 (1991).
19. D. W. Garvey, K. Zimmerman, P. Young, J. Tostenrude, J. S. Townsend, Z. Zhou, M. Lobel, M. Dayton, R. Wittorf, and M. G. Kuzyk, "Single-Mode Nonlinear-Optical Polymer Fibers," J. Opt. Soc. Am. B **13**, 2017–23 (1996).
20. D. W. Garvey, Q. Li, M. G. Kuzyk, C. W. Dirk, and S. Martinez, "Sagnac Interferometric Intensity-Dependent Refractive-Index Measurements of Polymer Optical Fiber," Opt. Lett. **21**, 104–06 (1996).
21. D. Bosc and C. Toinen, "Full Polymer Single-Mode Optical Fiber," IEEE Photonics Technology Letters **4**, 749–750 (1992).

22. D. Welker, Personal Communication (2000).
23. E. E. Havinga and P. van Pelt, "Electrochromism of Substituted Polyalkenes in Polymer Matrixes: Influence of Chain Length on Charge Transfer," Ber. Bunsenges. Phys. Chem. **83**, 816–821 (1979).
24. E. E. Havinga and P. van Pelt, "Intramolecular Charge Transfer, Studied by Electrochromism of Organic Molecules in Polymer Matrix," Mol. Cryst. Liq. Cryst. **52**, 449–459 (1979).
25. K. D. Singer, J. E. Sohn, and S. J. Lalama, "Second Harmonic Generation in Poled Polymer Films," Appl. Phys. Lett. **49**, 248–250 (1986).
26. K. D. Singer, M. G. Kuzyk, and J. E. Sohn, "Second-order Nonlinear-optical Processes in Orientationally Ordered Materials: Relationship Between Molecular and Macroscopic Properties," J. Opt. Soc. Am. **4**, 968–76 (1987).
27. J. I. Thackara, G. F. Lipscomb, M. A. Stiller, A. J. Ticknor, and R. Lytel, "Poled Electro-Optic Waveguide Formation in Thin-Film Organic Media," Appl. Phys. Lett. **52**, 1031–33 (1988).
28. D. Chen, D. Bhattacharya, A. Udupa, B. Tsap, H. R. Fetterman, H. R. Chen, S. S. Lee, J. Chen, W. H. Steier, and L. R. Dalton, "High-Frequency Polymer Modulators with Integrated Inline Transitions and Low V_π," IEEE Photonics Tech. Lett. **11**, 54 (1999).
29. D. J. Welker, J. Tostenrude, D. W. Garvey, B. K. Canfield, and M. G. Kuzyk, "Fabrication and Characterization of Single-Mode Electro-Optic Polymer Optical Fiber," Opt. Lett. **23**, 1826–1828 (1998).
30. A. Tagaya, Y. Koike, T. Kinoshita, E. Nihei, T. Yamamoto, and K. Sasaki, "Polymer Optical Fiber Amplifier," Appl. Phys. Lett. **63**, 883–884 (1993).
31. K. Kuriki, T. Kobayashi, N. Imai, T. Tamura, S. Nishihara, Y. Nishizawa, A. Tagaya, and Y. Koike, "High-Efficiency Organic Dye-Doped Polymer Optical Fiber Lasers," Appl. Phys. Lett. **77**, 331–333 (2000).
32. T. Kobayashi, S. Nakatsuka, T. Iwafuji, K. Kuriki, N. Imai, T. Nakamoto, C. D. Claude, K. Sasaki, and Y. Koike, "Fabrication and Superfluorescence of Rare-Earth Chelate-Doped Graded Index Polymer Optical Fibers," Appl. Phys. Lett. **71**, 2421–2423 (1997).
33. S. R. Vigil, Z. Zhou, B. K. Canfield, J. Tostenrude, and M. G. Kuzyk, "Dual-Core Single-Mode Polymer Fiber Coupler," J. Opt. Soc. Am. B **15**, 895–900 (1998).
34. R. Adler, "Waves on Inhomogeneous Cylindrical Structures," Proc. I.R.E. **40**, 339–348 (1952).
35. A. W. Snyder and J. D. Love, *Optical Waveguide Theory* (Chapman and Hall, London, 1983).
36. J. Jackson, *Classical Electrodynamics* (John Wiley and Sons, New York, 1982), 2nd ed.
37. M. Born and E. Wolf, *Principles of Optics* (Pergamon Press, Oxford, 1980), 6th ed.
38. J. W. Goodman, *McGraw-Hill Series in Electrical and Computer Engineering* (McGraw-Hill, New York, 1996). Introduction to Fourier Optics, Ch. 4, 2nd ed.
39. M. Born and E. Wolf, *Principles of Optics: Electromagnetic Theory of Propagation, Interference and Diffraction of Light* (Cambridge University Press, New York, 1999), 7th expanded ed.
40. A. W. Snyder, "Coupled-Mode Theory for Optical Fibers," J. Opt. Soc. Am. **62**, 1267–77 (1972).
41. R. Boyd, *Nonlinear Optics* (Academic Press, Boston, 1992).

42. R. W. Hellwarth, "Third-Order Optical Susceptibilities of Liquids and Solids," Prog. Quantum Electron. **5**, 1–68 (1977).
43. S. M. Jensen, "The Nonlinear Coherent Coupler," IEEE J. Quantum Electron. **QE-18**, 1580–83 (1982).
44. S. M. Jensen, "The Non-Linear Coherent Coupler: A New Optical Logic Device," in *Integrated and Guided-Wave Optics* (1980).
45. P. D. McIntyre and A. W. Snyder, "Power Transfer Between Optical Fibers," J. Opt. Soc. Am. **63**, 1518–27 (1973).
46. E. Caglioti, S. Trillo, S. Wabnitz, and G. I. Stegeman, "Limitations to all-optical Switching Using Nonlinear Couplers in the Presence of Linear and Nonlinear Absorption and Saturation," J. Opt. Soc. Am. B **5**, 472–82 (1988).
47. T. Kaino, K. Jinguji, and S. Nara, "Low Loss Poly(methyl methacrylate-d5) Core Fibers," Appl. Phys. Lett. **41**, 802–804 (1982).
48. M. G. Kuzyk, C. W. Dirk, and J. E. Sohn, "Mechanisms of Quadratic Electrooptic Modulation of Dye-Doped Polymer Systems," J. Opt. Soc. Am. B **5**, 842 (1990).
49. D. W. Garvey, *Nonlinear Optics in Polymer Fibers*, Ph.D. thesis, Dissertation, Washington State University (1999).
50. P. Avakian, W. Y. Hsu, P. Meakin, and H. L. Snyder, "Optical Absorption of Perdeuterated PMMA and Influence of Water," J. Polym. Sci. Polym. Ed. **22**, 1607 (1984).
51. "Fabrication and Mechanical Behavior of Dye-Doped Polymer Optical Fiber," J. Apl. Phys. **92**, 4–12 (2002).
52. C. W. Dirk, A. R. Nagarur, J. J. Lu, L. Zhang, P. Kalamegham, J. Fonseca, S. Gopalan, S. Townsend, G. Gonzalez, P. Craig, M. Rosales, L. Green, K. Chan, R. J. Twieg, S. P. Ermer, D. S. Leung, S. M. Lovejoy, S. Lacroix, N. Godbout, and E. Monette, "Molecular Studies and Plastic Optical Fiber Device Structures for Nonlinear Optical Applications," Proc. SPIE **2527**, 116–126 (1995).
53. G. D. Peng, P. L. Chu, Z. Xiong, T. W. Whitbread, and R. P. Chaplin, "Dye-Doped Step-Index Polymer Optical Fiber for Broadband Optical Amplification," J. Lightwave. Technol. **14**, 2215–2223 (1996).
54. Y. Ohtsuka, Y. Koike, and H. Yamazaki, "Studies on the Light-Focusing Plastic Rod. 6: The Photocopolymer Rod of Methyl Methacrylate with Vinyl Benzoate," Appl. Opt **20**, 280–285 (1981).
55. Y. Ohtsuka, Y. Koike, and H. Yamazaki, "Studies on the Light-Focusing Plastic Rod. 10: A Light-Focusing Plastic Fiber of Methyl Methacrylate-Vinyl Benzoate Copolymer," Appl. Opt **20**, 2726–2730 (1981).
56. K. Iga and N. Yamamoto, "Plastic Focusing Fiber for Imaging Applications," Appl. Opt **16**, 1305–1310 (1977).
57. T. Kaino, *Polymers for Lightwave and Integrated Optics* (Marcel Dekker, Inc., New York, 1992), Vol. 32 of Optical Engineering, chap. Polymer Optical Fibers, 1–38.
58. Y. Ohtsuka and Y. Shimizu, "Radial Distribution of the Refractive Index in Light-Focusing Rods: Determination Using Interphako Interference Microscopy," Appl. Opt **16**, 1050–1053 (1977).
59. Y. Ohtsuka, E. Nihei, and Y. Koike, "Graded-Index Optical Fibers of Methyl Methacrylate-Vinyl Benzoate Copolymer with Low Loss and High Bandwidth," Appl. Phys. Lett. **57**, 120–2 (1990).
60. Y. Koike, Y. Kimoto, and Y. Ohtsuka, "Studies on the Light-Focusing Plastic Rod. 12: The GRIN Fiber Lens of Methyl Methacrylate-Vinyl Phenylacetate Copolymer," Appl. Opt **21**, 1057–1062 (1982).

61. J. Zubia and J. Arrue, "Plastic Optical Fibers: An Introduction to Their Technological Processes and Applications," Opt. Fib. Tech. **7**, 101–140 (2001).
62. T. Flipsen, A. Pennings, and G. Hadziioannou, "Polymer Optical Fiber with High Thermal Stability and Low Optical Losses Based on Novel Densely Crosslinked Polycarbosiloxanes," J. Appl. Polym. Sci, **67**, 2223–2230 (1998).
63. A. A. Hamza and M. A. Kabeel, "Optical Anisotropy in Polypropylene Fibres as a Function of the Draw Ratio," J. Phys. D: Appl. Phys. **20**, 963–968 (1987).
64. P. Ji, D. Q. Li, and G. D. Peng, "Transverse Birefringence in Polymer Optical Fiber Introduced in Drawing Process," Proc. SPIE Int. Soc. Opt. Eng. **5212**, 108 (2003).
65. C. A. Buckley, E. P. Lautenschlager, and J. L. Gibert, "Deformation Processing of PMMA into High-Strength Fibers," J. Appl. Polym. Sci. **44**, 1321–1330 (1992).
66. M. G. Kuzyk and C. W. Dirk eds, *Characterization Techniques and Tabulations for Organic Nonlinear Optical Materials*, Vol. 60 of Optical Engineering (Marcel Dekker, New York, 1998).
67. K. Bløtekjær, "Strain Distribution and Optical Propogation in Tension-Coiled Fibers," Opt. Lett. **18**, 1059–1061 (1993).
68. N. I. Muskhelishvili, *Some Basic Problems of the Mathematical Theory of Elasticity* (Noordhoff, Groningen, Netherlands, 1953).
69. D. Marcuse, "Curvature Loss Formula for Optical Fibers," J. Opt. Soc. Am. **66**, 216–220 (1976).
70. J. J. Sakurai, *Advanced Quantum Mechanics*, Series in Advanced Physics (Addison-Wesley, Reading, MA, 1978).
71. J. D. Jackson, *Classical Electrodynamics* (Wiley, New York, 1996), 3rd ed.
72. M. G. Kuzyk and C. W. Dirk, *Characterization Techniques and Tabulations for Organic Nonlinear Optical Materials* (Marcel Dekker, 1998).
73. S. Huard, *Polarization of Light* (John Wiley and Sons, New York, 1997).
74. B. K. Canfield, C. S. Kwiatkowski, and M. G. Kuzyk, "Direct Deflection Method for Determining Refractive Index Profiles of Polymer Optical Fiber Preforms," Appl. Opt. **41**, 3404–3411 (2002).
75. M. Born and E. Wolf, *Principles of Optics* (Pergamon Press, 1980), 6th ed.
76. K. I. White, "Practical Application of the Refracted Near-field Technique for the Measurement of Optical Fibre Refractive Index Profiles," Opt. Quant. Electron. **11**, 185–96 (1979).
77. M. J. Saunders, "Optical Fiber Profiles Using the Refracted Near-Field Technique: A Comparison with Other Methods," Appl. Opt. **20**, 1645–51 (1981).
78. W. J. Stewart, "Optical Fiber and Preform Profiling Technology," IEEE J. Quan. Elec. **18**, 1451–1465 (1982).
79. C. Koeppen, R. F. Shi, W. Chen, and A. F. Garito, "Properties of Plastic Optical Fibers," J. Opt. Soc. Am. B **15**, 727 (1998).
80. R. F. Shi, C. Koeppen, G. Jiang, J. Wang, and A. F. Garito, "Origin of High Bandwidth Performance of Graded-Index Plastic Optical Fibers," Appl. Phys. Lett. **71**, 3625 (1997).
81. T. Kaino, "Absorption Losses of Low Loss Plastics Optical Fibers," Japanese Journal of Applied Physics **24**, 1661 (1985).
82. A. Skumanich, M. Jurich, and J. D. Swalen, "Absorption and Scattering in Nonlinear Optical Polymeric Systems," Appl. Phys. Lett. **62**, 446–448 (1993).
83. R. J. Kruhlak and M. G. Kuzyk, "Side-Illumination Fluorescence Spectroscopy. I. Principles," J. Opt. Soc. Am. B **16**, 1749–1755 (1999).

84. R. J. Kruhlak and M. G. Kuzyk, "Side-Illumination Fluorescence Spectroscopy. II. Applications to Squaraine-Dye-Doped Polymer Optical Fibers," J. Opt. Soc. Am. B **16**, 1756–1767 (1999).
85. A. C. Boccara, D. Fournier, W. B. Jackson, and N. M. Amer, "Sensitive Photothermal Deflection Technique for Measuring Absorption in Optically Thin Media," Opt. Lett. **5**, 377 (1980).
86. W. B. Jackson, N. M. Amer, A. C. Boccara, and D. Fournier, "Photothermal Deflection Spectroscopy and Detection," Appl. Opt. **20**, 1333–1344 (1981).
87. J. C. Murphy and L. C. Aamodt "Photothermal Spectrocopy using Optical Beam Probing: Mirage Effect," J. Appl. Phys. **51**, 4580–4588 (1980).
88. C. Pitois, A. Hult, and D. Wiesmann, "Absorption and Scattering in Low-Loss Polymer Optical Waveguides," J. Opt. Soc. Am. B **18**, 908–912 (2001).
89. G. Jiang, R. F. Shi, and A. F. Garito, "Mode Coupling and Equilibrium Mode Distribution Conditions in Plastic Optical Fibers," IEEE Photon. Technol. Lett. **9**, 1128–1130 (1997).
90. A. Garito, J. Wang, and R. Gao, "Effects of Random Perturbations in Plastic Optical Fibers," Science **281**, 962–967 (1998).
91. V. Koncar, "Optical Fiber Fabric Displays," Optics and Photonics News **16**, 40–44 (2005).
92. G. D. Peng, Z. Xiong, and P. L. Chu, "Fluorescence Decay and Recovery in Organic Dye-Doped Polymer Optical Fibers," J. Lightwave Technol. **16**, 2365–2372 (1998).
93. B. Howell and M. G. Kuzyk, "Amplified Spontaneous Emission and Recoverable Photodegradation in Disperse-Orange-11-Doped-Polymer," J. Opt. Soc. Am. B **19**, 1790 (2002).
94. B. Howell, *Transient Absorption and Stimulated Emission of the Organic Dye Disperse Orange 11*, Ph.D. thesis, Washington State University (2001).
95. B. Howell and M. G. Kuzyk, "Lasing Action and Photodegradation of Disperse Orange 11 Dye in Liquid Solution," Appl. Phys. Lett. **85**, 1901–1903 (2004).
96. M. A. Díaz-García, S. Fernández De Ávila, and M. G. Kuzyk, "Dye-Doped Polymers for Blue Organic Diode Lasers," App. Phys. Lett. **80**, 4486–4488 (2002).
97. T. Kobayashi, W. Blau, H. Tillman, and H.-H. Hörhold, "Blue Amplified Spontaneous Emission from a Stilbenoid-Compound-Doped Polymer Optical Fiber," Opt. Lett. **26**, 1952–1954 (2001).
98. A. Tagaya, Y. Koike, T. Kinoshita, E. Nihei, T. Yamamoto, and K. Sasaki, "Polymer Optical Fiber Amplifier," Appl. Phys. Lett. **63**, 883–4 (1993).
99. K. Kuriki, Y. Koike, and Y. Okamoto, "Plastic Optical Fiber Lasers and Amplifiers Containing Lanthanide Complexes," Chem. Rev. **102**, 2347–2356 (2002).
100. S. Xiaohong, M. Hai, D. Ning, X. Aifang, H. Jun, Z. Qijin, Y. Min, Z. Zebo, and X. Jianping, "Using Spectra Analysis and Scanning Near-Field Optical Microscopy to Study Eu Doped Polymer Fiber," Opt. Com. **208**, 111–115 (2002).
101. E. H. Huffman, "Stimulated Optical Emission of a Terbium Ion Chelate in a Vinylic Resin Matrix," Nature **200**, 158–159 (1963).
102. E. H. Huffman, "Additional Observations of Probable Stimulated Emission of a Terbium Ion Chelate in a Vinylic Resin Matrix," Nature **203**, 1372–1374 (1964).
103. M. A. Díaz-García, S. Fernández De Ávila, and M. G. Kuzyk, "Energy Transfer from Organics to Rare-Earth Complexes," Apl. Phys. Lett. **81**, 3924–3926 (2002).

104. R. Gao, C. Koeppen, G. Zheng, and A. F. Garito, "Effects of Chromophore Dissociation on the Optical Properties of Rare-Earth-Doped Polymers," Appl. Opt. **37**, 7100–7106 (1998).
105. T. Kobayashi, S. Nakatsuka, T. Iwafugi, K. Kuriki, I. N., T. Nakamoto, C. D. Claude, K. Sasaki, Y. Koike, and Y. Okamoto, "Fabrication and Superfluorescence of Rare-Earth Chelate-Doped Graded Index Polymer Optical Fibers," Appl. Phys. Lett. **71**, 2421–2423 (1997).
106. H. Liang, Q. Zhang, Z. Zheng, H. Ming, L. Zengchang, J. Xu, B. Chen, and H. Zhao, "Optical Amplification of Eu(DBM)$_3$Phen-Doped Polymer Optical Fiber," Opt. Lett. **29**, 477–479 (2004).
107. K. D. Singer, M. G. Kuzyk, R. B. Comizzoli, H. E. Katz, M. L. Schilling, J. E. Sohn, and S. J. Lalama, "Electrooptic Phase Modulation and Second Harmonic Generation in Corona-Poled Polymer Films," App. Phys. Lett. **53**, 1800 (1988).
108. M. G. Kuzyk and C. W. Dirk, "Quick and Simple Method to Measure Third-Order Nonlinear Optical Properties of Dye-Doped Polymer Films," Appl. Phys. Lett. **54**, 1628–30 (1989).
109. R. A. Norwood, M. G. Kuzyk, and R. A. Keosian, "Electro-Optic Tensor Ratio Determination of Side-Chain Copolymers with Electro-Optic Interferometry," J. Appl. Phys. **75**, 1869–74 (1994).
110. C. C. Teng and H. T. Man, "Simple Reflection Techniques for Measuring the Electro-Optic Coefficient of Poled Polymers," Appl. Phys. Lett. **56**, 1734–36 (1990).
111. M. G. Kuzyk and C. Poga, *Molecular Nonlinear Optics: Materials, Physics, and Devices*, Quantum Electronics — Principles and Applications (Academic Press, San Diego, 1994), chap. Quadratic Electro-Optics of Huest-Host Polymers, 299–337.
112. D. J. Welker, J. Tostenrude, D. W. Garvey, B. K. Canfield, and M. G. Kuzyk, "Fabrication and Characterization of Single-Mode Electro-Optic Polymer Optical Fiber," Opt. Lett. **23**, 1826 (1998).
113. K. Zimmerman, F. Ghebremichael, M. G. Kuzyk, and C. W. Dirk, "Electric-Field-Induced Polarization Current Studies in Guest-Host Polymers," J. Appl. Phys. **75**, 1267 (1994).
114. G. D. Peng, P. Ji, and P. L. Chu, "Electro-Optic Polymer Opical Fibres and Their Device Applications," SPIE Proc. **4459** (2002).
115. H. Ma, B. Chen, L. R. Takafumi, L. R. Dalton, and A. K. Y. Jen, "Highly Efficient and Thermally Stable Nonlinear Optical Dendrimer for Electro-Optics," J. Am. Chem. Soc. **123**, 986 (2001).
116. W. Wang, D. Chen, H. R. Fetterman, W. H. Steier, L. R. Dalton, and P. M. D. Chow, "Optical Heterodyne Detection of 60 *GHz* Electro-Optic Modulation From Polymer Waveguide Modulators," Appl. Phys. Lett. **67**, 1806–1808 (1995).
117. M. C. Gabriel, N. A. W. Jr., C. W. Dirk, M. G. Kuzyk, and M. Thakur, "Measurement of Ultrafast Optical Nonlinearities Using a Modified Sagnac Interferometer," Opt. Lett. **16**, 1334–1336 (1991).
118. H. Avramopoulos, P. M. W. French, M. C. Gabriel, H. H. Houh, J. N. A. Whitaker, and T. Morse, "Complete Switching in a Three-Terminal Sagnac Switch," IEEE Photonics Technology Letters **3**, 235–7 (1991).
119. H. Avramopoulos, P. M. W. French, M. C. Gabriel, D. J. DiGiovanni, R. E. LaMarche, H. M. Presby, and N. A. Whitaker, "All-Optical Arbitrary Demultiplexing at 2.5 *Gb/s* with Tolerance to Timing Jitter," CPDP **18** (1991).

120. N. A. J. Whitaker, M. C. Gabriel, H. Avramopolous, and A. Huang, "All-Optical, All-Fiber Circulating Shift Register with an Inverter," Opt. Lett. **24**, 1999 (1991).
121. Z. Xiong, G. D. Peng, and P. L. Chu, "Nonlinear Coupling and Optical Switching in a β-Carotene-Doped Twin-Core Polymer Optical Fiber," Opt. Eng. **39**, 624–7 (2000).
122. A. W. Snyder, "Couple-Mode Theory for Optical Fibers," J. Opt. Soc. Am. **62**, 1267–1277 (1972).
123. A. W. Snyder and Y. Chen, "Nonlinear Fiber Couplers: Switches and Polarization Beam Splitters," Opt. Lett. **14**, 517–519 (1989).
124. P. L. Chu, "Polymer Optical Fiber Bragg Gratings," Optics and Photonics News **16**, 53–56 (2005).
125. P. L. Knight, T. A. Birks, P. S. J. Russel, and D. M. Atkin, "All-Silica Single-Mode Optical Fiber with Photonic Crystal Cladding," Opt. Lett. **21**, 1547–1549 (1996).
126. M. A. van Eijkelenborg, C. J. Large, A. Argyros, J. Zagari, S. Manos, N. A. Issa, I. Bassett, S. Fleming, R. C. McPhedran, C. M. de Sterke, and N. A. P. Nicorovici, "Microstructured Polymer Optical Fibre," Opt. Express **9**, 319–327 (2001).
127. Y. Fink, D. J. Ripin, S. Fan, C. Chen, J. D. Joannopoulos, and E. L. Thomas, "Guiding Optical Light in Air Using an All-Dielectric Structure," L. Lightwave Technol. **17**, 2039–2041 (1999).
128. B. Temelkuran, S. D. Hart, G. Benoit, J. D. Joannopoulos, and Y. Fink, "Wavelength-Scalable Hollow Optical Fibres with Large Photonic Bandgaps for CO_2 Laser Transmission," Nature **420**, 650–653 (2002).
129. M. Dumont, Z. Sekkat, R. Loucif-Saibi, K. Nakatani, and J. A. Delaire, "Photoisomerization, Photoinduced Orientation and Orientational Relaxation of Azo Dyes in Polymeric Films," Nonlinear Optics (1992).
130. Z. Sekkat and M. Dumont, "Polarization Effects in Photoisomerization of Azo Dyes in Polymeric Films," Appl. Phys. B **53**, 121–123 (1991).
131. Z. Sekkat and M. Dumont, "Poling of Polymer Films by Photoisomerization of Azo Dye Chromophores," (1992).
132. Z. Sekkat and M. Dumont, "Photoassisted Poling of Azo Dye Doped Polymeric Films at Room Temperature," Appl. Phys. B **54**, 486–489 (1992).
133. Z. Sekkat and W. Knoll, "Creation of Second-Order Nonlinear Optical Effects by Photoisomerization of Polar Azo Dyes in Polymeric Films: Theoretical Study of Steady-State and Transient Properties," J. Opt. Soc. Am. B. **12**, 1855–67 (1995).
134. S. Bian, W. Zhang, and M. G. Kuzyk, "Erasable Holographic Recording in Photosensitive Polymer Optical Fibers," Opt. Lett. **28**, 929–931 (2003).
135. S. Bian and M. G. Kuzyk, "Phase Conjugation by Low-Power Continuous-Wave Degenerate Four-Wave Mixing in Nonlinear Polymer Optical Fibers," App. Phys. Lett. **84**, 858–860 (2004).
136. M. Froggatt, "Distributed Measurement of the Complex Modulation of a Photoinduced Bragg Grating in an Optical Fiber," Appl. Opt **35**, 5162–5164 (1996).
137. A. L. Jenkins, O. M. Uy, and G. M. Murray, "Polymer-Based Lanthanide Luminescent Sensor for Detection of the Hydrolysis Product of the Nerve Agent Soman in Water," Anal. Chem. **71**, 373–378 (1999).
138. B. R. Arnold, A. C. Euler, A. L. Jenkins, O. M. Uy, and G. M. Murray, "Progress in the Development of Molecularly Imprinted Polymer Sensorts," Johns Hopkins APL Tech. Digest **20**, 190–198 (1999).
139. A. L. Jenkins, R. Yin, and J. L. Jensen, "Molecularly Imprinted Polymer Sensor for Pesticide and Insecticide Detection in Water," Analyst **126**, 798–802 (2001).

140. K. Uchino and L. E. Cross, "Electrostriction and Its Interrelation with Other Anharmonic Properties of Materials," Japanese Journal of Applied Physics **19**, 171–173 (1980).
141. K. Uchino, "Photostrictive Actuator," Ultrasonics Symposium, 721–723 (1990).
142. K. Uchino, "Ceramic Actuators: Principles and Applications," MRS Bulletin, April **29**, 42 (1993).
143. M. Camacho-Lopez, H. Finkelmann, P. Palffy-Muhoray, and M. Shelley, "Fast Liquid-Crystal Elastomer Swims into the Dark," Nature Materials **3**, 307–310 (2004).
144. E. L. Buckland and R. W. Boyd, "Electrostrictive Contribution to the Intensity-Dependent Refractive Index of Optical Fibers," Optics Letters **21**, 1117–1119 (1996).
145. C. Poga and M. G. Kuzyk, "Quadratic Electroabsorption Studies of Third-Order Susceptibility Mechanisms in Dye-Doped Polymers," J. Opt. Soc. Am. B **11**, 80–91 (1994).
146. M. Kuzyk, R. C. Moore, and L. A. King, "Second-Harmonic-Generation Measurements of the Elastic Constant of a Molecule in Polymer Matrix," J. Opt. Soc. Am. B **7**, 64 (1990).
147. D. J. Welker and M. G. Kuzyk, "Photomechanical Stabilization in a Polymer Fiber-Based All-Optical Circuit," Appl. Phys. Lett. **64**, 809–811 (1994).
148. M. G. Kuzyk, D. J. Welker, and S. Zhou, "Photomechanical Effects in Polymer Optical Fibers," Nonlinear Optics **10**, 409–419 (1995).
149. D. J. Welker and M. G. Kuzyk, "Optical and Mechanical Multistability in a Dye-Doped Polymer Fiber Fabry-Perot Waveguide," Appl. Phys. Lett. **66**, 2792–2794 (1995).
150. D. J. Welker and M. G. Kuzyk, "All-Optical Switching in a Dye-Doped Polymer Fiber Fabry-Perot Waveguide," Appl. Phys. Lett. **69**, 1835–6 (1996).
151. M. Dumont and Z. Sekkat, "Dynamic Study of Photoinduced Anisotropy and Orientation Relaxation of Azo Dyes in Polymeric Films. Poling at Room Temperature." SPIE Proceeding **1774**, 1–12 (1992).
152. M. Dumont, G. Proc, and S. Hosotte, "Alignment and Orientation of Chromophores by Optical Pumping," Nonlinear Optics (1994).
153. M. Dumont, S. Hosotte, G. Froc, and Z. Sekkat, "Orientational Manipulation of Chromophores Through Photoisomerization," SPIE Proceedings **2042**, 1–12 (1993).
154. D. J. Welker and M. G. Kuzyk, "Suppressing Vibrations in a Sheet with a Fabry-Perot Photomechanical Device," Opt. Lett. **22**, 417–418 (1997).
155. S. Bian, D. Robinson, and M. G. Kuzyk, "Optically Activated Cantilever Using Photomechanical Effects in Dye-Doped Polymer Fibers," J. Opt. Soc. Am. B **23**, 697–708 (2006).
156. C.F. Bohren and D.R. Huffman, "Absorption and Scattering of Light by Small Particles," John Wiley & Sons (New York, 1983).
157. T. Kaino, M. Fujiki, S. Oikawa, and S. Nara, "Low Loss Plastic Fibers," Appl. Opt. 20, 2886–8 (1981).

Index

A

B

C

D

E

F

G

H

I

J

K

L

M

N

O

P

Q

R

S

T

U

X

Y

Z